APPLIED MECHANICS AND CIVIL ENGINEERING VI

PROCEEDINGS OF THE 6TH INTERNATIONAL CONFERENCE ON APPLIED MECHANICS AND CIVIL ENGINEERING (AMCE 2016), HONG KONG, 30–31 DECEMBER 2016

Applied Mechanics and Civil Engineering VI

Editor

Liquan Xie

Department of Hydraulic Engineering, Tongji University, China

CRC Press
Taylor & Francis Group
Boca Raton London New York Leiden

CRC Press is an imprint of the
Taylor & Francis Group, an **informa** business
A BALKEMA BOOK

CRC Press/Balkema is an imprint of the Taylor & Francis Group, an informa business

Typeset by V Publishing Solutions Pvt Ltd., Chennai, India
Printed and bound in Great Britain and USA

Published by: CRC Press/Balkema
P.O. Box 11320, 2301 EH Leiden, The Netherlands
e-mail: Pub.NL@taylorandfrancis.com
www.crcpress.com – www.taylorandfrancis.com

ISBN: 978-1-138-62631-7 (Hbk)
ISBN: 978-1-315-22907-2 (eBook)

Applied Mechanics and Civil Engineering VI – Xie (Ed.)
© 2017 Taylor & Francis Group, London, ISBN 978-1-138-62631-7

Table of contents

Applied Mechanics and Civil Engineering VI – Xie (Ed.)
© 2017 Taylor & Francis Group, London, ISBN 978-1-138-62631-7

Preface

Applied Mechanics and Civil Engineering VI includes the selected, peer reviewed papers that will be presented at the 6th International Conference on Applied Mechanics and Civil Engineering (AMCE 2016). AMCE 2016 showcases the exciting and challenging developments occurring in the area of applied mechanics, civil engineering and associated engineering practice, and serves as a major forum for researchers, engineers and manufacturers to share recent advances, discuss problems, and identify practical challenges associated with the engineering applications. The contributions from experts and world-renowned scientists cover a wide variety of topics:

- Applied mechanics and its applications in civil engineering;
- Bridge engineering;
- Underground engineering;
- Structural safety and reliability;
- Reinforced concrete (RC) structures;
- Rock mechanics and rock engineering;
- Geotechnical in-situ testing & monitoring;
- New construction materials and applications;
- Computational mechanics;
- Natural hazards and risk, and
- Water and hydraulic engineering.

Applied Mechanics and Civil Engineering VI will appeal to professionals and academics involved in the above mentioned areas. Although these papers represent only modest advances toward overcoming major scientific problems in civil engineering, some of the technologies might be key factors in the success of future engineering advances. It is expected that this book will stimulate new ideas, methods and applications in ongoing civil engineering advances.

We would like to express our deep gratitude to all authors, reviewers for their excellent work, and Léon Bijnsdorp, Lukas Goosen and other editors from Taylor & Francis Group for their wonderful work.

Applied Mechanics and Civil Engineering VI – Xie (Ed.)
© 2017 Taylor & Francis Group, London, ISBN 978-1-138-62631-7

Behavior of large deformation and supporting measures in soft rock tunnel

Kang Bian
State Key Laboratory of Geomechanics and Geotechnical Engineering, Institute of Rock and Soil Mechanics, Chinese Academy of Science, Wuhan, Hubei Province, P.R. China

Xiaoqing Zheng
State Key Laboratory of Geomechanics and Geotechnical Engineering, Institute of Rock and Soil Mechanics, Chinese Academy of Science, Wuhan, Hubei Province, P.R. China
University of Chinese Academy of Sciences, Beijing, P.R. China

Yuwei He
Hubei Wuxue Yangtze River Highway Bridge Co. Ltd., Wuxue, Hubei Province, P.R. China

Jian Liu
State Key Laboratory of Geomechanics and Geotechnical Engineering, Institute of Rock and Soil Mechanics, Chinese Academy of Science, Wuhan, Hubei Province, P.R. China

ABSTRACT: Typical large-deformation phenomena of soft rock occurred frequently during tunneling in Huangjiazhai Tunnel. The monitoring results indicated that the deformation had the characteristics of large magnitude, high velocity, long duration and inhomogeneity of spatial distribution. The results of in-situ stress test showed that, as a result of high tectonic stress and lower rock strength, the drilling test area of tunnel was regarded as extremely-high stress area, even if the maximum buried depth (270 m) of tunnel was not too large. Based on the New Austrian Tunneling Method, new comprehensive supporting measures were utilized in Huangjiazhai Tunnel, which were consisted of increase of the reserved deformation, bolt and grouting supporting, accelerating the construction of inverted arch and secondary lining and adjusting the spacing and type of steel arch. It was proved in the subsequent construction that the above supporting design was effective for controlling the large deformation in Huangjiazhai Tunnel.

1 INTRODUCTION

A large number of highways run across the mountains in the central and western regions of China, therefore, tunnel engineering plays a significant role in the highway construction. Due to the complex topographic and geological conditions, large deformation phenomena in soft rock tunnels are frequently encountered, which induce different degrees of damage of primary support and secondary lining. More seriously, severe extrusion induced by large deformation commonly leads to instability of tunnel, resulting in significant difficulties in tunnel design and construction. Consequently, the proper supporting measures applied in large-deformation tunnels are of importance to complete the constructions safely.

Studies on deformation behavior and supporting measures in soft rock tunnel have been conducted by a large number of researchers (He Manchao et al., 2002, 2013; Li Hongbo et al., 2011; Liu Gao et al., 2005; Dai Yonghao et al., 2015; Karmen F.B. et al., 2004). He Manchao et al. (2002, 2013) proposed an initiative supporting technology system which was based on constant resistance with large deformation coupling support. Liu Gao et al. (2005) researched the intense deformation and serious failure in Muzhailing tunnel, and proposed the mechanism of large deformation in this tunnel. Li Hongbo et al. (2011) studied the monitoring results of deformation and structural stress for Xiakou soft rock tunnel. Dai Yonghao et al. (2015) presented comprehensive support measures in Daliang tunnel which suffered large deformation. Nevertheless, the existing researches are lacking of combining analysis of the behavior of large deformation, the geostress results and supporting measures in soft rock tunnel. In this paper, taking an example of Huangjiazhai Tunnel in Northwest Hubei Province, the large-deformation characteristics are summarized. And then, combining the in-situ stress test and laboratory experiments, the cause of large deformation is analyzed. Finally, new comprehensive supporting measures were proposed and utilized, which were proved effectively for controlling the large deformation in the subsequent construction in Huangjiazhai Tunnel.

2 PROJECT OVERVIEW

Huangjiazhai Tunnel is one of 29 tunnels of Macheng-Zhuxi Highway, which is located at north-west of Hubei province in Central China. Designed as a long separated tunnel, Huangjiazhai Tunnel has the maximum buried depth of 270 m, with the whole length of 1437 m. Both of the two tunnels are of 12.4 m width and 9.9 m height, and the end-wall portal form is chosen. The tunnel sites are mainly hilly areas, also including small amounts of valleys and terraces. Both the entrances and exits of the tunnel are situated at the halfway up the hills, and the slope angles are mostly between 15 degree and 20 degree.

The principal strata locating at the construction site of Huangjiazhai Tunnel is Silurian shale belonging to the Sllm Group (S_1l). Quaternary residual gravelly soil (Q_4^{el+dl}) composes the covering layer of this area. Geological explorations via boreholes drilled at the tunnel route show that three layers could be divided from the earth's surface to the bottom: gravelly soil, intense weathering shale and intermediary weathered shale. In general, the rock masses present mostly fragmentized, jointed and pelitic, with thin-layered structure.

The New Austrian Tunnelling Method (NATM) is applied in construction of Huangjiazhai Tunnel. At the regions where the rock masses are considered as grade V, the ring cut method is used; at the other parts, the bench cut method is utilized. The initial supports in the tunnel are combinative technologies, including rock bolt, steel arch support, reinforcing mesh and shotcrete. After that, the reinforced concrete is applied as the secondary lining.

3 LARGE-DEFORMATION PHENOMENA DURING TUNNELLING

During tunneling, typical large-deformation phenomena of soft rock occurred frequently in Huangjiazhai Tunnel, some of the examples are described as follows.

Fig. 1 shows the typical squeezing phenomenon of the sidewall. Affected by the collapse at the section between K60 + 733 and K60 + 734.2, severe deformation occurred at the section between K60 + 745 and K60 + 753 of the right tunnel. The primary support cracked and peeled off, with steel arch at the left sidewall distorted. More seriously, the primary support intruded into the outline of the secondary lining, the maximum intrusion value even reached 105.3 cm at the section K60 + 745. As a result, the primary support at the intrusion section should be chiseled off and replaced.

Collapse happened at the section between K60 + 733 and K60 + 734.2, as shown in Fig. 2. When the primary supports were chiseled off at the section between K60 + 733 and K60 + 734.2, some of the rock fragments slipped from the left sidewall and crown; and as time passed by, the amount of rock fragments increased rapidly, and collapse was induced.

Fig. 3 shows the cracking of secondary lining at the sidewall. The surface concrete gradually peeled off, and the reinforcement meshes were exposed at the section between ZK60 + 690.5 and ZK60 + 707. The above phenomena aggravated when the continuous raining days came.

The shotcrete cracking occurred at the entrance slope of right tunnel after continuous heavy rain, as shown in Fig. 4. Through uninterrupted observation, the width and extended length of the crack increasingly expanded as time went on, and the maximum values reached 40 cm and 15 m respectively. Finally, a through crack appeared, and the adjacent parts were at risk from

Figure 1. Squeezing of the sidewall.

Figure 2. Collapse inside the tunnel.

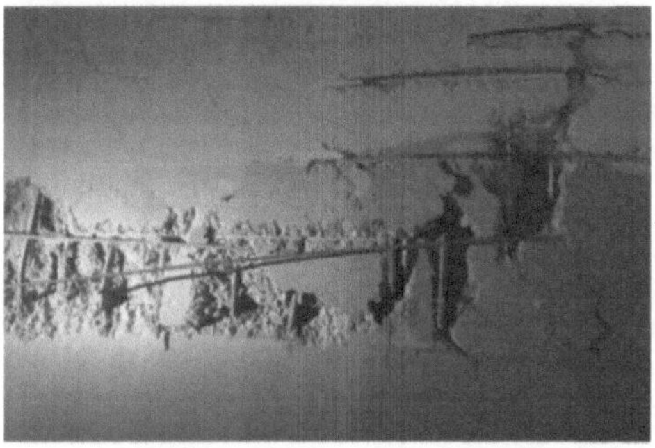

Figure 3. Cracking of secondary lining.

Figure 4. Shotcrete cracking at the entrance slope.

landslides. According to the monitoring results, the maximum velocity of surface subsidence at the entrance slope exceeded 16 mm/d, which accelerated the expansion of shotcrete crack.

4 MONITORING SITUATION

The monitoring results could present quantitative data to analyze the real situation during tunneling process at Huangjiazhai Tunnel. Fig. 5 and Fig. 6 show the monitoring results of horizontal convergence and crown settlement of four typical sections in Huangjiazhai Tunnel. Fig. 7 shows the monitoring results of surface subsidence at the entrance slope of right tunnel. In general, several characteristics of deformation in Huangjiazhai Tunnel could be summarized as follows.

1. The large accumulative deformation. According to the monitoring results, the accumulative horizontal displacement and crown settlement of all four sections exceeded 100 mm. At the zones where the surrounding rocks were loose and fragmentized, e.g. Section ZK60 + 645, the maximum horizontal displacement even exceeded 350 mm within 50 days, and the maximum crown settlement reached 314 mm. Besides, the accumulative surface subsidence at the entrance slope of right tunnel also reached 183 mm within 23 days.
2. The large deformation velocity, especially at the initial period. In the initial several days, the deformation velocity inside the tunnel remained at a high level; as time gone by, after the monitoring lasting for 3 to 5 days, the deformation velocity would decrease. The maximum velocity of horizontal convergence exceeded 24 mm/d at Section ZK60 + 750 when the monitoring lasted for three days, and the maximum velocity of crown settlement was 18 mm/d at the fourth day after monitoring.
3. The long duration of deformation. During the whole monitoring period (50 days), both the horizontal and vertical deformations of monitoring sections had no convergent tendency. In general, the monitoring results of deformation at Huangjiazhai Tunnel represented typical rheologic characteristics of surrounding rock.

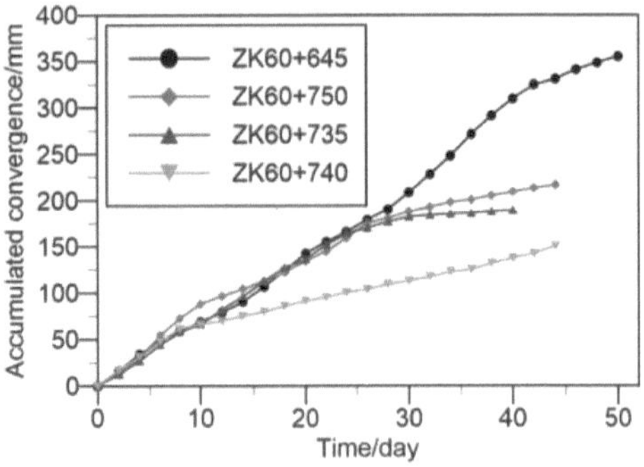

Figure 5. Curve between horizontal convergence and time of four sections in tunnel.

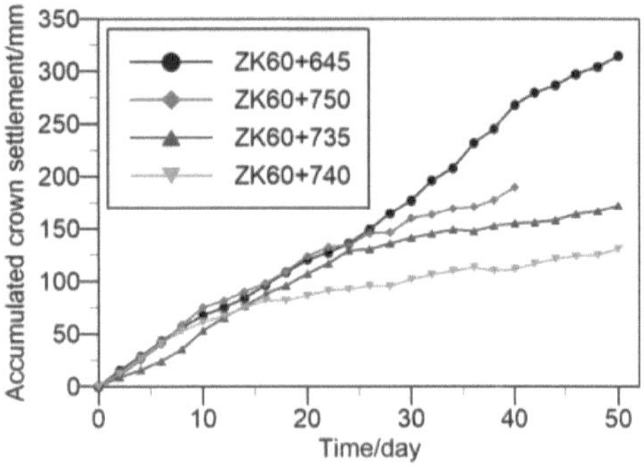

Figure 6. Curve between crown settlement and time of four sections in tunnel.

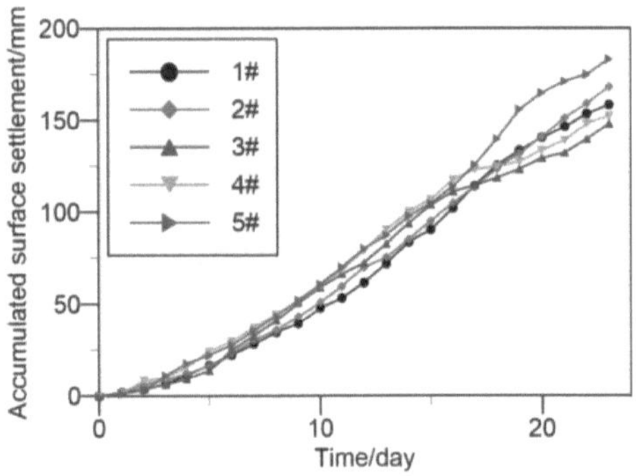

Figure 7. Curve between surface subsidence and time at the entrance slope of right tunnel.

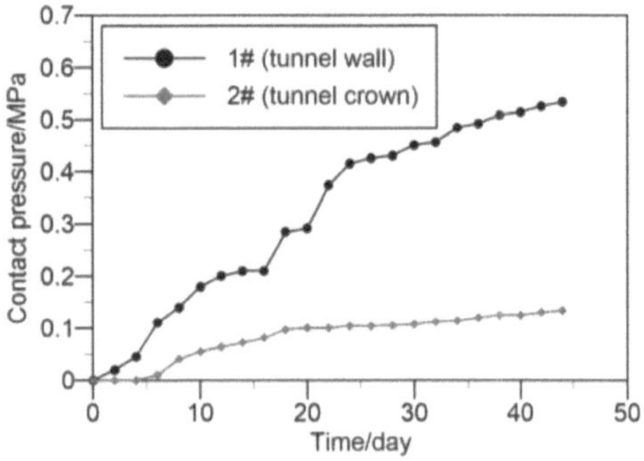

Figure 8. Curve between contact pressure and time at stake YK61+666.

4. The inhomogeneity of deformation distribution. Even though the displacements at Huangjiazhai Tunnel were generally large, the tunnel deformations were obviously much bigger at the fractured zones and the groundwater-developed zones than the other positions. Moreover, the horizontal displacements at most of monitoring sections were larger than the values of crown subsidence.

Fig. 8 shows the monitoring results of contact pressure between primary support and surrounding rock at stake YK61 + 666 in Huangjiazhai Tunnel. It was seen that during the observation period, the contact pressure kept a continuous increasing trend and had no convergent tendency within 43 days. Moreover, sudden increase of contact pressure occurred at the 18th and 22th day after monitoring, as a result of blasting disturbance at adjacent sections. From the point of view of magnitude, the pressure at tunnel wall was generally bigger than the value at arch crown; and the maximum contact pressure between primary support and surrounding rock reached 0.534 MPa.

5 GROUND STRESS TEST

The ground stress test was utilized to assess the geostress level at Huangjiazhai Tunnel. The vertical borehole for geostress measurement, with a depth of 50 m, was located at the stake YK61 + 195 in the right tunnel, and the buried depth of measurement part was 260 m. The hydraulic fracturing method was applied in the geostress test, and the results are listed in Table 1.

Table 1. Results of geostress test in vertical borehole at Huangjiazhai Tunnel.

Depth m	Maximum horizontal principal stress MPa	Minimum horizontal principal stress MPa	Vertical stress MPa	Direction of maximum horizontal principal stress
18	2.44	2.14	7.23	/
22	6.41	4.30	7.33	/
26	13.20	7.01	7.44	/
30	12.61	6.56	7.54	/
37	11.90	6.56	7.72	/
39	14.33	7.85	7.77	NE8°
41	14.60	7.86	7.83	NE10°

As shown in Table 1, within the depth range of geostress measurement, the maximum horizontal principal stress at the borehole is 14.60 MPa, and the direction of maximum horizontal principal stress is NE10°. The results indicate that the maximum principal stress and the second principal stress are both horizontal stresses, and only the minimum principal stress is vertical stress. Furthermore, the lateral pressure coefficient of geostress at the borehole, defined as the ratio of the maximum horizontal principal stress and the vertical stress, is nearly 1.8. Therefore, the tectonic stress is the primary composition of the geostress at the tunnel site. Consequently, a conclusion could be drawn that the region where Huangjiazhai Tunnel locates had experienced intense tectonic movements in the geological history, which could be also verified by the littery attitude of tunnel rock.

The grade of rock mass stresses could be defined by the ratio of uniaxial compressive strength to maximum principal stress, and it could be considered as extremely-high stress area when the ratio is less than 4. In fact, according to laboratory experiments, the uniaxial compressive strength of shale specimen taking from Huangjiazhai Tunnel is 23.3 MPa. Therefore, the ratio is only 1.6 at the borehole, which indicates that although the maximum buried depth (270 m) of tunnel is not too large, as a result of high tectonic stress and lower rock strength, the drilling test area of Huangjiazhai Tunnel is regarded as extremely-high stress area.

6 SUPPORTING MEASURES AGAINST LARGE DEFORMATION

Based on the above analysis, the squeezing deformation in Huangjiazhai tunnel may be due to the rheology of shale under extremely-high geostress, and the low strength and integrity of the rock masses constitute the basic conditions of large deformation.

According to the monitoring results, the deformation and contact pressure of surrounding rock in Huangjiazhai Tunnel changed typically over time, and the existing supporting design could not satisfy the demands of tunnel stability. Therefore, adjusted supporting design should be presented to adapt the special geological condition in Huangjiazhai Tunnel. Based on the New Austrian Tunneling Method, in the design of the primary support, a certain extent of deformation of surrounding rock induced by excavation is allowable, and then sufficient support should be applied in order to prevent the tunnel from buckling failure. The specific measures were listed as follows.

1. Reinforcement at the large deformation and cracking parts. The temporary steel support was utilized at the tunnel wall of large deformation parts, in order to provide resistance of creep deformation. To the cracking parts of primary support, grouting reinforcement in surrounding rock with perforated steel pipes was used.
2. Increase of the reserved deformation. The original reserved deformation was chosen as 12 cm, which was obviously less than the actual deformation monitored in Huangjiazhai Tunnel. To prevent the primary support intruding into the outline of the secondary

Table 2. Parameters of advance ductile grouting in the adjusted supporting design.

Diameter mm	Length m	Circumferential spacing cm	External angle °	Grouting pressure MPa
42	3.5	80~100	5~10	0.8

lining, as well as reduce the surrounding rock pressure borne by the secondary lining, the adjusted reserved deformation was chosen as 30 cm.
3. Bolt and grouting supporting. Since the rock mass exposing at the working face in Huangjiazhai Tunnel was more fragmentized and jointed than design, in order to increase the integrity and deformation resistance of surrounding rock, bolt and grouting supporting was necessary. At the fractured zones, the advance ductile grouting was utilized at the whole excavation section; and the proportion of cement and water was chosen as 1:1. The parameters of ductile grouting were shown in Table 2.
4. Accelerating the construction of inverted arch and secondary lining. To the squeezing tunnel, the secondary lining should not be applied until the deformation of primary support was completely stable. The inverted arch would be constructed as early as possible, and the construction of secondary lining would be carried out in time, immediately after the velocity of deformation reducing to 3~5 mm/d, in order to form the enclosed load-bearing ring and improve the structure stress of tunnel.
5. Adjusting the spacing and type of steel arch. To improve the capability of deformation resistance of steel support, the steel arch was more densely distributed with spacing of 50 mm instead of 60 mm; and moreover, the I20b I-steel was used, taking the place of the original type I18.

After the adjusted supporting design was carried out, the magnitude and velocity of tunnel deformation reduced significantly, according to the subsequent monitoring results. Furthermore, the incidence of large deformation phenomena, including squeezing of the sidewall and damage of the primary support, was decreasing in the following days. It was proved that the adjusted supporting design was effective for controlling the large deformation in Huangjiazhai Tunnel.

7 CONCLUSIONS

During tunneling, typical large-deformation phenomena of soft rock occurred frequently in Huangjiazhai Tunnel, e.g. squeezing of the sidewall, collapse inside the tunnel, severe cracking of secondary lining and shotcrete cracking at the entrance slope.

The monitoring results indicate that the deformation in Huangjiazhai Tunnel has the characteristics of large magnitude, high velocity, long duration and inhomogeneity of spatial distribution.

Based on the results of geostress test and laboratory experiment, although the maximum buried depth (270 m) of tunnel is not big enough, as a result of high tectonic stress and lower rock strength, the drilling test area of Huangjiazhai Tunnel is regarded as extremely-high stress area.

Adjusted support measures were represented and utilized in Huangjiazhai Tunnel, including the temporary steel support utilized at the tunnel wall of large deformation parts; increase of the reserved deformation; bolt and grouting supporting; accelerating the construction of inverted arch and secondary lining; adjusting the spacing and type of steel arch. It was proved in the subsequent construction that the above supporting design was effective for controlling the large deformation in Huangjiazhai Tunnel. It also provides feasible support measures against large deformation in squeezing tunnels, which could be referred in similar projects.

ACKNOWLEDGEMENT

The authors gratefully acknowledge the support by the National Key Research and Development Project of China (Grant No. 2016YFC0401802), the State Key Program of National Natural Science of China (Grant No. 51539002) and the National Natural Science Foundation of China (Nos. 51209198 and 51204158).

REFERENCES

Dai Yonghao, Chen Weizhong, Tian Hongming, et al. 2015. Study of Large Deformation and Support Measures of Daliang Tunnel with Soft Surrounding Rockmass. Chinese Journal of Rock Mechanics and Engineering 34(Supp.2): 4149–4156.

He Manchao, Jing Haihe, Sun Xiaoming. 2002. Engineering Mechanics of Soft Rock. Beijing: Science Press.

He Manchao, Yuan Yue, Wang Xiaolei, et al. 2013. Control technology for large deformation of Mesozoic compound soft rock in Xinjiang and its application. Chinese Journal of Rock Mechanics and Engineering 32(3): 433–441.

Karmen F.B., Borut P. 2004. Displacement analysis of tunnel support in soft rock around a shallow highway tunnel at Golovec. Engineering Geology 75: 89–106.

Li Hongbo, Dai Yonghao, Song Jihong, et al. 2011. Construction monitoring and support design of Xiakou tunnel under high stress with soft rock surrounding. Rock and Soil Mechanics 32(Supp.2): 496–501.

Liu Gao, Zhang Fanyu, Li Xinzhao, et al. 2005. Research on large deformation and its mechanism of Muzhailing tunnel. Chinese Journal of Rock Mechanics and Engineering 24(Supp.2): 5521–5526.

Applied Mechanics and Civil Engineering VI – Xie (Ed.)
 ISBN 978-1-138-62631-7

Integrating *in situ* tests on the excavation damaged zone of underground caverns

F. Ai & J. Liu
State Key Laboratory of Geomechanics and Geotechnical Engineering, Institute of Rock and Soil Mechanics, Chinese Academy of Science, Wuhan, Hubei Province, P.R. China

N.S. Chen
Northwest Engineering Corporation Limited, Power China, Xian, Shanxi Province, P.R. China

H.B. Xiao
Huaneng Lancang River Hydropower Corporation Limited, Kunming, Yunnan Province, P.R. China

G.H. Shao
Wulonglong and Lidi Hydropower Construction Authority, Huaneng Lancang River Hydropower Corporation Limited, Kunming, Yunnan Province, P.R. China

ABSTRACT: This paper presents a comprehensive study of the unloading relaxation, damage fracture, and degradation of the mechanical parameters of surrounding rocks in underground caverns during excavation using acoustic velocity test, borehole camera test, and borehole elastic modulus test. First, the range of excavation damaged zone is determined by the results of acoustic velocity test. Then, from borehole camera images, information of the distribution and evolution of fissures and cracks during blasting excavation is obtained. Finally, the relationship between elastic modulus and acoustic velocity is established by the borehole elastic modulus tests and acoustic velocity results. The mechanical parameters of the excavation damaged zone and surrounding undisturbed rocks of underground caverns are directly determined by *in situ* geophysical and geotechnical tests. It provides a direct basis for feedback analysis of displacement monitoring and advanced deformation forecasting of surrounding rocks in underground caverns.

1 INTRODUCTION

Excavation damaged zone of the surrounding rock is a key science question on rock mechanics and construction site design. Through *in situ* geophysical and geotechnical tests, unloading relaxation, damage fracture, and degradation of the mechanical parameters of surrounding rocks during excavation can be obtained directly and efficiently. Scientists have made several achievements on *in situ* geophysical and geotechnical tests on the excavation damaged zone.

Anderson et al. (2009) carried out several experiments in the Hard Rock Laboratory on the Sweden Äspö Island. Yan et al. (2009) studied the damage characteristics of surrounding rocks for tunnels constructed using TBM and drill and blast. Li et al. (2010) conducted tests on the formation and evolution of TBM excavation damaged zone in deep-buried tunnel based on the digital panoramic borehole camera technique. Liu et al. (2011) detected and analyzed the excavation damaged zone of deep tunnel by means of acoustic test, drilling TV, and numerical methods. Zhang et al. (2011) analyzed a large number of velocity–depth curves and proposed an acoustic velocity fitting of the excavation damaged zone of surrounding rock in underground powerhouse. Dai et al. (2015) used acoustic wave test, borehole TV, and micro-seismic monitoring to study the characteristics of damaged zones due to excavation in deep underground powerhouse at Houziyan hydropower station.

Previous research works are mainly based on a single method, such as acoustic testing, micro-seismic monitoring, borehole elasticity modulus, and borehole camera, to evaluate the evolution of excavation damaged zone of underground engineering surrounding rock. It rarely combines with different types of testing methods. In this paper, a hydroelectric station underground power house is chosen for the case study. *In situ* geophysical and geotechnical tests of surrounding rock on the construction phase are conducted. Unloading relaxation, damage fracture, and degradation of the mechanical parameters of surrounding rocks during excavation are studied. It provides a direct basis for feedback analysis of displacement monitoring and advanced deformation forecasting of surrounding rocks in underground caverns.

2 METHODS

2.1 *Acoustic velocity test*

Single-borehole acoustic test uses single-emitting with double-receiving energy converters in the borehole. The outside equipment is composed of the pulse signal source and recording device, as shown in Figure 1. Water is used as coupling agent and test point spacing is 20 cm. Rock acoustic velocity V_P is calculated as follows:

$$V_P = \frac{L}{(t_2 - t_1)} \tag{1}$$

where L is the distance between two receiving energy converters and t_2 and t_1 are the arrival times of the second and first waves, respectively.

2.2 *Borehole camera test*

Digital borehole camera is composed of hardware part and image processing system. The hardware part includes winch, panoramic probe, control box, equipment box, camcorder, desktop computers, and other components that receive real-time information from the downhole probe. According to image orientation information, an image processing system expands each image in the order of NESW and then stitches together to a full-borehole expanded-view image, as shown in Figure 1b. The circular accuracy of the digital borehole camera is up to 0.1–0.2 mm.

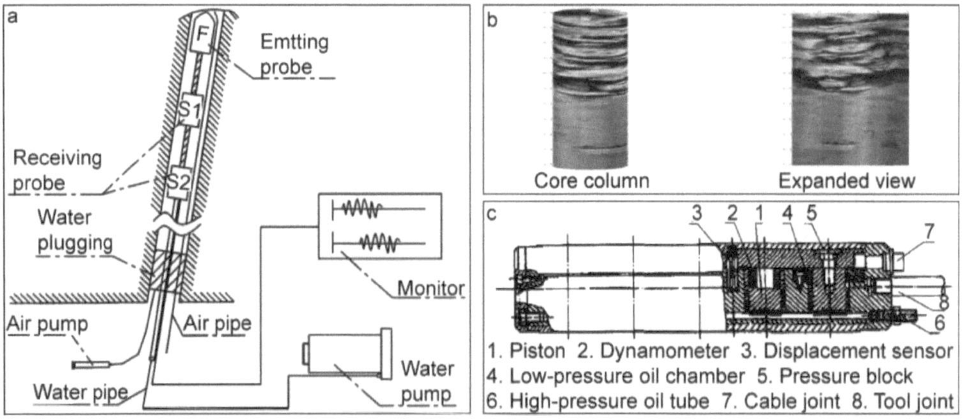

Figure 1. Schematic diagram of single-borehole acoustic test; b. Images of borehole camera; c. Schematic diagram of borehole dilatometer.

2.3 *Borehole elastic modulus tests*

Borehole elastic modulus test uses four internal pistons pushing two rigid bearing plates in borehole dilatometer, which applies symmetrical strip loads on the borehole wall rock, as shown in Figure 1c. The elastic modulus (E) of test site rock is calculated as follows:

$$E = A \times H \times D \times T(\nu, \beta) \times \frac{\Delta Q}{\Delta D} \tag{2}$$

where A is the influence coefficient of two-dimensional calculation formula considering three-dimensional problem; H is the hydraulic correction coefficient of dilatometer; D is the borehole diameter; T(ν, β) is the coefficient determined by rock Poisson ratio and angle of circumference of bearing plate; ΔQ is load increment; and ΔD is deformed increment.

2.4 *Determination of the excavation damaged zone*

At present, the excavation damaged zone of rock has no uniform definition and measuring method. There are two methods to determine the excavation damaged zone of rock in engineering practice:

1. The area of borehole acoustic velocity is significantly reduced, thereby determining the range of the excavation damaged zone.
2. Determining the area of significant change in the physical and geotechnical properties of rock, including rupture, stress redistribution, and elastic modulus.

In this paper, acoustic velocity and borehole camera methods are comprehensively used to determine the excavation damaged zone of rock in underground caverns.

3 CASE STUDY

3.1 *Engineering background*

The hydroelectric station underground caverns used as case study include main power house, transformer chamber, and tailrace surge chamber, which are arranged in parallel and in the axial direction of NE50° with spacing of 41.5 m and 35.5 m. The outline dimensions of the three main caverns are 189 m × 26.7 m × 70.25 m (length × width × height), 143.7 m × 18.0 m × 34.3 m, and 116.50 m × 20.0 m × 65.9 m, as shown in Figure 2. The vertical burial depth is in the range of 129–331 m and the lateral burial depth is 124–154 m. The lithology of the underground cavern is layered or laminated sandstone and slate.

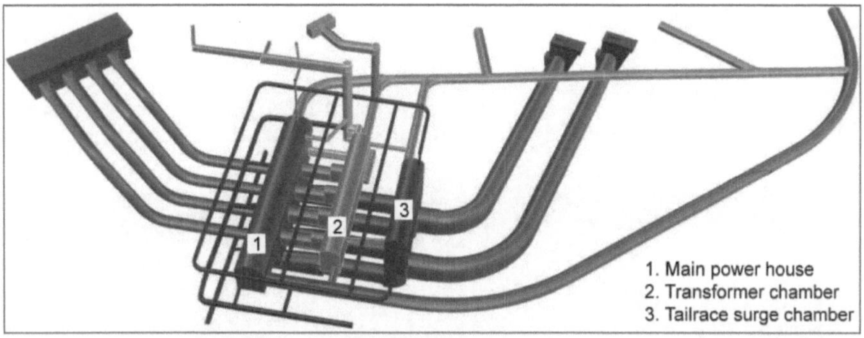

Figure 2. Layout of the hydroelectric station underground caverns.

3.2 *Monitoring scheme*

In order to reflect the relaxation processes and structure revolution of rocks in different lithologies in the process of excavation, boreholes are respectively drilled in sandstone and slate. Along the longitudinal axis of the three underground caverns, four rows of boreholes are arranged. Figure 3 shows the second row of physical boreholes. The diameter of the physical borehole is 76 mm and the depths of the boreholes are in the range of 25–48 m.

3.3 *The range of excavation damaged zones of caverns*

With acoustic velocity as the evaluation criterion for excavation damaged zone, the ranges of excavation damaged zone of the three underground caverns are obtained, as shown in Figures 4 and 5. Table 1 shows the results of geophysical and geotechnical tests of main power house. The range of average acoustic velocity is 4.31–4.74 km/s. The average elastic modulus ranges from 10.38 to 13.2 GPa, while the average deformation modulus ranges from 8.5 to 10.23 GPa. Comparing with other caverns, the main power house has the largest excavation span with 26.7 m. The downstream sidewall of main power has the deepest excavation damaged zone ranging from 1.2 to 4.6 m. The lengths of anchor in the support system of underground caverns are respectively 6 and 9 m, which can effectively limit the deformation of the surrounding rock.

3.4 *Relationships between the results of geophysical and geotechnical tests and depth*

Figure 5 shows the change of borehole acoustic velocity, elastic modulus, deformation modulus, and cracks with depth in the borehole C2-2. The fluctuation of physical properties has good correspondence with borehole cracks. This shows that physical properties in the crack area abruptly decrease. Furthermore, the deterioration of physical properties is higher in large crack.

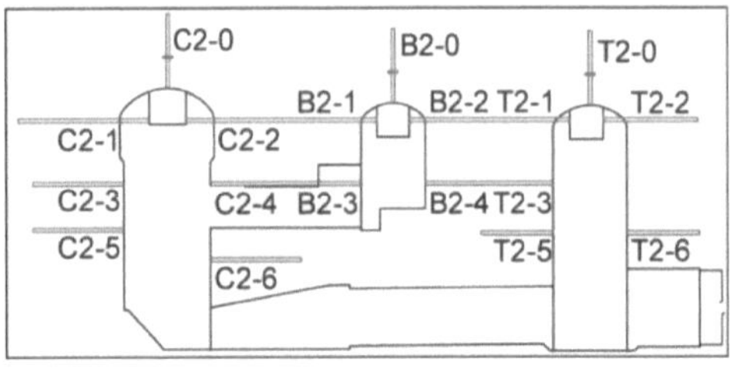

Figure 3. Layout of the physical boreholes.

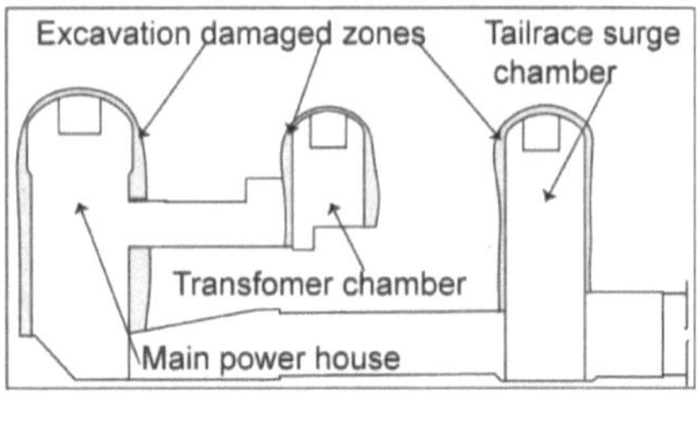

Figure 4. Excavation damaged zones of caverns.

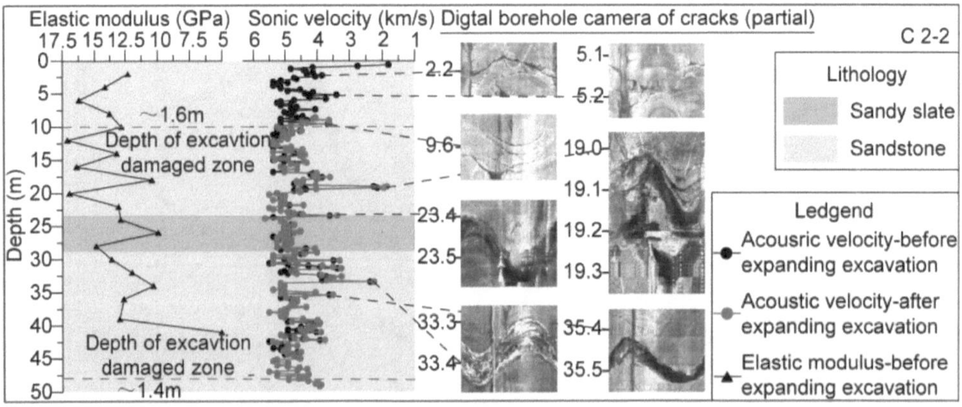

Figure 5. Geophysical test results of borehole C2-2.

Table 1. Geophysical and geotechnical test results of the main power house.

Location	Core No.	Acoustic velocity (km/s)	Elastic modulus (GPa)	Deformation modulus (GPa)	Excavation damaged zone (m)
Main power house	C2-0	/	12.05	9.72	/
	C2-1	4.56	11.23	8.7	1.2
	C2-2	4.74	13.02	8.97	1.6
	C2-3	4.38	11.45	9.37	3.4
	C2-4	4.39	11.9	9.47	4.6
	C2-5	4.67	10.38	8.5	3
	C2-6	4.31	13.2	10.23	1.6

Acoustic velocity, elastic modulus, and deformation are average values.

Table 2. Comparison of acoustic velocity between excavation damaged zone and undisturbed area.

Core No.	Undisturbed area (km/s)				Excavation damaged zone (km/s)				Decrement (km/s)	Attenuation ratio (%)
	Max.	Min.	Ave.	SD	Max.	Min.	Ave.	SD		
C1-2	5.64	2.27	4.81	0.60	5.26	1.92	4.03	0.91	0.78	16.28
C2-1	5.64	1.81	4.57	0.63	4.21	3.73	4.05	0.19	0.52	11.42
C2-2	5.64	2.16	4.78	0.56	5.14	1.81	4.09	0.86	0.69	14.44
C4-1	5.78	2.67	4.48	0.58	4.56	2.16	3.82	0.71	0.66	14.67

Max., maximum value; Min., minimum value; Ave., average value; SD, standard deviation.

3.5 *Comparison of acoustic velocity between excavation damaged zone and undisturbed area*

The results of acoustic velocity between excavation damaged zone and undisturbed area are shown in Table 2. The average acoustic velocity of undisturbed area ranges from 4.21 to 5.14 km/s, whereas the average acoustic velocity of excavation damaged zone ranges from 2.94 to 5.05 km/s. The average attenuation ratio of excavation damaged zone ranges from 11.42 to 16.28%.

4 DISCUSSION

4.1 *Influence of expanding excavation on the physical properties of rock*

Before and after the first-layer expanding excavation, borehole acoustic velocity test and camera test were conducted on four boreholes. The results of comparison of acoustic velocity between before and after first-layer expanding excavation are shown in Table 3. The results of the acoustic velocity test show that the attenuation ratio of acoustic velocity ranges from 1.66% to 2.51% in the impact of the first-layer blasting expanding excavation.

Figure 6 shows the crack evolution processes before and after expanding excavation. A number of irregular slight cracks are developed in the 0.00–0.10 m part of borehole C2-1 located at the upstream sidewall of main power house after first-layer expanding excavation. A similar situation is found in the 0.00–0.10 m and 1.0–1.1 m parts of borehole C4-1 at the upstream sidewall of the main power house. No new cracks are found in the remaining parts of boreholes due to the impact of the first-layer blasting expanding excavation. In general, the effect of the first-layer blast expanding excavation on crack growth and the physical properties of underground cavern rock is not significant.

4.2 *Relationship between acoustic velocity and elastic modulus*

Similarly to acoustic velocity, elastic modulus of rock is influenced by the crack development characteristics. Thus, these two parameters have good correspondence.

Exponential fitting between acoustic velocity and elastic velocity is established using the least squares method. The relationships between acoustic velocity and elastic modulus of sandstone and slate are obtained, as shown in Figure 7.

The results indicate that the relationships between elastic modulus and acoustic wave velocity have low correlation coefficient, in the range of 0.45–0.48. The potential reason is that a number of cracks and fissures exist on the wall of boreholes. On the one hand, when high dip angle cracks exist, the direction of propagation of acoustic wave is approximately parallel to the direction of cracks. The acoustic velocity decreases slightly, while the elastic modulus is more susceptible to the high dip angle cracks. Therefore, high acoustic velocity is achieved with low elastic modulus. On the other hand, when low dip angle cracks exist, the direction of propagation of acoustic wave is approximately perpendicular to the direction of cracks. The acoustic velocity decreases sharply, while the elastic modulus is less susceptible to the low dip angle cracks. Therefore, low acoustic velocity is achieved with high elastic modulus. In general, the direction of crack has a direct effect on the correlation between acoustic velocity and elastic modulus.

Table 3. Comparison of acoustic velocity between before and after first-layer expanding excavation.

Core No.	Before expanding excavation (km/s)				After expanding excavation (km/s)				Decrement (km/s)	Attenuation ratio (%)
	Max.	Min.	Ave.	SD	Max.	Min.	Ave.	SD		
C1-2	5.64	2.27	4.81	0.58	5.64	1.98	4.71	0.64	0.10	2.02
C2-1	5.64	1.81	4.59	0.65	5.44	1.71	4.48	0.64	0.11	2.37
C2-2	5.64	2.16	4.79	0.57	5.64	1.87	4.67	0.62	0.12	2.51
C4-1	5.78	2.67	4.48	0.60	5.78	2.63	4.40	0.58	0.08	1.66

Max., maximum value; Min., minimum value; Ave., average value; SD, standard deviation.

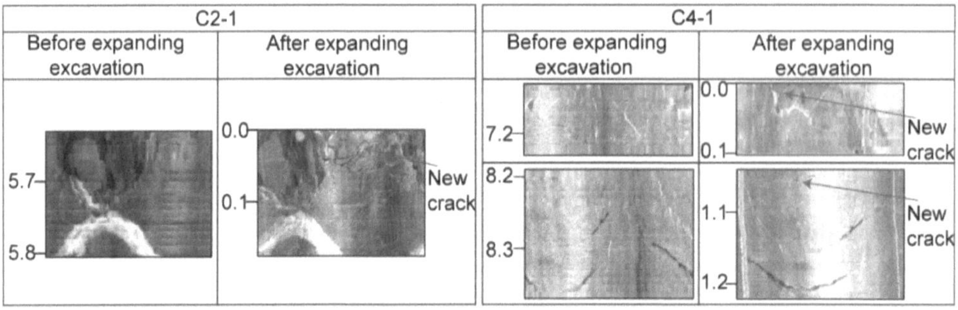

Figure 6. Borehole camera images of crack evolution process before and after expanding excavation.

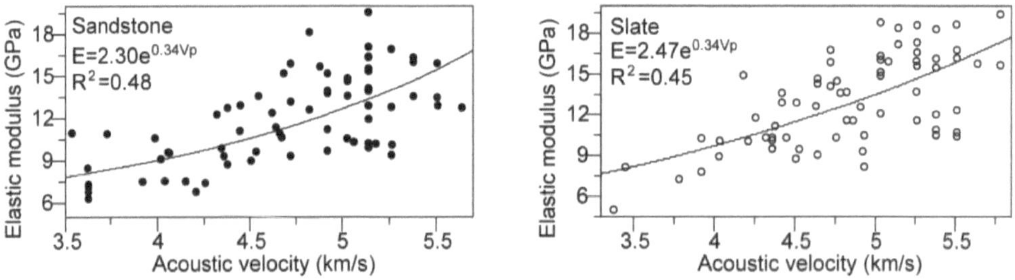

Figure 7. Relationships between acoustic velocity and elastic modulus of sandstone and slate.

5 CONCLUSIONS

In summary, integrating acoustic velocity test, borehole camera test, and borehole elastic modulus test have been used to study the unloading relaxation, damage fracture, and degradation of the mechanical parameters of surrounding rocks in underground caverns. The following conclusions can be drawn:

- The fluctuation of physical properties has good correspondence with borehole cracks. Physical properties abruptly deteriorate in the crack region.
- The damaged zone of underground caverns ranges from 1.2 to 4.6 m. The anchor support system can effectively limit the deformation of the surrounding rock.
- The average attenuation ratio of acoustic velocity in excavation damaged zone ranges from 11.42% to 16.28%.
- The effect of the first-layer blast expanding excavation on the crack growth and physical properties of underground cavern rock is not significant.
- The direction of crack has a direct effect on the correlation between acoustic velocity and elastic modulus.

ACKNOWLEDGMENTS

The authors gratefully acknowledge the support by the National Key Research and Development Project of China (Grant No. 2016YFC0401802), the State Key Program of National Natural Science of China (Grant No. 51539002), and the National Natural Science Foundation of China (Nos. 51209198 and 51204158).

REFERENCES

Anderson, J.C., C.D. Martin. 2009. The Aspo pillar stability experiment: Part I-Experiment design. *International Journal of Rock Mechanics and Mining Sciences* 46(5): 865–878.

Dai et al. 2015. Characteristics of damaged zones due to excavation in deep underground powerhouse at Houziyan hydropower station. *Chinese Journal of Rock Mechanics and Engineering* 34(4): 735–736.

Jianhai Zhang et al. 2011. Acoustic velocity fitting and monitoring feedback analysis of surrounding rock loosing zone in underground powerhouse. *Chinese Journal of Rock Mechanics and Engineering* 30(6): 1191–1197.

Ning Liu, Chunsheng Zhang, Weijiang Zhu. 2011. Detection and analysis of excavation damage zone of deep tunnel. *Rock and Soil Mechanics* 32(Suppl.2): 526–532.

Peng Yan et al. 2009. Study of the damage characteristics of surrounding rocks for tunnels constructed using TBM and drill-and-blast. *China Civil Engineering Journal* 42(11): 121–128.

Shaojun Li et al. 2010. Testing on formation and evolution of TBM excavation damaged zone in deep-buried tunnel based on digital panoramic borehole camera technique. *Chinese Journal of Rock Mechanics and Engineering* 29(6): 1106–1112.

Applied Mechanics and Civil Engineering VI – Xie (Ed.)
 ISBN 978-1-138-62631-7

Screening of *Lactobacillus rhamnosus* capable of yielding high levels of Conjugated Linoleic Acid by UV mutagenesis

Li Fang, Ming-zhu Zheng, Zhi-rui Zhang, Peng Wang, Zu-hao Wang & Wei-hong Min
College of Food Science and Engineering, Jilin Agricultural University, Changchun, China

ABSTRACT: This study, we took the strain A53.2 of *Lactobacillus rhamnosus* that was capable of yielding Conjugated Linoleic Acid (CLA) for UV mutagenesis experiment to obtain high-yielding CLA strain. The result showed that the optimum concentration gradient of mutagenesis was 10^{-5}, the optimum mutagenesis time was 40 s; and through three rounds of screening by UV mutagenesis, a high-yielding mutated strain UV3-4 was obtained, whose highest yield was 31.57 μg/mL, and which showed a good hereditary stability through five passages.

1 INTRODUCTION

Conjugated Linoleic Acid (CLA) is the generic terms of a group of positional and conformational isomers of the essential amino-acid, linoleic acid. The isomers cis-9, trans-11CLA and trans-10, cis-12CLA are have a strong physiological activity (Pariza et al., 2001). Recently, CLA, a functional fatty acid, has gained popularity due to its physiological functions such as e anti-cancer (Agatha et al., 2004; Cho et al., 2006; Choi et al., 2002), anti-atherosclerosis (Eder et al., 2010), and weight loss (Gaullier et al., 2007). CLA has already been used in clinical application (McGowan et al., 2013; Bhattacharya et al., 2006; Yang et al., 2015), healthcare, food addictives, and so on. However several studies showed that the content of CLA isomer having a physiological activity in the milk and flesh of rumen animal was extremely low, and it has not been found yet in the product of natural plant (Adolf et al., 1998). Thus, mass production of CLA having physiological functions has become the focus of scientists worldwide.

At present, the CLA is produced by either chemical method or biosynthesis method. The product synthesized by chemical method is a mixture of multiple isomers, and it is very difficult to separate the active isomer. The isomer produced by biosynthesis method has a simple composition, which is similar to that of the CLA isomer in the natural food, thus, it draws the attention of researchers. Studies showed that most *Lactobacillus* species could convert LA into CLA (Ogawa et al., 2005; Hosseini et al., 2015), and the linoleate isomerase contained in it had a strong selectivity to the substrate, therefore, the production of CLA based on *Lactobacillus* has a good research prospect.

This experiment used the *Lactobacillus rhamnosus* A53.2 capable of yielding conjugated linoleic acid by UV mutagenesis. Through the determination of optimum mutagenesis concentration gradient, optimum mutagenesis time and three successive UV mutagenesis and other experiments, the goal of obtaining high-yielding CLA strain has been achieved.

2 MATERIALS AND METHODS

2.1 *Materials*

Lactobacillus rhamnosus A53.2 was obtained by screening. Linoleic Acid (LA, purity ≥ 90%) was purchased from Aladdin Co., Ltd. Conjugated linoleic acid (purity ≥ 99%) was purchased from Sigma-Aldrich Co., St. Louis. Biochemical reagents include beef extract, peptone, yeast extract, and agar. Normal hexane, anhydrous sodium sulfate, maltose, glucose, dipotassium phosphate, sodium acetate, ammonium citrate, magnesium sulfate, Tween 80, and anhydrous sodium sulfate are analytically pure.

Basal culture medium: beef extract 1 g, peptone 1 g, yeast extract 0.5 g, glucose 2 g, ammonium citrate diammonium hydrogen 0.2 g, sodium acetate 0.5 g, dipotassium phosphate 0.2 g, magnesium sulfate 0.058 g, manganese sulfate 0.025 g, Tween 80 0.1 mL, distilled water was added till 100 mL, pH was adjusted to be 6.2–6.4, temperature 121°C, and sterilization 20 min. A solid medium was prepared by adding 1.5–2.0 g of agar.

Fermentation medium: beef extract 1 g, peptone 1 g, yeast extract 0.5 g, glucose 2 g, ammonium citrate diammonium hydrogen 0.2 g, sodium acetate 0.5 g, dipotassium phosphate 0.2 g, magnesium sulfate 0.058 g, manganese sulfate 0.025 g, Tween 80 0.1 mL, distilled water was added until 100 mL, pH was adjusted to be 6.2–6.4, temperature 121°C, and sterilization for 20 min. LA concentration was 0.4 mg/mL.

2.2 *Growth curve drawing*

A number of test tubes (18 mm × 180 mm) were used; the prepared culture medium was put into the test tubes accordingly with a loaded liquid volume of 10 ml. An inoculation was conducted with an inoculum size of 2%, then a static cultivation was conducted under 37°C. A sampling was made every 2 h to measure the OD value under 600 nm. The empty fluid medium was taken as a control, the growth curve of strain was drawn with the time of cultivation of thallus as horizontal coordinates and OD600 value as vertical coordinates.

2.3 *Determination of condition of UV mutagenesis*

The temperature of 37°C was used to cultivate the bacterial suspension to logarithmic phase, which was diluted by sterile saline solution, then successively diluted at a ratio of 10^{-1} gradient until the dilution ratio was 10^{-7}, and 100 μL of bacterial suspension of each dilution was drawn to respectively smear in the sterilized basal solid medium panels, uniformly smeared. Preheated by UV lamp for 20 min, the ultraviolet radiation distance was fixed at 20 cm (the following UV-induced mutations were all radiated with this distance); After 30 s of radiation, two parallel samples were set for each gradient, which received a light-free culture for 48 h, a change in clump count was observed, and the fatality rate was calculated to determine the concentration gradient of UV mutagenesis of optimum fatality rate. All the above experiments used culture panel free from ultraviolet radiation as control.

2.4 *Determination of optimum radiation time of UV mutagenesis*

Logarithmic phase bacterial suspension (100 μL) with an optimum dilution was taken and uniformly smeared in the differential medium; the ultraviolet radiation lasted respectively for 0, 10, 20, 30, 40, 50, and 60 s. Two parallels were set for each gradient, which were cultivated for 48 h under 37°C and dark condition, the post-inducement clump count had been observed and calculated, a mean value was taken, and the UV mutagenesis time with a optimum fatality rate was determined. All the above experiments used the culture panel free from ultraviolet radiation as control.

2.5 *Three rounds of experiment of UV mutagenesis and strain hereditary stability*

Logarithmic phase bacterial suspension (100 μL) with an optimum dilution was drawn uniformly smeared on the basal solid medium, which was radiated by the ultraviolet source for an optimum radiation time, then cultivated in a dark box for 48 h under 37°C. A fermentation experiment was conducted in the strain obtained from UV mutagenesis to measure the yield of CLA. The mutant strain with a higher yield was picked out, and then underwent five serial passages and fermented to measure its hereditary stability. In the above experiments, three parallel samples were used for each group.

2.6 *Production of standard curve of CLA*

CLA standard substance (2.5 mg) was taken for constant volume in a 25 ml volumetric flask with the normal hexane as a solvent, a 1 mg/mL CLA solution was obtained from uniform dissolution. The solution was drawn respectively by 0.05, 0.10, 0.15, 0.20, 0.25, 0.30, 0.35, 0.40, 0.45 and 0.50 mL and put into the volumetric flask, in which the solution had a constant volume of 10 mL. Using normal hexane, CLA standard samples of different concentration had been obtained, whose absorption value at wavelength 233 nm was measured with normal hexane as a reference. The standard curve of CLA ultraviolet absorption was drawn with CLA concentration as horizontal coordinates and light absorption value as vertical coordinates.

2.7 *Measurement of CLA yield*

The activated bacterial strain was inoculated at an inoculum size of 2% into the test tube loaded with medium beforehand, the substrate LA was added, the substrate concentration was made to reach 0.4 mg/mL, static cultivation was conducted for 24 h under 37°C.

After the fermentation was finished, the fermentation broth was transferred into 60 mL separating funnels, each of which was added with 20 mL of normal hexane, followed by a vibrational extraction for 5–10 min and static standing for 20 min. After a sufficient extraction, the aqueous-phase liquid at the lower layer was discharged after delamination, 20 mL of distilled water was added for vibrational washing, followed by a static standing for 10 min. Then, the aqueous-phase liquid at the lower layer was discharged, followed by two times of washing with water. The extract liquor was transferred into a 50 mL small beaker after washing, and a specific amount of anhydrous sodium sulfate was added for dehydration until the emulsion layer disappeared. The dehydrated normal hexane extract liquor was transferred into a 25 mL volumetric flask, where the normal hexane was added to achieve a constant volume of 25 mL, followed by a constant stirring. The normal hexane extract liquor obtained from the same steps and under the same conditions with the exception of inoculation of bacterial strain was taken as a control. Ultraviolet spectrophotometer was used to measure the light absorption value of CLA at a wavelength of 233 nm, and the CLA yield was determined by standard curve equation.

3 RESULTS AND DISCUSSION

3.1 *Growth curve of Lactobacillus rhamnosus*

Figure 1 shows the growth curve of strain A53.2. This strain from seventh hour entered the logarithmic growth phase, the 16th hour entered the stationary phase, and the 20th hour entered the lag phase.

3.2 *Standard curve of CLA*

The CLA standard curve was drawn as shown in Figure 2, whose linear regression equation was: $y = 0.1165x + 0.0032$, where y represents the absorbance of CLA standard sample at 233 nm, x represents the mass concentration of CLA standard sample (μg/mL), and the degree of fitting is $R^2 = 0.9986$.

3.3 *Determination of concentration gradient of optimum ultraviolet induced mutation*

All of panel radiated by ultraviolet grew into a bacterial lawn-like shape when the dilution degree of bacterial solution was 10^{-1}~10^{-2}, which made it impossible to calculate the fatality rate. The clump count when the concentration gradient was 10^{-3}–10^{-7} is shown in Table 1, indicating that, it went against the screening of bacterial strain when the dilution was too high or too low. A large number of studies showed that, the possibility of forward mutation

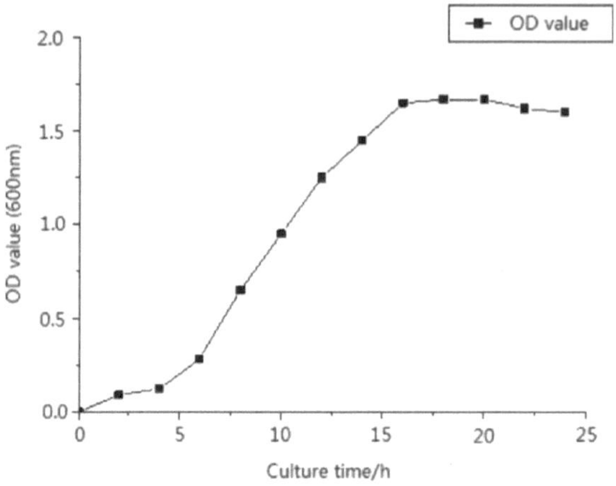

Figure 1. Growth curve of A53.2.

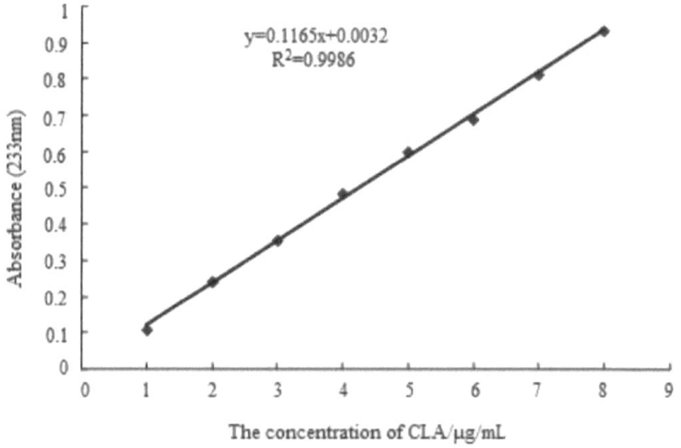

Figure 2. Standard curve of conjugated linoleic acid.

Table 1. Determination of change in the germ liquid dilution degree for UV mutagenesis.

Dilution degree	10^{-1}	10^{-2}	10^{-3}	10^{-4}	10^{-5}	10^{-6}	10^{-7}
0 s	Bacterial lawn	Bacterial lawn	1125	547	206	7	4
30 s	Bacterial lawn	Bacterial lawn	619	173	44	1	0
Mortality rate			45%	68%	79%	86%	100%

was maximum when the mortality rate of microorganism was between 70% and 80%. Thus, we selected 10^{-5} as the optimum concentration gradient.

3.4 *Determination of optimum UV mutagenesis time*

The panels underwent a treatment of different radiation duration in the context of fixed radiation distance and same power of UV lamp, followed by an observation of clump count of the panels and calculation of fatality rate. As shown in Table 2, the fatality rate increased with the extension of mutagenesis time; when the mutagenesis time was 40 s, the mutant strain had a mortality rate of 78%, and thus 40 s was selected as the optimum mutagenesis time.

3.5 *Result of three rounds of UV mutagenesis and the experiment of strain hereditary stability*

Through the first round of UV radiation treatment, the original strain A53.2 was measured for each mutant strain yield after initial screening and secondary screening and fermentation, as shown in Table 3, where the mutant strain UV1-16 had the highest yield, which was 20.57 μg/mL.

The mutant strain UV1-25 that had the highest yield in the first UV mutagenesis was taken as the original strain in the second round: the yield measured by fermentation after mutagenesis is shown in Table 4, where the mutant strain UV2-9 had the highest yield, that is, 26.51 μg/mL.

The mutant strain UV2-9 that had the highest yield in the second UV mutagenesis was taken as the original strain in the third round: the yield measured by fermentation after mutagenesis is shown in Table 5, where the mutant strain UV3-4 had the highest yield, that is, 31.47 μg/mL.

After the third UV treatment, the mutant strain UV3-4 had the highest yield of CLA, which was selected for the fifth UV treatment. As shown in Table 6, the CLA yield was measured by fermentation as about 31.5 μg/mL, which was stable.

Table 2. Determination of UV mutagenesis time.

Mutagenesis time	0 s	10 s	30 s	40 s	50 s	60 s
Number of colonies	258	169	101	57	36	19
Mortality rate		35%	60%	78%	86%	92%

Table 3. Quantum of CLA after fermentation of the first UV mutagenesis of mutation strains.

Strain	A53.2	UV1-3	UV1-7	UV1-14	UV1-16	UV1-25	UV1-26
CLA (μg/mL)	14.65	18.59	16.66	14.86	20.12	17.71	14.21
	13.86	18.16	16.31	14.73	20.65	17.26	14.68
	15.11	18.76	15.79	14.75	20.95	17.54	14.98
Average value	14.54	18.50	16.25	14.78	20.57	17.51	14.63

Table 4. Quantum of CLA after fermentation of the second UV mutagenesis of mutation strains.

Strain	UV1-16	UV2-1	UV2-2	UV2-9	UV2-13	UV2-17
CLA (μg/mL)	20.83	22.79	22.33	26.26	25.12	23.71
	20.36	22.13	22.31	26.73	25.65	23.26
	20.41	22.06	21.79	27.55	25.95	23.54
Average value	20.53	22.33	22.25	26.51	25.57	23.50

Table 5. Quantum of CLA after fermentation of the third UV mutagenesis of mutation strains.

Strain	UV2-9	UV3-2	UV3-4	UV3-5	UV3-11	UV3-12	UV3-16
CLA (μg/mL)	26.52	29.19	31.46	26.75	27.12	30.91	28.81
	26.61	29.34	31.36	26.43	26.95	31.26	29.08
	26.47	29.09	31.60	26.72	26..94	31.14	28.94
Average value	26.61	29.21	31.47	26.63	27.01	31.10	28.94

Table 6. Quantum of CLA after fermentation of going down to five posterity alternation of strain UV3-4.

Times	1	2	3	4	5
CLA (μg/mL)	31.33	31.25	31.54	31.47	31.48
	31.66	31.69	31.56	31.65	31.65
	31.41	31.38	31.62	31.62	31.52
Average value	31.46	31.44	31.57	31.47	31.55

4 CONCLUSION

This study was aimed at improving the CLA yield and studying the UV mutagenesis of *Lactobacillus rhamnosus* A53.2. From the experiment, the optimum mutagenesis concentration gradient was determined as 10^{-5} and the optimum mutagenesis time was 40 s. The mutant strain UV3-4 with a yield of 31.57 μg/mL was obtained from three rounds of serial UV mutagenesis screening, with its yield nearly doubled. Through five serial passages, CLA yield was stable, presenting a good hereditary stability.

REFERENCES

Adolf. R. 1998. Fractionation of cis-oleic and trans-oleic, linoleic and conjugated linoleic fatty acid methyl esters by silver ion high performance liquid chromatography. J. Chromator A. 799: 329–332.

Agatha, G., A. Voigt, E. Kauf, and F. Zintl. 2004. Conjugated linoleic acid modulation of cell membrane in leukemia cells. Cancer Lett. 209: 87–103.

Bhattacharya, A., Banu, J., Rahman, M., Causey, J., Fernandes, G., 2006. Biological effects of conjugated linoleic acids in health and disease. J. Nutr. Biochem. 17: 789–810.

Choi, Y., Y. Park, J. M. Storkson, M. W. Pariza, and J. M. Ntambi. 2002. Inhibition of stearoyl-CoA desaturase activity by the cis-9, trans-11 isomer and the trans-10, cis-12 isomer of conjugated linoleic acid in MDA-MB-231 and MCF-7 human breast cancer cells. Biochem. Biophys. Res. Commun. 294: 785–790.

Cho, H. J., Kim, E. J., Lim, S. S., Kim, M. K., Sung, M. K., Kim, J. S., & Park, J. H. Y. 2006. Trans-10, cis-12, not cis-9, trans-11, conjugated linoleic acid inhibits G1-S progression in HT-29 human colon cancer cells. Journal of Nutrition, 136(4), 893–898.

Gaullier, J. M., Halse, J., Hoivik, H. O., Hoye, K., Syvertsen, C., Nurminiemi, M., Hassfeld, C., Einerhand, A., O'Shea, M., & Gudmundsen, O. 2007. Six months supplementation with conjugated linoleic acid induces regional-specific fat mass decreases in overweight and obese. British Journal of Nutrition. 97(03), 550–560.

Hosseini, E., Kermanshahi, R., Hosseinkhani, S., Shojaosadati, S., & Nazari, M. 2015. Conjugated linoleic acid production from various substrates by probiotic Lactobacillus. Annals of Microbiology. 65, 27–32.

McGowan, M., Eisenberg, B., Lewis, L., Froehlich, H., Wells, W., Eastman, A., Kuemmerle, N., Rosenkrantz, K., Barth, R., Jr., Schwartz, G., Li, Z., Tosteson, T., Beaulieu, B., Jr., & Kinlaw, W. 2013. A proof of principle clinical trial to determine whether conjugated linoleic acid modulates lipogenic pathway in human breast cancer tissue. Breast Cancer Research and Treatment, 138(1), 175–183.

Ogawa, J., Kishino, S., Ando, A., Sugimoto, S., Mihara, K., & Shimizu, S. 2005. Production of conjugated fatty acids by lactic acid bacteria. Journal of Bioscience and Bioengineering. 100(4), 355–364.

Pariza, M. W., Park, Y., & Cook, M. E. 2001. The biologically active isomers of conjugated linoleic acid. Progress in Lipid Research, 40(4), 283–298.

Applied Mechanics and Civil Engineering VI – Xie (Ed.)
 ISBN 978-1-138-62631-7

A method for intelligent safety monitoring and hazard warning of highway dangerous rock

Zongling Yan
China Merchants Chongqing Communications Research and Design Institute Co, Ltd., Chongqing, China

Hejun Chai
National Local Joint Engineering Laboratory for Road Engineering and Disaster Prevention and Reduction Technology in Mountainous Areas, Chongqing, China

ABSTRACT: Because of special topographical and geological conditions, numerous dangerous rocks result from the construction of mountainous highways. Rockfall and collapse hazard caused by dangerous rock have a decisive effect on highway traffic safety and operation efficiency. Conducting safety monitoring, predicting hazard, reducing rockfall and collapse disaster, and improving highway traffic safety are becoming extremely difficult problems. Highway dangerous rock hazard has devastating characteristics such as large quantity, small scale, small deformation, sudden occurrence, and high hazard-forming ability. The safety monitoring of dangerous rock should be cheap, durable, and highly precise, and accurate. A systematic safety monitoring system involving a non-contact surface displacement monitoring method based on the laser ranging method, the crack and macroscopic deformation monitoring method, and the dislocation of dangerous rock monitoring method based on the Radio Frequency Identification (RFID) technique is put forward to meet the safety monitoring requirements of highway dangerous rock. A sound and light alarm device is used to alert vehicles about the rockfall and collapse disaster, in order to avoid secondary accidents. The proposed method for safety monitoring and hazard warning of dangerous rock provides technical support for decision making in the reinforcement treatment of dangerous rock and the early warning of rockfall and collapse hazard.

1 INTRODUCTION

Inevitable excavation of mountains for highway construction leads to a series of high slopes. Some rock mass containing fractures and structure plane distribution on steep slopes or cliffs will change from the stable state to the ultimate equilibrium state or failure due to external agents such as gravity, rainfall, underground water, and earthquake, which will cause rockfall and collapse hazard. Some rock mass will have the main requirement condition of instability or forerunning phenomena of rockfall or collapse such as deformation and cracks. These rock masses are called dangerous rock, irrespective of their state of stability.

The safety of dangerous rock has a decisive effect on the efficiency of highway operation. The rock mass and protective structures are aging and under accumulated damage because of the complicated mountainous hydrogeology, exploration of long-term field environment, and repeated loading or vibration of vehicles; thus, rockfall or collapse hazard is inevitable. For example, the collapse of Taiwan Formosa express in Xizhi, Keelung section, on 25 April 2010, causing the burial of two-way tines and complete traffic disruption, was due to the breakage of Dapu Bridge.

Rockfall and collapse of dangerous rock in mountainous highways can destroy pavements, block traffic, or cause casualty.

Safety monitoring and hazard warning of dangerous rock can avoid or prevent rockfall or serious collapse efficiently and decrease casualty and economic loss. Hence, it is important for the construction and operation of mountainous highways.

2 CHARACTERISTICS OF SAFETY MONITORING OF DANGEROUS ROCK

2.1 *Features of highway dangerous rock hazard*

Highway rockfall and collapse hazard has the following features in time and space distribution, whose mechanism of occurrence is different from landslide and debris flow.

1. Wide range and large quantity
Highway is a linearly stretched construction, which ranges from 20–30 km to over 100 km. Massive dangerous rocks distribute randomly along the highway due to geological and terrain conditions.

2. Small scale
The scale of individual highway dangerous rock is small, just dozens or even several stere.

3. Sudden occurrence
Rockfall and collapse hazard is suddenly occurred compared with landslide, and precursory is not obvious, which leads to difficult for hazard warning.

4. Small deformation
The displacement of dangerous rock from stabilization to ultimate equilibrium or unstable failure is one or two orders of magnitude smaller than landslide.

5. High hazard-forming ability
Rockfall and collapse hazard caused by dangerous rock is suddenly occurred and difficult to monitor and alert. Furthermore, it is very difficult for vehicles to prevent collision in narrow space of highway channel when the hazard occurs due to the large traffic volume and high speed of vehicles.

2.2 *Requirements of safety monitoring of highway dangerous rocks*

According to the requirements of highway operation safety, dangerous rock, and the characteristics of its hazard, the following are the demands of safety monitoring of dangerous rock:

1. High precision and accuracy
The requirements of precision and accuracy of the monitoring device is higher than landslide safety monitoring, because the precursory of sudden occurred rockfall and collapse hazard is not obvious, and the deformation is very slight.

2. Low cost and high durability
The design service life of highway is 50 years, so it requires that the safety monitoring device has the characteristics of high durability, low maintenance cost, and easy replacement. Furthermore, the expensive monitoring device is not suitable for application due to wide range and large quantity and small scale of highway dangerous rock.

3. All-weather working
The safety monitoring system of dangerous rock must have all-weather continuous working. Night monitoring ability is quite important because the drivers cannot find the rockfall on the highway easily due to poor vision at high speed.

4. Real-time performance
The alarm information must be given to drivers immediately by the monitoring system of dangerous rock during rockfall and collapse, in order to avoid casualties.

3 INTELLIGENT SAFETY MONITORING OF HIGHWAY DANGEROUS ROCK

Both the monitoring techniques and device of slide slope can be used in the monitoring of dangerous rock, such as geodetic method, instrument measure method, measurement robot technique, GPS monitoring technique, and some other new monitoring techniques, including optical fiber monitoring technique and three-dimensional laser scanning technique. Some proprietary techniques are developed based the failure characteristics of rock mass applied in the monitoring of dangerous rock successfully, such as time-domain reflectometry, acoustic emission technique.

The direct monitoring method of dangerous rock is displacement monitoring, because the displacement of dangerous rock is the most perceptual intuition for the safety management of highway operation. Displacement is the most direct forewarning of rockfall and collapse. Therefore, safety monitoring of highway dangerous rock should be focused on displacement monitoring.

3.1 *Non-contact surface displacement monitoring method based on laser ranging method*

Dangerous rock is often located in cut slope ranging from several meters to several tens of meters where the monitoring instrument and device are difficult to install. The field installation on dangerous rock can be avoided by adopting a wide range of laser finder in non-contact surface displacement monitoring. Laser range finder is often composed of laser transmitter module, laser receiver module, and power supply. Laser range finder should be installed far away from the rockfall or collapse-affected region. Pulsed laser inside laser range finder emits single-pass laser pulse or a series of laser pulse to dangerous rock. A single laser pulse or laser pulse train is emitted to dangerous rock by laser range finder, and the laser receiver module can capture the laser pulse reflected by the dangerous rock surface. The distance between the laser range finder and the dangerous rock can be calculated through measuring the roundtrip time of laser pulse by counter module of laser range finder, as shown in Figure 1.

The basic framework of laser monitoring system constitutes of phase-shift laser range finder, data collection module, data transmission module, power supply system, and remote control center. The first step is transforming the signal of laser sensor into standard electrical signal. The second step is delivering the data from data collection module to data transmission module. Finally, the data will be transmitted to remote control center by mobile network or satellite communication.

3.2 *Surface crack monitoring of dangerous rock*

Automatic crack gauge and displacement meters can be installed on the crack to monitoring its development on the surface of dangerous rock.

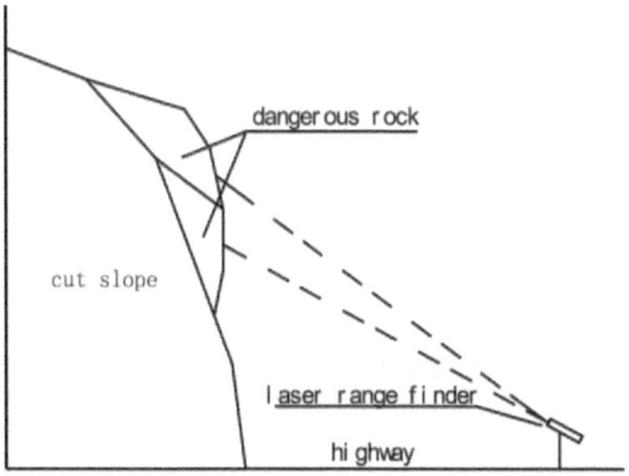

Figure 1. Surface displacement monitoring by laser ranging.

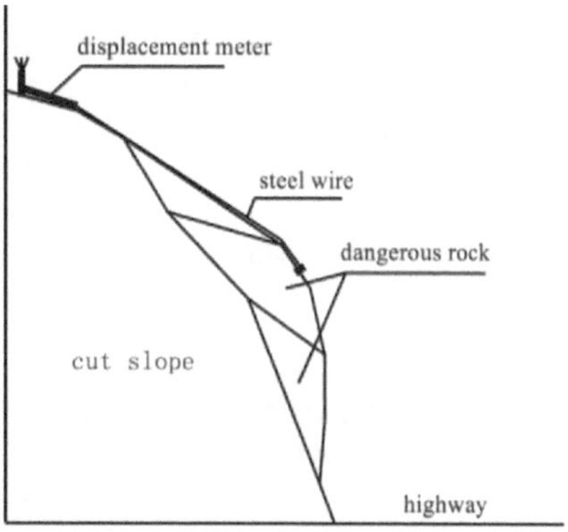

Figure 2. Macroscopic deformation monitoring system.

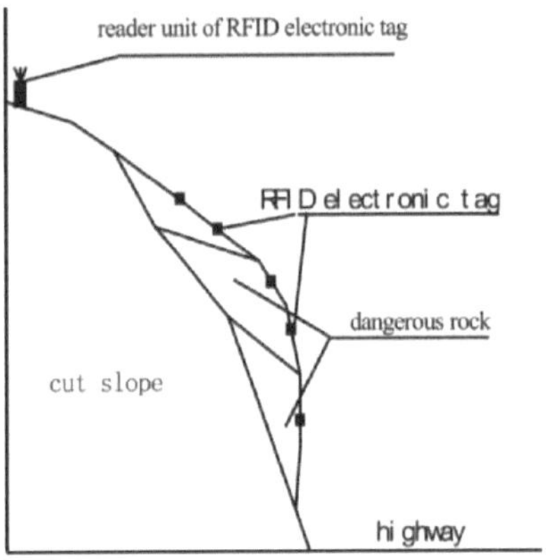

Figure 3. Rockfall and collapse monitoring by RFID.

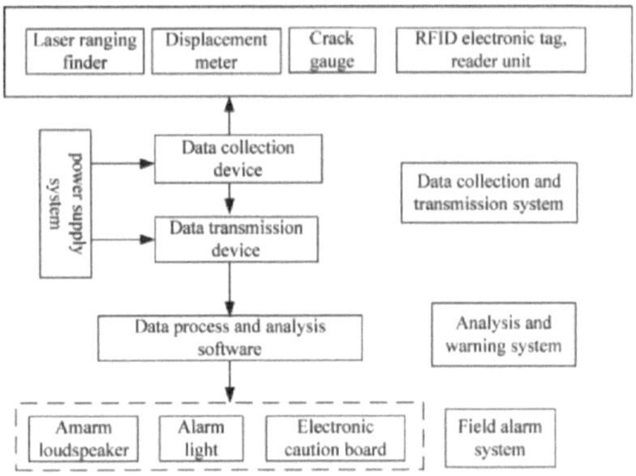

Figure 4. Safety monitoring and hazard alarm system.

3.3 *Macroscopic deformation monitoring of dangerous rock*

Macroscopic deformation of dangerous rock can be monitored for real-time evaluation of the rockfall or collapse hazard process for early hazard warning.

Select monitoring point of the representative regions on dangerous rock, connect the monitoring point and stable parent rock by steel wire, and fix the steel wire on stable parent rocks through displacement meter. Deformation of dangerous rock can be monitored in real time by displacement meter. The alarm device will be triggered while deformation of dangerous rock begins accelerating or reaches certain threshold value. Zero value on the displacement meter indicates that the steel wire has been snapped and dangerous rock has broken away from its original position or stable parent rocks. This indicates the occurrence of rockfall and collapse hazard, and the linkage warning device on the hazard field will be activated automatically.

3.4 *Hazard monitoring of dangerous rock based on the radio frequency identification technique*

RFID, which stands for radio frequency identification technique, represents automatic identification technology of Internet of Things. Complete RFID system is composed of information collection side (RFID electronic tag, reader unit, RF antenna) and Internet application system in the background. RFID can identify different pieces of information through transmitting and receiving radio frequency signal.

Rockfall and collapse hazard can be monitored by the information perception function of RFID through installing RFID electronic tag on the dangerous rock and putting the reader unit on the stable region far away from the dangerous rock.

The dangerous rock with RFID electronic tag will tumble below the slope when rockfall or collapse occurs. The reader unit on the stable rock cannot trace the electronic tag or the reader unit at toe of slope traces the electronic tag, which indicates that rockfall or collapse hazard has occurred. Then, the monitoring system sends out warning automatically, as shown in Figure 4.

4 MONITORING INFORMATION MANAGEMENT AND LOCAL HAZARD WARNING

Monitor data of surface displacement, crack, macroscopic deformation monitoring, and the change of dangerous rock location are transmitted to monitoring information management center by data transmission system.

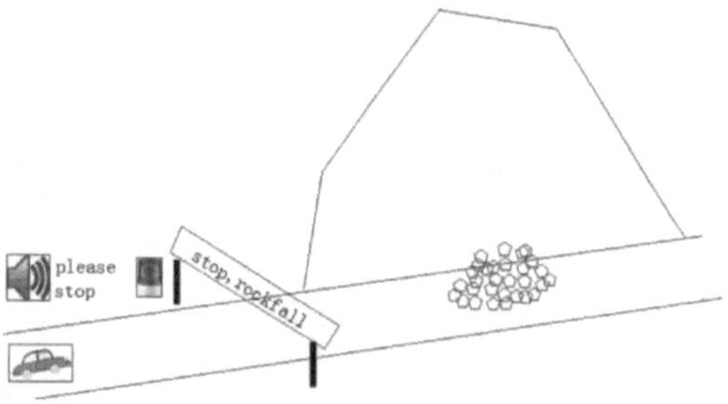

Figure 5. Field alarming devices alarm system for dangerous rock hazard.

Monitor data are analyzed to evaluate the safety state of dangerous rock, and hazard warning is produced according to the results of evaluation. Field hazard warning devices can be activated immediately once hazard warning occurs, as shown in Figure 4.

Field alarm system includes wireless warning information reception module and communicates safety monitoring information with monitoring system. The field alarm system will inform passing vehicles of hazard information by means of caution light, loudspeaker, warning sign, and so on, so that drivers can stop or slow down to avoid traffic accidents.

The field alarm system equipped with wireless warning information reception module is installed at 200–300 m on both sides of dangerous rock. Electronic caution board, caution light, and loudspeaker will be opened on receiving the command sent by monitoring information management center, as shown in Figure 5.

5 CONCLUSIONS

Highway dangerous rock and its hazard have features of wide range, large quantity, small scale, sudden occurrence, small deformation, high hazard-forming ability. Therefore, the characteristics of dangerous rock must consider these features during safety monitoring.

The safety monitoring system of dangerous rock should have high precision and accuracy, low cost, and high durability, all-weather working ability, and real-time performance.

According to the development and monitoring characteristics of highway dangerous rock, a safety monitoring system of dangerous rock and field warning method composed of non-contact surface displacement monitoring, crack monitoring, macroscopic deformation monitoring, rockfall and collapse monitoring is put forward. Furthermore, the safety monitoring information management system and the field alarm system of dangerous rock are established.

ACKNOWLEDGMENTS

The authors gratefully acknowledge the financial supports from the National Science and Technology Support Plan under Grant Nos. 2015BAK09B01 and 2015BAK09B02, the Integrated Demonstration Project of Chongqing under Grant No. cstc2014jcsf30001, the Application Development Project of Chongqing under Grant No. cstc2014yykfb30001.

REFERENCES

Fairley, Mike. RFID smart labels—A challenge for converters [J]. AP Australia Print Magazine, 2005(4).

Li Tianwen, Wu lin, Li Jiaquan. Application in Landslide Monitoring with GPS technique [J]. Journal of Mountain Science, 2004, 22(6): 713–718.

Liu Wenlong, Zhao Xiaoping. Study on the Application of 3D Laser Scanning Technology in Landslide Monitoring [J]. Metal Mine, 2009(2): 131–133.
Liu Yongli. Application research on distributed on optical fiber sensing technology for monitoring of slope engineering [D]. Zhe Jiang University, 2011.
Mei Wensheng. Research on Application in Deformation Monitoring with Georobot. Dam & Safety, 2002(5): 33–35.
Ning Huansheng, Xu Quanyu. Research on Global Internet of Things Developments and It's Lonstruction in China [J]. Acta Electronica Sinica, 2010(11): 31–33.
Shinomiya T, Nakanishi Y, Morishima H, et a1. Damage diagnosis technique for brick structures using acoustic emission [J]. Journal of Acoustic Emission, 2003(20): 145–152.
Wang Yousheng. Realization of Laser Ran e Finding Based on Phase [J]. Journal of Beijing University of Technology, 2003, 29(4): 424–426.
Wei Xuefeng. Study on high precision phase determination technology in handheld laser range finder. National University of Defense Technology [D], 2013.
Xu ShaoQuan, Yang Zhi qiang, Zhang Huahai. GPS Measurement Principle and Application [M]. Wu Han, Wu Han University Publishing House, 1998.
Yan Zongling, Huang He, Li Cong, Chai He Jun. Development and application of intelligent warning and monitoring system in mountainous highway landslide [C]. Academic conference collected papers in 2015 about Chinese academic seminar of International Road Federation and China Highway and Transportation Society, 2015(9): 103–110.
You Zhanqing, Liu Kesheng. Planning and implementation of wireless Radio Frequency Identification technique [M]. Bei Jing: Publishing House of Electronics Industry, 2005: 49–53.
Zhang Qin, Shi Yanxing. The Landslide Monitoring System Based on TDR [J]. Chinese Journal of Scientific Instrument, 2005, 26(11): 1199–1202.

Applied Mechanics and Civil Engineering VI – Xie (Ed.)

Studies on the mechanism of ecological restoration of eutrophic water in a tropical shallow urban lake

Guangrong Chen
GuangDong Construction Polytechnic, Guanghua, GuangZhou, China

ABSTRACT: Sampling and detection of the stable carbon and nitrogen isotopes of submerged plant, periphyton, seston and fish were conducted in the demonstration area and unrestored area of a tropical shallow urban lake. The results indicated that in the demonstration area, the average $\delta^{13}C$ values of the submerged plant, periphyton, seston and tilapia were −20.1‰, −23.0‰, −28.0‰ and −23.8‰, respectively, and their average $\delta^{15}N$ values were 5.9‰, 5.7‰, 5.5‰ and 11.0‰, respectively. By using a $\delta^{13}C$ mass balance model, the contributions of periphyton to the diet of tilapia were estimated to be 80%, which indicated that the submerged plant not only provided shelter but also increased the food source for fish indirectly.

1 INTRODUCTION

In the study of water ecology, food-web structure research has been crucial in the restoration of the lake ecological system and the evaluation of the lake ecosystem response to environmental changes. The stable isotopic technique has become an important means of the ecosystem's food-web structure and the relationship between nutrition and its dynamic change. Studies have shown that there exist certain differences among the carbon stable isotope ratios ($\delta^{13}C$) of different sources of organic matter. The $\delta^{13}C$ of consumers is almost the same as that of the digested and assimilated food, with the difference between changes being mainly in the range of 0–1‰. The nitrogen stable isotope ratio ($\delta^{15}N$) will produce enrichment in the process of animal feed. The $\delta^{15}N$ of consumers is about 3.4 ± 1.1‰ higher than that of the food in the process of the trophic level transfer. Both $\delta^{13}C$ and $\delta^{15}N$, respectively, reflect the food source of biology and the nutrition level position in the ecosystem's food web, and therefore, it is possible to determine the source of natural or man-made carbon and nitrogen by carbon and nitrogen isotope technology, and track the act or process of assimilation and transfer way in the ecosystem's food web.

In the past few decades, the stable isotope technology has been widely applied to the study of freshwater ecosystems. Recently, stable isotope techniques (especially the carbon and nitrogen stable isotopes) have been applied to study the structure and change of the ecosystem's food web and the relationship of the biological nutrient level. However, there are not many studies conducted on freshwater ecosystems in China. Using a stable isotope technique for the ecological restoration of food-web structure in a tropical shallow urban lake, this paper aims at revealing the underlying mechanism of substance flow and energy transfer based on $\delta^{13}C$ and $\delta^{15}N$.

2 MATERIALS AND METHODS

The sample included submerged plant, periphyton, seston (phytoplankton, zooplankton, organic debris, etc.) and fish in the demonstration area and unrestored area of a tropical shallow urban lake. The leaves of the dominant submerged plant were collected and washed with water; the epiphytic algae was gathered from the surface of aquatic higher plants and washed with water. The water samples were filtered by screen mesh to get the seston, using vacuum suction to the pre-burned glass fiber filter (Whatman GF/C). Then, the seston was wrapped in aluminum foil paper and put into a polyethylene sealing bag, which should be taken back to the lab in a sampling box with ice cubes and acidized in time. The fish was captured from the demonstration area and unrestored area and its back muscles were obtained by cryopreservation. The above samples were brought back to the laboratory for drying at 60 degrees to constant weight, and then they were ground into uniform powder using a mortar and placed in a dryer for isotope detection ([a]$\delta^{13}C$ vs. PDB; [b]$\delta^{15}N$ vs. atmospheric nitrogen).

3 RESULTS AND ANALYSIS

3.1 *The $\delta^{13}C$ and $\delta^{15}N$ values of the submerged plant and plankton in the demonstration area and unrestored area*

The dominant species of the submerged plant in the demonstration area were *Vallisneria* sp. and *Myriophyllum spicatum*, whose $\delta^{13}C$ values were between −22.1‰ and −18.9‰ with an average of (−20.1 ± 0.3) ‰. The $\delta^{13}C$ value of *Vallisneria* sp. was (−19.8 ± 0.3) ‰, higher than the average of Myriophyllum which was (–20.6 ± 0.8) ‰. The $\delta^{15}N$ values of the submerged plant were in the range of 4.4‰ to 7.0‰ with an average of (5.9 ± 0.3) ‰. The $\delta^{15}N$ value of *Vallisneria* sp. was (6.1 ± 0.3) ‰, slightly higher than the average of Myriophyllum which was (5.5 ± 0.5) ‰ (see Figure 1). Compared to the submerged plant, the $\delta^{13}C$ values of periphyton and seston were lower, which were −23.0‰ and −28.0‰, respectively, and the $\delta^{15}N$ values of periphyton and seston were 5.7‰ and 5.5‰.

There was no submerged plant in the unrestored area, and hence the detection of the stable carbon and nitrogen isotopes of the submerged plant and periphyton was not possible. Compared to the demonstration area, the $\delta^{13}C$ value of seston was similar, which was (−28.0 ± 0.1) ‰, but the $\delta^{15}N$ value of seston was higher, which was (9.4 ± 0.5) ‰.

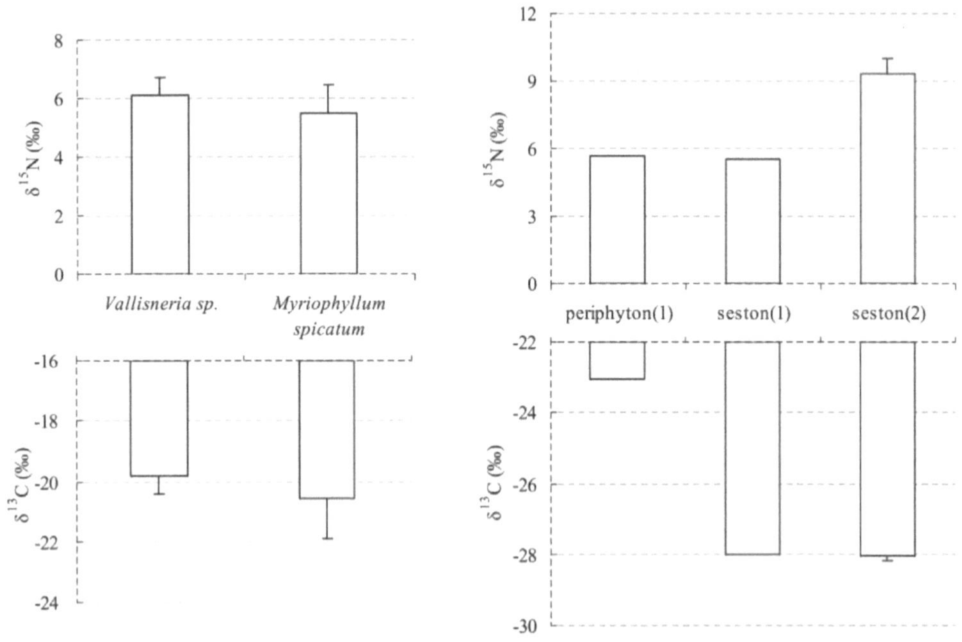

Figure 1. The $\delta^{13}C$ and $\delta^{15}N$ values of submerged macrophytes and plankton in the demonstration area (1) and unrestored area (2).

3.2 *The $\delta^{13}C$ and $\delta^{15}N$ values of fish in the demonstration area and unrestored area*

The dominant species of fish in the demonstration area was tilapia (*Oreochomis* sp.), which is omnivorous and its food sources include phytoplanktons, zooplanktons and organic detritus. The $\delta^{13}C$ values of tilapia were between −21.2‰ and −24.6‰ with an average of (−23.8 ± 0.6) ‰, which is almost equal to the periphyton value but had a difference with the seston value. The $\delta^{15}N$ values of tilapia were in the range of 10.7‰ to 11.4‰ with an average of (11.0 ± 0.1) ‰.

The dominant species of fish in the unrestored area were tilapia (*Oreochomis* sp.), crucian carp (*Carassius auratus*) and bighead (*Aristichthys nobilis*). The average values of $\delta^{13}C$ and $\delta^{15}N$ of tilapia in the unrestored area were (−25.9 ± 0.0) ‰ and (14.3 ± 0.1) ‰, the crucian carp's were −27.8‰ and 15.8‰, and the bighead's were (−27.1 ± 0.1) ‰ and (17.3 ± 0.0) ‰. Compared to the demonstration area, the $\delta^{13}C$ values of fish were lower and almost equal to those of seston, but the $\delta^{15}N$ values of fish were higher (see Figure 2).

3.3 *Analysis of the food-web structure in the demonstration area by using $\delta^{13}C$ and $\delta^{15}N$*

The average value of $\delta^{13}C$ of tilapia in the demonstration area was −23.8‰, which mainly feeds on periphyton (see Figure 3). The average value of $\delta^{13}C$ of fish in the unrestored area was −26.9‰, which mainly feeds on seston (phytoplanktons, zooplanktons, organic debris, etc.).

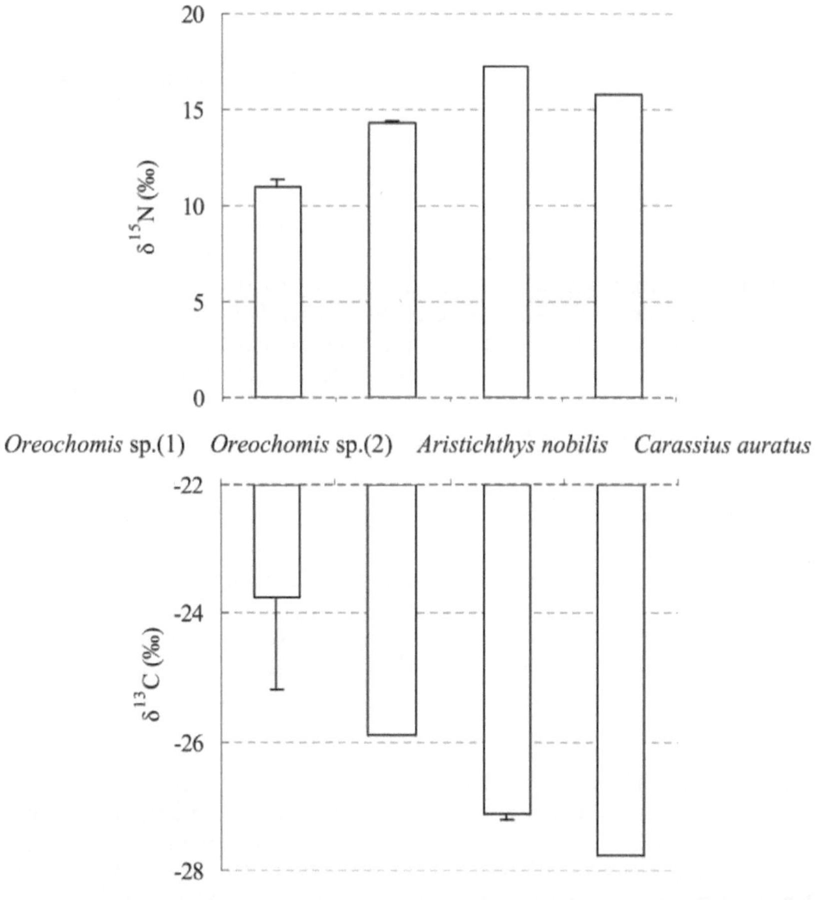

Figure 2. Differences in $\delta^{13}C$ and $\delta^{15}N$ of fish in the demonstration area (1) and unrestored area (2).

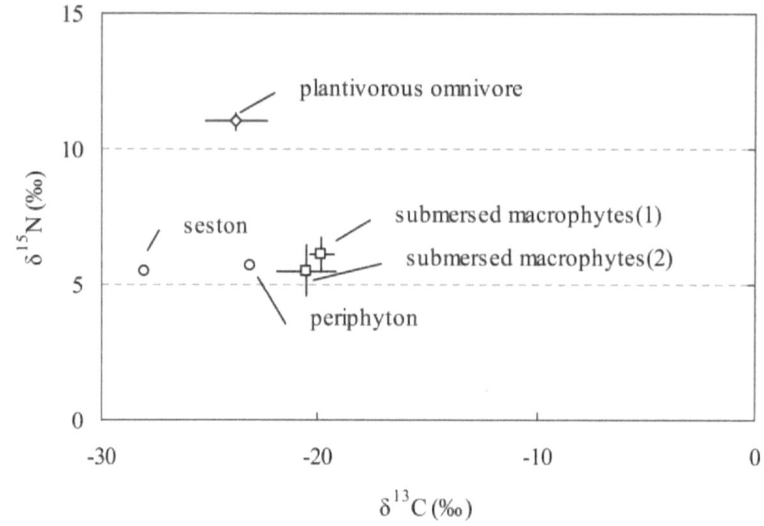

Figure 3. The trophic structure based on the mean stable carbon and nitrogen isotope values (± standard deviation) in the demonstration area. Submerged macrophytes: (1) *Vallisneria* sp. and (2) *Myriophyllum spicatum*.

The values were different because the primary producers were different in the demonstration area and unrestored area. The water was clear in the demonstration area and the high content and biomass of seston were very low, and hence omnivorous tilapia could only feed on periphyton. The abundance and biomass of seston were high in the unrestored area, and therefore filter-feeding fish could mainly feed on seston.

By using the mass balance model, the ratio of periphyton as the diet of tilapia was found to be 80% in the demonstration area, which indicates that the submerged plants not only provided biological habitat and breeding ground, but also provided a biological source of food indirectly. The average values of $\delta^{15}N$ of fish were higher than the average values of seston and periphyton, which are 5.5 and 5.4, respectively, slightly higher than a trophic level range of (3.4 ± 1.1) ‰, which may relate with feeding some zooplanktons, organic debris and so on. The submerged plants were the main primary producers in the demonstration area, which lowers the concentration of dissolved CO_2 and inorganic nitrogen sources, and therefore reducing the values of $\delta^{13}C$ and $\delta^{15}N$ of seston.

4 CONCLUSIONS

In the demonstration area, the $\delta^{13}C$ values had the following order: submerged plant <periphyton> seston and the $\delta^{15}N$ values were similar. Compared to the submerged plant, the $\delta^{13}C$ values of periphyton and seston were lower. The $\delta^{15}N$ values of fish were higher than their food source such as seston or periphyton. The $\delta^{13}C$ values of fish in the demonstration area were higher than that in the unrestored area, but the $\delta^{15}N$ values of fish were lower in the demonstration area.

The average value of $\delta^{13}C$ of tilapia in the demonstration area was −23.8‰, which mainly feed on periphyton. The average value of $\delta^{13}C$ of fish in the unrestored area was −26.9‰, which mainly feed on seston (phytoplanktons, zooplanktons, organic debris, etc.). By using the mass balance model, the ratio of periphyton as the diet of tilapia was estimated to be 80% in the demonstration area. The water was clear in the demonstration area and the abundance and biomass of seston were very low, and therefore zooplanktons fed on phytoplanktons similar to omnivorous tilapia. The submerged plant provided shelter for zooplanktons, so that omnivorous tilapia could only feed on periphyton. This indicated that submerged plants not only provided biological habitat and breeding ground, but also provided a biological source of food indirectly.

ACKNOWLEDGMENTS

This work was supported by the Guangdong Natural Science Foundation (No. 10412741004261).

REFERENCES

Aberle N, Hillebrand H, Grey J, Wiltshire K.H. 2005. Selectivity and competitive interactions between two benthic invertebrate grazers (*A. aquaticus* and *P. antipodarum*)—an experimental study using ^{13}C- and ^{15}N-labelled diatoms [J]. *Freshw Biol*, 50: 369–379.

Chen G. 2012. Study on biomanipulation and its effect in restoration of eutrophic water in tropical shallow urban lake [J]. *Environmental science & management*, (9): 141–143.

Cole M.L, Valiela I, Kroeger K.D, et al. 2004. Assessment of a $\delta^{15}N$ isotopic method to indicate anthropogenic eutrophication in aquatic ecosystems [J]. *J Environ Qual*, 33: 124–132.

Gu B, Schelske C.L, Hodell D.A. 2004. Extreme ^{13}C enrichments in a shallow hypereutrophic lake: Implications for carbon cycling [J]. *Limnol Oceanogr*, 49: 1152–1159.

Jepsen D.R, Winemiller K.O. 2002. Structure of tropical river food webs revealed by stable isotope ratios [J]. *Oikos*, 96: 46–55.

Post D.M. 2002. Using stable isotopes to estimate trophic position: modals, methods, and assumptions [J]. *Ecology*, 83: 703–718.

Vanderklift M.A, Ponsard S. 2003. Sources of variation in consumer-diet $\delta^{15}N$ enrichment: a meta-analysis [J]. *Oecologia*, 136: 169–182.

Woodward G, Hildrew A.G. 2002. Food web structure in riverine landscapes [J]. *Freshw Biol*, 47: 777–798.

Applied Mechanics and Civil Engineering VI – Xie (Ed.)
© 2017 Taylor & Francis Group, London, ISBN 978-1-138-62631-7

Development of a modal identification method using the random decrement technique combined with HHT

Wen Qiao
College of Water Conservancy and Civil Engineering, South China Agricultural University, Guangzhou, Guangdong, China

Guoming Liu
College of Civil Engineering, Fuzhou University, Fuzhou, Fujian, China

Renkun Li
CIECC Engineering Company Limited, Beijing, China

ABSTRACT: In this paper, a method was developed to link the band-pass filtering technique with the time and frequency-domain modal identification method by using the random decrement technique in combination with Hilbert–Huang transform. The proposed method improved the accuracy and reliability of parameter identification by effectively avoiding mixed frequencies that occurred during signal decomposition. The validity of this method was verified by identifying the frequency and damping ratio of the IASC-ASCE structural health monitoring benchmark structure. The results indicated that the identification error can be reduced by combining the RDT with EMD to process the time sequences that resulted from filtering the specific ranges of frequencies of structural earthquake time histories by using the HHT method. Therefore, the proposed method can be applied to other modal identification techniques.

1 INTRODUCTION

The use of vibration signals in structural engineering to identity the modal parameters of a structure is one of the important means for structural health monitoring and damage detection. Therefore, this approach has become one of the many research focuses worldwide (Jianping Han, 2011). The current parameter identification methods include the frequency domain method, the time domain method, the time–frequency domain method and the simulated evolution-based method. Over the recent years, the time–frequency domain method has been the rapidly developed method, and up to over 10 such transform methods have been introduced by research scholars (Yanjun Li, 2010). Of these methods, Hilbert–Huang Transform (HHT) has strong adaptability, which has attracted much attention. This characteristic is attributed to the mode decomposition of HHT that is based on the characteristics of data itself without an *a priori* base.

For the past few years, both the fields of academia and engineering in China and abroad have made extensive studies on the theory and application of structural modal parameter identification by using the HHT method. In China, J. Chen et al explored the application of the HHT method to structures with dense frequency damping identification (Juan Chen, 2003). X. B. Duan et al. (2006) combined the HHT method with state variable analysis to identity the system parameters of nonlinear multi-degree-of-freedom structures. A study by Lihang Kou (2008) discussed the combination of HHT with NEXT (Natural EXcitation Technique) that effectively identified the modal parameters of an arch dam model and improved the accuracy of parameter identification. Another study by Hongping Zhu (2011) indicated that the HHT method identified the variation of the structure's physical parameters by monitoring the vibration performance of the structure for the defection of structure damage. Research in this field started relatively earlier outside China. In one paper (Cheraghi N, 2005), EMD was used for the detection of structural faults. Moreover, a study by Lin Silian (2005) explored the combination of HHT with NEXT for structural health monitoring and obtained satisfactory results.

In this article, a modal parameter identification method that was more suitable for a large-scale structure was developed by combining the random decrement technique with Hilbert–Huang transform. The proposed method was then used to identify the modal frequency and damping ratio of the IASC-ASCE structural health monitoring benchmark structure. Based on the comparative analysis of the identification results obtained with theoretical values and the results reported in S. Lin & J. N. Yang (2005), the newly developed method was proved to be not only valid but also superior to other existing methods.

2 RANDOM DECREMENT TECHNIQUE

Random Decrement Technique (RDT) is a processing method that extracts the free attenuation vibration signal from the random vibration response signal of a structure. The idea is to distinguish the measured vibration response signal that contains both the determinate signal and the random signal by taking advantage of the zero mean value of stationary random vibration signals. Then, the free attenuation vibration signal can be obtained by subtracting the determinate signal.

For a linear system, the forced vibration response of a measure point under any excitation can be expressed as:

$$v(t) = v(0)D(t) + \dot{v}(0)V(t) + \int_0^t h(t-\tau)f(\tau)d\tau \tag{1}$$

where $D(t)$ is the free vibration response with initial displacement at 1 and initial velocity at 0; V(t) is the free vibration response with initial displacement at 0 and initial velocity at 1; $v(0), \dot{v}(0)$ are the initial displacement and the initial velocity, respectively; $h(t)$ is the unit pulse response function of the system; and $h(t)$ is the external excitation.

To truncate a measured random vibration response signal of a structure with an appropriate range, a series of different intersection points $t_i(i=1,2,\cdots,n)$ can be obtained. The response $v(t-t_i)$ starting from t_i can be taken as the linear superposition of the following three components:

1. a free vibration response resulting from the initial displacement since t_i;
2. a free vibration response resulting from the initial velocity since t_i;
3. a forced vibration response resulting from a random excitation $f(t)$ since t_i.

Therefore:

$$v(t-t_i) = v(t_i)D(t-t_i) + \dot{v}(t_i)V(t-t_i) + \int_{t_i}^t h(t-\tau)f(\tau)d\tau \tag{2}$$

Since $f(t)$ is stable, the random characteristics are independent of the starting point of time. Thus, the series of starting points t_i of $v(t-t_i)$ are moved to the origin of coordinates, to obtain a series of subsample functions $x_i(t)(1=1,2,\cdots,n)$ for this random response process:

$$x_i(t) = AD(t) + \dot{v}(0)V(t) + \int_0^t h(t-\tau)f(\tau)d\tau \tag{3}$$

By taking the statistical average of $x_i(t)$:

$$
\begin{aligned}
x(t) &= \frac{1}{n}\sum_{i=1}^{n} x_i(t) \approx E\left[AD(t) + \dot{v}(t_i)V(t) + \int_0^t h(t-\tau)f(\tau)d\tau\right] \\
&= AD(t) + E\left[\dot{v}(t_i)\right]V(t) + \int_0^t h(t-\tau)E\left[f(\tau)\right]d\tau
\end{aligned}
\tag{4}
$$

since excitation $f(t)$ is a stationary random process with a mean value being 0, and the system response is also a zero mean value stationary random vibration, that is $E\left[f(t)\right]=0,\ E\left[\dot{v}(t_i)\right]=0,$ thus:

$$
x(t) \approx AD(t) \tag{5}
$$

A free vibration response with initial displacement at A and initial velocity at 0 is obtained, whose modal parameters can be identified using various time-domain identification methods.

3 A MODAL PARAMETER IDENTIFICATION METHOD WITH THE COMBINATION OF RDT AND HHT

For a given signal sequence, the first n order inherent frequencies $(\omega_1, \omega_2, \cdots, \omega_n)$ can be obtained by preliminarily estimating the general range of the main frequencies based on the auto-power spectrum of the signal. Then, a band-pass filter is applied on $X(t)$ based on the assumption that the range of the first-order inherent frequency is between $\omega_{1M} \sim \omega_{1N}$. The frequency scope of the filter is selected between the range of $\omega_{jM} \sim \omega_{jN}$ to obtain the j-th order modal response. Similarly, n time sequence signals are obtained, denoted by $x_j(t)(j=1,2,3,\cdots,n)$. The cut-off frequency of the filter should be properly selected. The narrower the scope is, the better it is. However, the interest frequency must be ensured to be in the filter frequency range. Empirical mode decomposition is then used for every time sequence $x_j(t)$. The true intrinsic mode function is distinguished according to the correlation coefficient of the intrinsic mode function and the original signal. So far, the complex time sequence has been decomposed into several orders of basic time series from a high frequency to a low frequency (Juan Chen, 2003). Finally, after transforming every intrinsic mode function into the free vibration response using the RDT, the modal frequency and damping ratio of the structure can be identified using Hilbert–Huang transform and the least squares fitting method.

Let us assume that the free attenuation signal is $X_j(t)$ and after applying the random decrement technique on the time sequence $x_j(t)$, it can be expressed as:

$$
X_j(t) = U_{0j}e^{-\xi_j\omega_{nj}}\cos\left(\omega_{dj}t - \alpha_j\right) \tag{6}
$$

where ω_{nj} is the undamped circular frequency of the j-th modal; ω_{dj} is the damped circular frequency of the j-th modal; and ξ_j is the damping ratio of the j-th modal.

After $X_j(t)$ is Hilbert transformed:

$$
Y_j(t) = HT\left[X_j(t)\right] = \frac{1}{\pi}P\int_{-\infty}^{+\infty}\frac{X_j(\tau)}{t-\tau}d\tau \tag{7}
$$

where P is the Cauchy principal component.

An analytical equation can be written as:

$$
Z_j(t) = X_j(t) + iY_j(t) = U_j(t)e^{i\theta_j(t)} \tag{8}
$$

where amplitude $U_j(t)$ and phase $\theta_j(t)$ can be further expressed as:

$$U_j(t) = U_{0j}e^{-\xi_j\omega_{nj}t} \tag{9}$$

$$\theta_j(t) = \omega_{dj}t - a_j \tag{10}$$

By applying logarithm and differential to the amplitude and phase, we obtain:

$$\ln U_j(t) = -\xi_j\omega_{nj}t + \ln U_{0j} \tag{11}$$

$$\frac{d\theta_j(t)}{dt} = \omega_{dj} \tag{12}$$

From Equations (11) and (12), the damped inherent frequency ω_{di} can be obtained from the slope of $\theta_i(t)$, and the damping ratio ξ_i can be obtained from the slope of the relation curve of the logarithmic amplitude and time. In actual processing, the error resulting from the extracted free attenuation function and the numerical calculation will cause the logarithmic amplitude and time curve not to be strictly linear. Therefore, the damping ratio is further determined by fitting the logarithmic amplitude to a straight line using the least squares fitting method.

4 NUMERICAL EXAMPLE VERIFICATION OF THE RDT + HHT METHOD

To explore the reliability of the RDT + HHT modal parameter identification method, an experiment was performed to identify the frequency and damping ratio of the IASE-ASCE structural health monitoring benchmark structure (Johnson E A, 2004).

As a research platform for the comparative analysis of various damage identification methods, the benchmark structure was developed by the IASE (International Association for Structural Control) and the ASCE (American Society of Civil Engineers). In this paper, the modal parameters of the benchmark structure are first identified under initial conditions. Figure 2 shows the acceleration response on the fourth floor of the structure during noise excitation. The acceleration is composed of the measured signal and 10% white noise.

Figure 3 shows the power spectrum of the acceleration signal. The four peaks shown in Figure 4 correspond to four orders of modal frequency. To improve the accuracy of parameter identification, a preliminary estimation of the general frequency scope was made according to the spectral distribution of the signal auto-power spectrum. It can be seen from the figure that the first-order frequency lies in the range of 8.6 ~ 10.5 Hz, the second-order frequency lies in the range of 25.2 ~ 25.8 Hz, the third-order frequency lies in the range of 37.5 ~ 39.0 Hz and the fourth-order frequency lies in the range of 47.5 ~ 48.5 Hz.

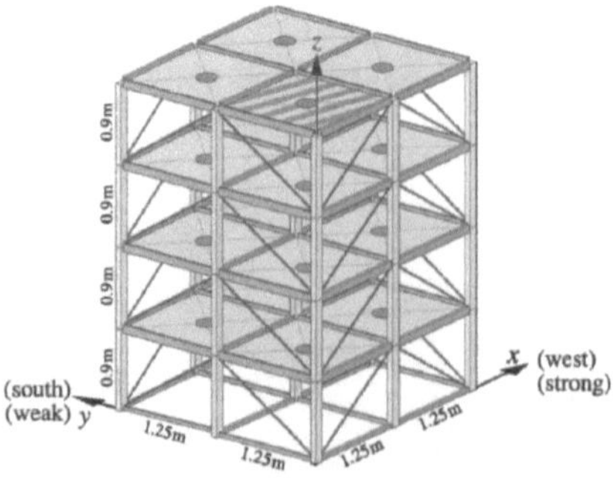

Figure 1. IASC-ASCE structural health monitoring benchmark structure numerical analysis model.

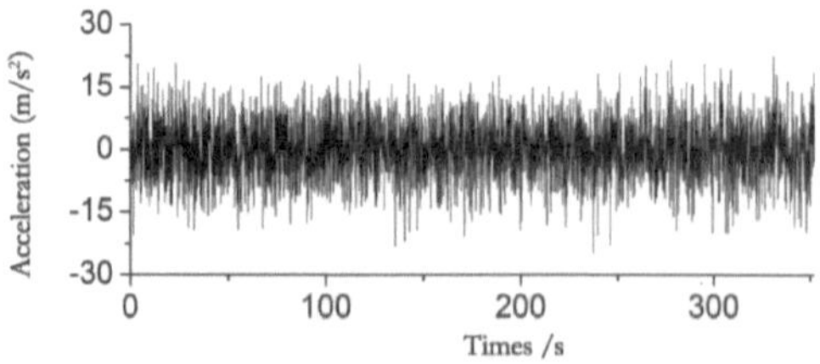

Figure 2. The acceleration on the fourth floor.

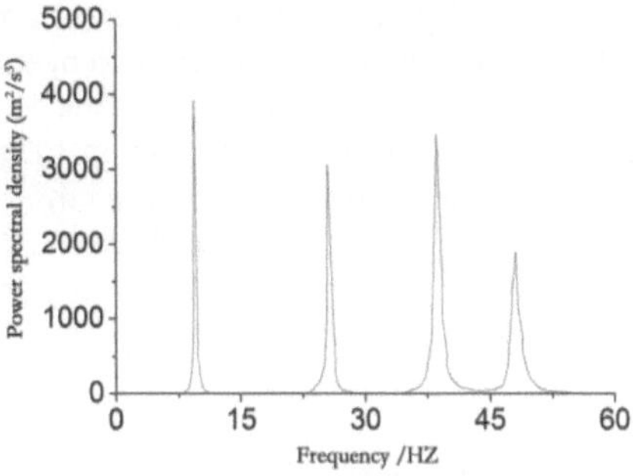

Figure 3. Power spectrum of the acceleration signal.

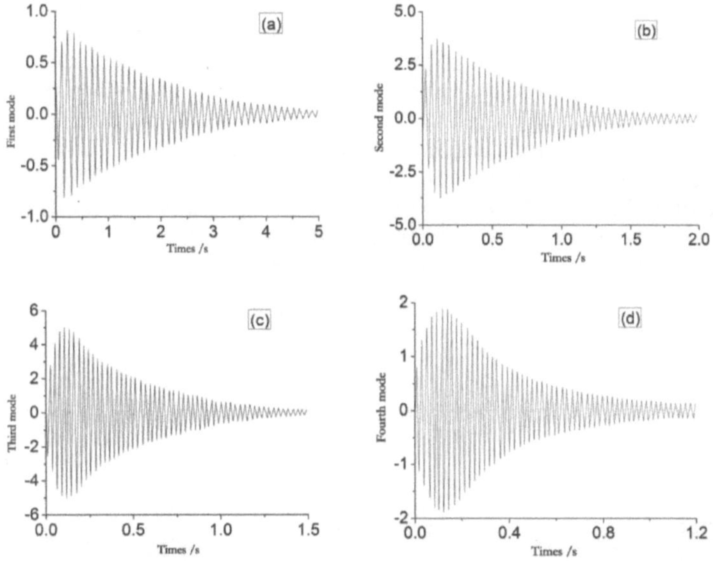

Figure 4. Free vibration response of each order modal obtained using the RDT [unit: m/s^2].

Table 1. Results of the modal parameter identification of the IASE-ASCE benchmark structure.

	Frequency [Hz]			Damping ratio [%]		
Order	Theoretical value	Reference*	Our results	Theoretical value	Reference*	Our results
1	9.41	9.39	9.40	1.00	1.09	1.07
2	25.54	25.41	25.45	1.00	1.16	1.12
3	38.66	38.59	38.60	1.00	1.05	1.04
4	48.01	48.13	48.10	1.00	1.03	1.03

*S. Lin & J. N. Yang (2005).

Then, the band-pass filters were used to process the signal, with the frequency band of 8 ~ 11 Hz, 25 ~ 26 Hz, 37 ~ 39 Hz and 47 ~ 49 Hz, respectively. Thereafter, the free attenuation curves, shown in Figure 4, were obtained by using the RDT to process IMF1 ~ IMF4, which resulted from the EMD of the filtered data. The figures show that the structure exhibits an obvious free attenuation tendency. Lastly, from Equations (11) and (12), the natural vibration frequency and damping ratio was obtained by using Hilbert–Huang transform and the least squares fitting method. The results are summarized in Table 1, which also includes the identification results from another study (S. Lin & J. N. Yang 2005). By comparison, it is obvious that the accuracy of parameter identification can be improved by using the proposed method in this paper, i.e. using a band-pass filter to process the signal based on the frequency distribution and then carrying out the analysis based on the filtered results. At the same time, the validity and reliability can be verified by using the RDT–HHT modal identification method proposed in this paper and the corresponding MATLAB program. Because the frequency and damping ratio of the structure is determined by the mathematical model established only by measuring the output signals of the system, this method is more suitable for modal parameter identifications in large-scale engineering structures.

5 CONCLUSION

This paper proposed to apply the band-pass filter technique to the modal identification method by using the RDT in combination with Hilbert–Huang transform. The reliability of parameter identification was improved by effectively avoiding mixed frequencies that occurred during signal decomposition, which was obtained by using the RDT to transform the intrinsic modal functions into free vibration responses, where the intrinsic modal functions resulted from the modal decomposition of the filtered signal. The accuracy of parameter identification was improved by using the least squares method when determining the modal frequency and damping ratio. The reliability of this RDT-HHT method was verified by identifying the frequency and damping ratio of the IASC-ASCE structural health monitoring benchmark structure. It proved that the identification error could be reduced by the combined use of RDT and EMD to process the time sequences that resulted from filtering the specific frequency ranges of structural earthquake time histories by using the HHT method. Therefore, this method can be applied to other modal identification techniques.

ACKNOWLEDGEMENTS

This work was financially supported by the Fujian Province Natural Science Foundation (2011J01309).

REFERENCES

Cheraghi N., Riley M.J., Taheri F., 2005. A novel approach for detection of damage in adhesively bonded joints in plastic pipes based on vibration method using piezoelectric sensors [M]: IEEE: 3472–3478.

Hongping Zhu, Jing Yu, Junbing Zhang. 2011. A summary review and advantages of vibration-based damage identification methods in structural health monitoring. Engineering mechanics, Vol. 28: 1–11.

Jianping Han, Lin Li, Hongtao Wang, et al. 2011. Modal parameter identification based on Hilbert-Huang transform and random decrement technique. World earthquake engineering, Vol. 27: 72–77.

Johnson E.A., Lam H.F., Katafygiotis. 2004. Phase I IASC-ASCE structural health monitoring benchmark problem using simulated data. [J]. Journal of Engineering Mechanics-Asce, Vol. 1: 3–15.

Juan Chen, Youlin Xu, Jie Li. 2003. Hilbert-Huang transform for damping ratio identification of structures with closely spaced modes of vibration. Earthquake engineering and engineering vibration, Vol. 23: 34–42.

Juan Chen, Youlin Xu. 2003. Application of HHT for modal parameter identification to civil structures. Journal of vibration engineering, Vol. 16: 383–388.

Lihang Kou, Feng Jin. 2008. Modal parameter identification for engineering structures based on HHT method. Advances in science and technology of water resourses, Vol. 28: 45–49.

Lin Silian, Yang Jann N., Zhou Li. 2005. Damage identification of a benchmark building for structural health monitoring [J]. Smart materials and structures. 14(3): S162.

Lin S., J.N. Yang. 2005. Damage identification of a benchmark building for structural health monitoring. [J]. Smart Materials and Structures, Vol. 3: 162–169.

Xibo Duan, Jiguang Zhang, Yong Yan. 2006. Identification of modal parameters of multiple degree-of-freedom system based on Hilbert-Huang transform. Journal of Shanghai university (natural science), Vol. 12: 403–407.

Yanjun Li, Xiangyun Hu, Zhanxiang He. 2010. Application situation and prospects of Hilbert-Huang transform in geophysics. Chinese journal of engineering geophysics, Vol. 7: 537–543.

Applied Mechanics and Civil Engineering VI – Xie (Ed.)
 ISBN 978-1-138-62631-7

Study on the creep characteristics and constitutive model of schist in natural and saturated states

S.G. Liu
State Key Laboratory of Geomechanics and Geotechnical Engineering, Institute of Rock and Soil Mechanics, Chinese Academy of Sciences, Wuhan, China

A.J. Guan
Hubei Provincial Communications Investment Group Co. Ltd., China

J. Liu & Z.P. Liu
State Key Laboratory of Geomechanics and Geotechnical Engineering, Institute of Rock and Soil Mechanics, Chinese Academy of Sciences, Wuhan, China

ABSTRACT: This study explores the creep behaviors of schist in different water contents and establishes the creep model and its constitutive equations. First, the *in situ* displacement tests were conducted and the time-dependent deformation regularity of surrounding rock was studied based on the sericite schist tunnel construction in the northwest area of Hubei Province of China. Then, rock samples were taken and shear creep tests were conducted along the direction of schistosity plane in natural and saturated states. On the basis of the results of *in situ* and laboratory tests, a viscoelastic–plastic creep model was constructed by model identification and introducing a nonlinear rheological component. Finally, parameters of the creep model were obtained by multiple regression and back analysis and the rationality of the creep model was verified by comparing it with the experimental data.

1 INTRODUCTION

Schist is a metamorphic rock with obvious lamellar structure that can often be revealed in underground engineering construction. For example, in the construction of highway tunnels in the northwest area of Hubei Province of China, soft schist is widely distributed. Because that the distance between schistosity planes is small and the binding power is weak, the compressive and shear strength of schist mass is low. Because of the effect of chlorite, montmorillonite, and other clay minerals, schist can be easily weathered and softened when containing water. The stability of surrounding rock is poor and the deformation of rock mass is large in the process of tunnel excavation, especially under the condition of high groundwater level. Therefore, disasters with large deformation of surrounding rock and tunnel collapse happen frequently. The long-term stability of engineering rock mass is affected significantly by its rheological property. Therefore, it is of great significance to realize the characteristics and regularities of the creep behaviors of schist and its creep mechanism in order to control the deformation, prevent disasters, and ensure safety of the tunnel operation.

Schist is water-sensitive, and hence, water content has significant effects on the strength and deformation parameters of rock mass. Some studies indicate that both the peak strength and residual strength of schist decreased significantly in saturated than in dry condition.

Reduction of the strength of schist influences its rheological deformation characteristics in response. When the stress increased to the stage of yield, the plastic deformation increases with the increase of loading, thereby inducing the acceleration of the breakdown of rock. Recently, some researchers have attempted to take the water content into account in creep test of the rocks. Therefore, it is necessary to consider the influence of water content in studying the creep characteristics of schist.

This study is based on the sericite schist tunnel construction in the northwest area of Hubei Province of China. *In situ* displacement tests are carried out to realize the time-dependent deformation regularity of surrounding rock. Then, shear creep tests are conducted along the direction of schistosity plane in natural and saturated sates. On the basis of the results of *in situ* and laboratory tests, a creep model and its constitutive equation are constructed. Finally, the creep mechanical parameters are obtained and the rationality of the model is verified.

2 DEFORMATION REGULARITY OF SCHIST TUNNEL

2.1 *In situ deformation test*

Sericite schist is widely spread in the northwest area of Hubei Province. Therefore, a typical sericite schist tunnel was selected as the research object. According to the requirement of the New Austrian Tunneling Method, both the horizontal and vertical displacements of surrounding rock were monitored every day during the process of tunnel construction to realize the dynamic development regularity of the deformation of tunnel rock and supporting structure.

The stress field was readjusted after the excavation. The process of stress redistribution goes with the release of surrounding rock pressure. Different release of the surrounding rock pressure induces different corresponding deformation. Figure 1 shows the horizontal and vertical displacements of schist tunnel of different sections.

2.2 *Time-dependent deformation regularity*

According to the statistical analysis of the monitoring data of schist tunnels, the process of deformation development of surrounding rock can be divided into three stages, including the initial stage of rapid growth, the stage of slow growth, and the stage of steady-state growth.

The initial stage of rapid growth. This stage occurs mostly during the first or second week after the excavation and primary support of tunnels. In this stage, the deformation rate is relatively high, the deformation curve increases linearly, and the deformation mainly includes the elastic deformation and plastic deformation of the surrounding rock.

The stage of slow growth. This stage occurs mostly during 15–30 days after the excavation and primary support of tunnels. The deformation rate decreases gradually and the

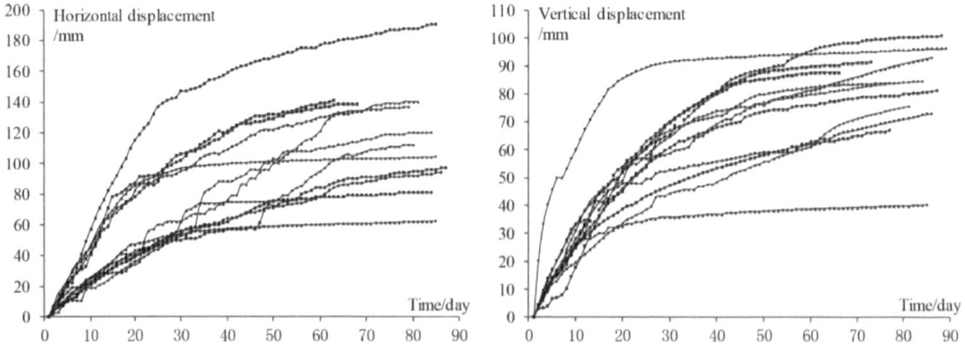

Figure 1. Horizontal and vertical displacements of surrounding rock of different sections.

deformation curve increases exponentially. The deformation rate is easily affected by construction disturbance and the deformation curve has a step increase. The deformation of the surrounding rock mainly includes rheological deformation and a certain amount of plastic deformation. In this stage, the quality of the surrounding rock degrades further. Cracks will occur if the initial supporting strength is insufficient and the deformation rate will accelerate suddenly if the reinforcement is not timely enough.

The stage of steady-state growth. When the time after initial supporting reaches a certain value, the deformation rate decreases further and reaches a steady state. The displacement of surrounding rock is mainly induced by the rheological deformation of rock mass. When the deformation reaches a certain limit, the rock mass will fail and disaster may happen if there is no further reinforcement or secondary lining.

3 CREEP CHARACTERISTICS OF SCHIST

3.1 *Creep test*

In order to further study the creep characteristics of the schist, rock samples were taken from tunnel and shear creep tests were conducted along the direction of schistosity plane in natural and saturated states. The cylindrical test specimens of sericite schist are shown in Figure 2. The test used stepwise shear loading method when the normal loading is kept constant. First, normal loading was applied to the predicted value and kept constant. Then, stepwise shear loading was applied when the normal deformation tended to be stable. Each step of loading was needed to persist more than 6 days and the water content must be kept constant. The instantaneous deformation was measured when applying each step of shear loading and the shear deformation was measured at 5, 10, and 30 min and 1, 2, 4, 8, 12, 16, and 24 h after loading. Then, the shear deformation was measured daily. The shear loading values were 20, 32, 48, 64, and 80 MPa. The creep curves of the tests of sericite schist are shown in Figure 3.

3.2 *Creep curve analysis*

It can be seen from Figure 3 that during the application of shear loading, the schist specimen generated elastic strain, which increased with the increase of horizontal stress. The stage from elastic deformation to creep deformation persisted only for a moment. When the shear stress is small, the deformation induced by each step of shear loading was instantaneous while the creep deformation was small and the creep rate decreased rapidly. With the increase of shear stress, the creep rate decreased to a steady value and persisted until the next step of loading. The steady creep rates were different depending on the shear stress. The shape of the creep curve, especially the decelerating creep stage, was influenced by the shear stress and the curve changed from approximately straight line to notch curve with the increase of the shear stress. The viscosity resistance decreased rapidly and the specimen trended to fail when the shear deformation of the specimen increased to a

Figure 2. Cylindrical test specimens of sericite schist.

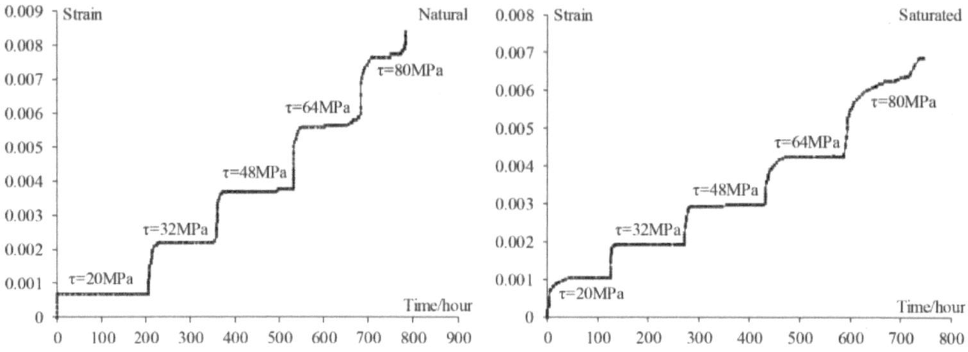

Figure 3. Creep curves of sericite schist in both natural and saturated states under different loading.

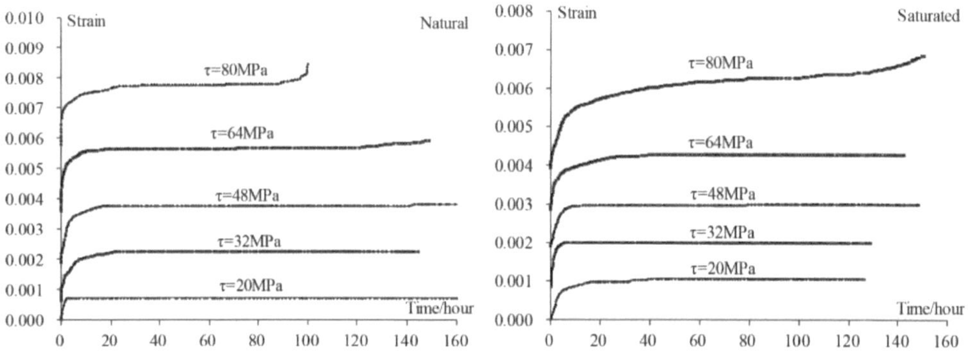

Figure 4. Separated creep curves of sericite schist in both natural and saturated states under different loading dislodged the effect of loading history.

certain value while the shear stress remaining at 80 MPa. Obviously, the regularity of the creep deformation of schist specimen was similar to the regularity of the deformation of schist tunnel rock mass.

Because that the creep test used stepwise loading method, the plastic deformation caused by last step of loading would have an effect on the shear deformation generated by the next step of loading and, on the contrary, the shear creep deformation might decrease when the stress increased. This result was against the creep behaviors of rock mass. Therefore, the effect of loading history on shear deformation must be taken into consideration in processing the test data. The creep curves separated from the whole loading process that dislodged the influence of loading history are shown in Figure 4.

4 CREEP MODEL AND THE CONSTITUTIVE EQUATION

4.1 *Creep model*

Considering both the regularities of creep curve of laboratory tests and the *in situ* deformation tests, a modified viscoelastic–plastic model was chosen to describe the creep characteristics of schist by model identification. The generalized Kelvin model was selected to describe the steady creep stage when the stress was low, while a nonlinear rheological component must be introduced to describe the accelerating creep stage of schist when the stress is higher than a certain value. The relationship between the strain and time of the nonlinear rheological component was indicated as follows:

$$e(t)=\frac{\sigma}{G}\left(1-e^{-\left(\frac{H(t-t_2)}{t_c-t_2}\right)^n}\right) \quad (1)$$

$$H(t-t_2)=\begin{cases}0 & t\le t_2\\ t-t_2 & t>t_2\end{cases} \quad (2)$$

where G and n are rheological parameters; t_2 is the time from steady creep to accelerating stage; and t is the time of shear failure.

4.2 *Constitutive equations*

On the basis of the creep model of schist composed of six basic models, the constitutive equations of the nonlinear viscoelastic–plastic creep model of schist constructed by model identification are given in Equations 3–5.

If $\sigma\le\sigma_s$,

$$e(t)=\frac{s}{E_1}+\frac{s}{E_2}\left(1-e^{-\frac{E_2}{\eta_2}t}\right) \quad (3)$$

If $\sigma>\sigma_s$,

$$t\le t_2,\ \varepsilon(t)=\frac{\sigma}{E_1}+\frac{\sigma}{E_2}\left(1-\varepsilon^{-\frac{E_2}{\eta_2}t}\right)+\frac{\sigma-\sigma_s}{\eta_s}t \quad (4)$$

$$t>t_2,\ \varepsilon(t)=\frac{\sigma}{E_1}+\frac{\sigma}{E_2}\left(1-\varepsilon^{-\frac{E_2}{\eta_2}t}\right)+\frac{\sigma-\sigma_s}{\eta_s}t+\frac{\sigma}{G}\left(1-\varepsilon^{-\left(\frac{H(t-t_2)}{t_c-t_2}\right)^n}\right) \quad (5)$$

where E_1 = instantaneous shear modulus, E_2 = viscoelasticity shear modulus, σ_s = frictional resistance.

The follow equations are used to describe the shear creep characteristics of schist in high stress.

$$t\le t_2,\ \varepsilon(t)=\frac{\tau}{E_1}+\frac{\tau}{E_2}\left(1-\varepsilon^{-\frac{E_2}{\eta_2}t}\right)+\frac{\tau-\tau_s}{\eta_s}t \quad (6)$$

$$t>t_2,\ \varepsilon(t)=\frac{\tau}{E_1}+\frac{\tau}{E_2}\left(1-\varepsilon^{-\frac{E_2}{\eta_2}t}\right)+\frac{\tau-\tau_s}{\eta_s}t+\frac{\tau}{G}\left(1-\varepsilon^{-\left(\frac{H(t-t_2)}{t_c-t_2}\right)^n}\right) \quad (7)$$

4.3 *Calculating of parameters*

Multiple regression analysis was used to obtain the parameters of the creep constitutive equations based on the test data processed with Boltzmann superposition principle to analyze the regularities of these parameters. Then, these parameters were plugged into Equations 6 and 7 and the optimal parameters were obtained by back analysis.

Parameters of the creep constitutive equations of schist in both natural and saturated states by back analysis are shown in Table 1. The creep hook face was obtained by plugging the parameters into Equations 6 and 7, and its rationality was verified by comparing it with the experimental curves.

Therefore, the creep model shown in Figure 5 and the constitutive equations can be used to analyze the time-dependent deformation and damage mechanism of schist rock mass in the construction of tunnel or other underground engineering processes.

Table 1. Parameters of the creep constitutive equations of schist in both natural and saturated states by back analysis.

Parameters	τ MPa	E_1 GPa	E_2 GPa	μ_2 GPa · h	μ_2 GPa · h	G GPa	n	R
Natural	18	5.34	13.56	32.45	–	–	–	1.02e-5
	32	6.75	24.20	30.56	–	–	–	4.56e-6
	48	9.82	30.25	78.25	–	–	–	3.25e-5
	64	9.87	21.23	50.56	–	–	–	4.40e-5
	80	10.02	35.24	65.45	75.26			3.25e-6
	80	10.89	26.47	60.47	58.90	16.25	17	2.60e-5
Saturated	18	6.68	11.56	26.16	–	–	–	3.69e-5
	32	7.35	23.77	46.50	–	–	–	1.90e-5
	48	6.95	50.46	38.29	–	–	–	6.20e-5
	64	8.56	31.35	68.44	–	–	–	7.90e-5
	80	9.11	45.60	74.28	62.33	–	–	8.25e-5
	80	9.45	36.66	68.10	35.40	10.10	24	4.56e-5

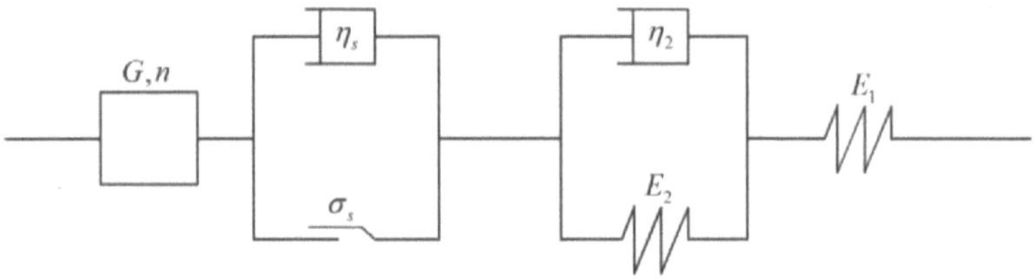

Figure 5. Nonlinear viscoelastic–plastic creep model of schist containing six components.

5 CONCLUSIONS

In situ displacement and experimental creep tests were conducted to study the characteristics of schist in both natural and saturated states, and the creep model and its constitutive equations were established. The main conclusions were drawn as below.

The process of deformation development of schist tunnel rock can be divided into three stages, including the initial stage of rapid growth, the stage of slow growth, and the stage of steady-state growth.

The viscosity resistance decreased rapidly and the specimen trended to fail when the shear deformation of the specimen increased to a certain value while the shear stress remaining at 80 MPa.

The regularity of creep deformation of the schist specimen was similar to that of the schist tunnel rock mass.

The modified viscoelastic–plastic model constructed by model identification and introducing a nonlinear rheological component was sufficient to describe the creep characteristics of schist.

ACKNOWLEDGMENT

The authors gratefully acknowledge the support of the National Key Research and Development Project of China (Grant No. 2016YFC0401802), the State Key Program of National Natural Science of China (Grant No. 51539002), and the National Natural Science Foundation of China (Nos. 51209198 and 51204158).

REFERENCES

Bikong C., Hoxha D., Shao J.F. 2015. A micro-macro model for time-dependent behavior of clayey rocks due to anisotropic propagation of microcracks. *International Journal of Plasticity* 69: 73–88.

Khandelwal M., Singh T.N. 2009. Predicting elastic properties of schistose rocks from unconfined strength using intelligent approach. *Arabian Journal of Geosciences* 4(3–4): 435–442.

Lajtai E.Z., Schmidtke R.H., Bielus L.P. 1987. The effect of water on the time-dependent deformation and fracture of a granite. *International Journal of Rock Mechanics and Mining Sciences & Geomechanics Abstracts* 24(4): 247–255.

Liu L., Wang G-m., Chen J-h., Yang S. 2013. Creep experiment and rheological model of deep saturated rock. *Transactions of Nonferrous Metals Society of China* 23(2): 478–483.

Read S.A.L., Perrin N.D., Brown I.R. 1987. Measurement and analysis of laboratory strength and deformability characteristics of schistose rocks. *6th ISRM Congress. Montreal: International Society for Rock Mechanics.*

Xiong L., Li T., Yang L. 2014. Biaxial compression creep test on green-schist considering the effects of water content and anisotropy. *KSCE Journal of Civil Engineering* 18(1): 103–112.

Yang S.Q., Jing H.W., Cheng L. 2014. Influences of pore pressure on short-term and creep mechanical behavior of red sandstone. *Engineering Geology* 179: 10–23.

Applied Mechanics and Civil Engineering VI – Xie (Ed.)

Effect of Nano-$CaCO_3$ on performance of cement and concrete

X. Chen, H. Cao, L. Weng & J. Qin
Shenzhen Academy of Aerospace Technology, Shenzhen, P.R. China

ABSTRACT: The effect of Nano-$CaCO_3$ (NC) on the hydration, durability and strength development of the Portland cement with a high volume of fly ash is studied. The compressive strength, porosity, hydration products, carbonation depth and diffusion coefficient of chloride of the binder are characterized. The results show that the addition of NC can improve greatly the compressive strength of the materials both at the early and later stages. It also refines the porosity, promotes the hydration process of the binder and improves the concrete durability.

1 INTRODUCTION

The concretes with high volume fly ash have been gradually accepted in applications due to the benefit of energy conservation and CO_2 emission reduction. However, the addition of high volume fly ash generally reduces the concrete early strength, which results in some difficulties in curing process (Shaikh, 2015). The nanotechnology could overcome these disadvantages because the concrete is a combination that consists of nanocrystals, micrometer crystals, amorphous phases and irreducible water, as studied by Sanchez and Sobolev (Sanchez, 2010). They and others also investigated the relationship between the performance and the particle size of raw materials for preparation of high performance concrete (Sobolev, 2005). These studies provided useful information for the practical applications of novel concrete technologies (Hou, 2012 and Bjornstrom, 2004). On the other hand, Bentz et al. (Bentz, 2012) and Kakali et al. (Kakali, 2000) showed that limestone powder can accelerate early-age hydration, and provide nucleation sites for calcium silicate hydrates (C–S–H), and also react with calcium aluminates and tricalcium silicate to produce calcium carbo-aluminates and calcium carbo-silicate hydrates. However, the excess addition of limestone powder deteriorates the concrete performance. Up to date, most of the studies have focused on micro-sized limestone powders, while Nano-limestone (NC) has not been thoroughly investigated. In fact, the presence of NC could increase the rate of C_3A hydration reaction to form a carbo-aluminate complex, consequently increasing the amount of hydration products and mechanical strength (Camileti, 2013). Meanwhile it also reacts with tricalcium silicate and accelerates setting time and early strength development (Pera, 1999).

In the paper the effect of Nano-$CaCO_3$ on the hydration, durability and strength development of the Portland cement with high volume fly ash is studied to further the previous studies.

2 EXPERIMENTAL

2.1 *Materials*

The properties of the P.O 42.5 cement (from China Resources Cement Holdings Limited), ground granulated blast furnace slag and fly ash are listed in Table 1, and the Nano-$CaCO_3$

Table 1. Properties of cement, slag and fly ash.

Materials	Type	Density [g/cm³]	Surface area [m²/kg]	Setting time [min]	
				Initial	Final
Cement	42.5	3.02	361	183	388
Fly ash	–	2.31	–	–	–
Slag	S95	2.91	419	–	–

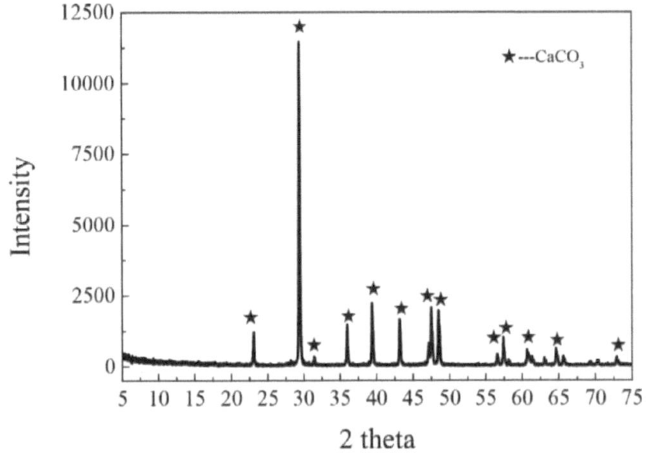

Figure 1. XRD pattern of Nano-$CaCO_3$.

powder is from a commercial source with a purity of 95% (Figure 1) and a surface area of 37.6 m^2/g. The 5–30 mm crushed stones and the river sand with a module of 2.70 are used as aggregates. The water reducer is a 20% aqueous solution of poly-carboxylic acid.

2.2 *Process*

1. Mix design

The samples of paste and mortar are prepared according to the mix design as shown in Table 2, and the concrete samples as Table 3.

2. Testing methods

The workability of mortar and concrete with different NC additions are tested. The $40 \times 40 \times 160$ mm^3 mortar specimens, Φ100 mm cylinder and $100 \times 100 \times 100$ mm^3 cube concrete specimens are prepared and cured at 20 ± 3°C, 60 ± 5% RH for 24 hours according to the standard process. The compressive strength of the concrete and mortar is tested at 3d, 7d and 28d respectively. The specimens for the carbonation tests were cured till 26d, dried in a oven at 60°C for 48 h, and then carbonated in a carbonation chamber (T = 20 ± 3°C, CO_2 = 20 ± 3%, RH = 70 ± 5%). After being carbonated for 7d and 28d, the specimens were cut and the cross section was sprayed with phenolphthalein solution to identify the carbonation depth. The concrete cylinders were prepared for testing the diffusion coefficient of chloride by RCM method when cured in 28d. The samples for SEM, MIP, XRD and hydration heat examination are extracted at 7d and 28d respectively.

Table 2. Mix design of the paste and mortar samples.

	No.	Cement	Fly ash	Nano-$CaCO_3$	Water/Bind ratio
Paste	P0	100	0	0	0.45
	P2	98	0	2	0.45
	PF0	70	30	0	0.45
	PF2	69	29	2	0.45
Motar	M0	100	0	0	0.5
	M2	98	0	2	0.5
	M3	97	0	3	0.5
	MF0	30	70	0	0.5
	MF2	29	69	2	0.5
	MF3	29	68	3	0.5

Table 3. Mix design of concrete.

	Binder[kg/m³]								
No.	Cement	Slag	FA	SF	NC	Aggregates [kg/m³]	W/B	WR[%]	Sand rate
C0	147	113.5	147	12.6	0	1080	0.38	0.30	0.40
C1	145	113.5	144.5	12.6	4.2	1080	0.38	0.37	0.40
C3	142.8	110	142	12.6	12.6	1080	0.38	0.46	0.40

3 RESULTS AND DISCUSSION

3.1 *Workability*

The fluidity and slumps of the mortar and the concrete are tested and the results are as follows:

The fluidity results of the mortar and the concrete show that the fluidity and slump decrease gradually with the amount of NC. When NC is introduced to the concrete, the fluidity can be adjusted if the water reducer is added at a suitable quantity.

3.2 *Compressive strength of the mortar and concrete*

The strengths of the mortar and the concrete with a variable amount of NC at the different curing time are showed in Table 4.

It can be seen that the NC can enhance the strength of mortar and concrete at different age, the enhancement for the early age is the most, giving the 3d compressive strength increase by almost 20% to the mortar samples doped with 2% and 3% of NC, and the 28d compressive strength increases by 10% to the mortar samples involving high volume fly ash. The 3d and 28d compressive strengths of the concrete increase by more than 30% and 20% respectively.

3.3 *Carbonation depth*

The carbonation depth of the mortar after 7d accelerated carbonation test and concrete after 28d accelerated carbonation test are measured, and the results are shown in Figure 3. It is obvious that if NC is added, the carbonation depth is greatly reduced compared to those without NC addition. After the 7d accelerated carbonation, the mortar having high volume fly ash without NC carbonized completely, while doped with 3% of NC, the carbonation depth after 7d accelerated carbonation is 15 mm. The carbonation depth of reference concrete after 28d is 19 mm, and for concrete doped with 1% and 3% NC, the carbonation depth decrease by 25% and 38% respectively.

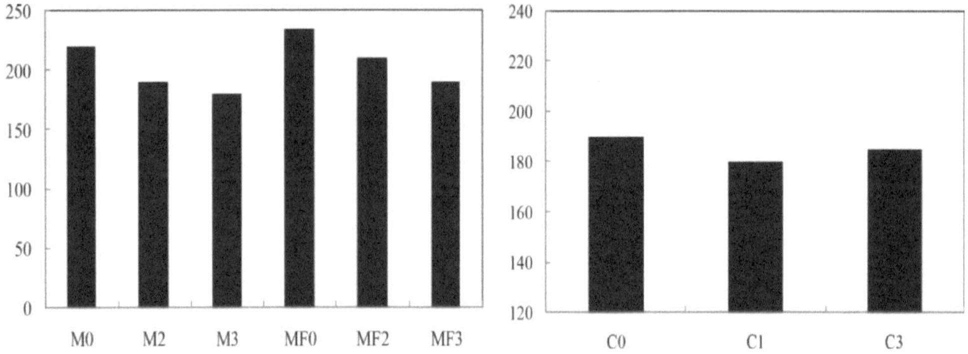

Figure 2. Workability tests of the mortar and the concrete.

Table 4. Compressive strength of the mortar and the concrete / [MPa].

Sample	3d	7d	28d
M0	21.6	33.6	40.3
M2	26.4	34.0	41.2
M3	29.7	46.0	47.5
MF0	5.1	11.1	23.4
MF2	5.6	11.9	24.9
MF3	6.1	12.6	25.2
C0	16.6	23.4	34.6
C1	18.8	25.6	38.5
C3	22.1	28.1	42.9

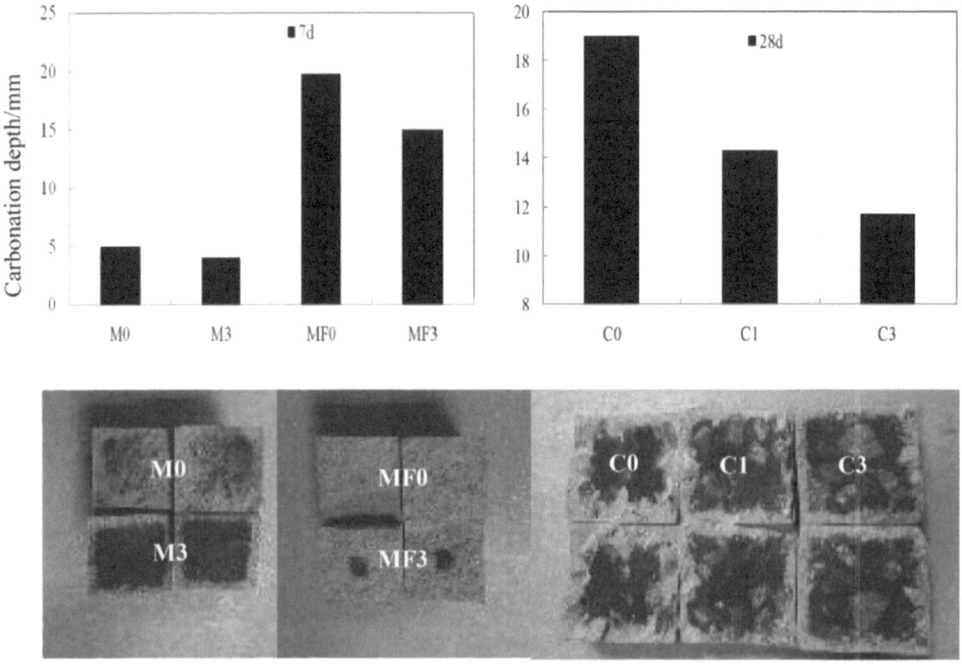

Mortar after 7d acceleration

Concrete after 28d acceleration

Figure 3. Carbonation depth of the mortar and the concrete.

3.4 *Chloride diffusivity coefficient*

The chloride diffusivity coefficients of the concrete with and without NC are listed in Table 5. The results show that the concrete with 1% and 3% NC can greatly decrease the chloride diffusivity coefficient by 19.1% and 36% respectively.

3.5 *Hydration heat evolution*

Fig. 5(a) shows the heat flow rate of the pastes with or without NC. The results show that the two curves of heat flow rate are almost coincides, which could be due to the heat flow rate of the plain cement paste is high and the addition of NC cannot increase the early hydration

Table 5. Chloride diffusivity coefficient of the concrete.

NO.	T0 [°C]	Tf [°C]	H [cm]	L [cm]	Dfucc/ $\times 10^{-12}$
C0	19.4	19.7	5.03	0.45	1.83
C1	19.3	18.6	4.94	0.39	1.48
C3	19.6	18.2	5.08	0.31	1.18

*Note: U = 30V, Time: 24 h.

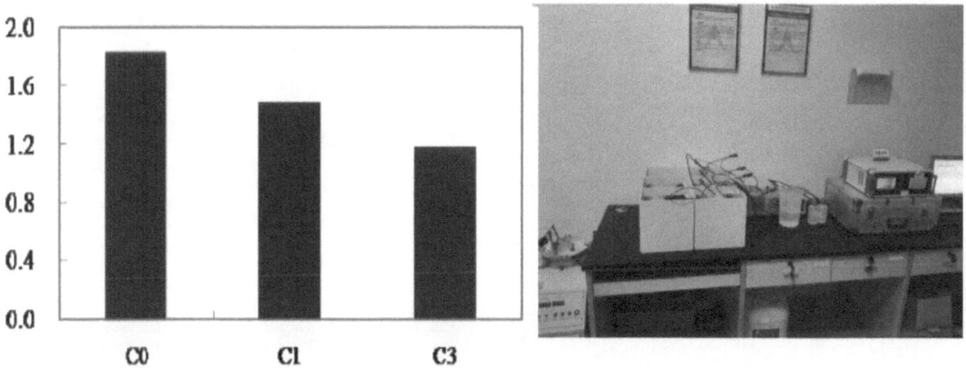

a b

Figure 4. Chloride diffusion coefficient of the concrete.

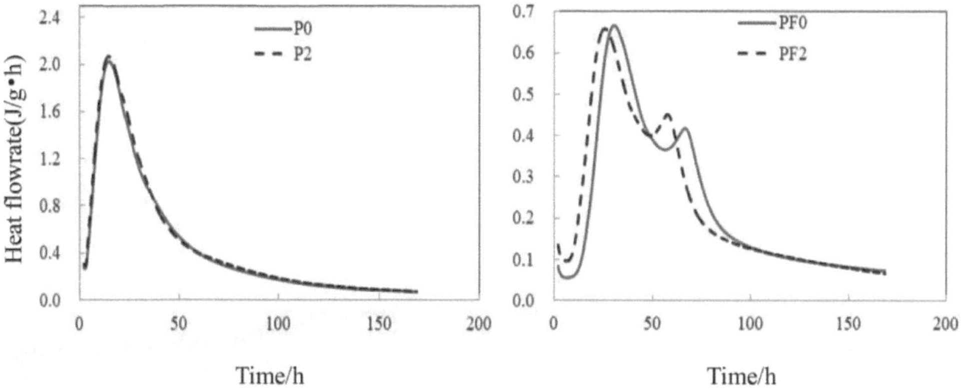

a. Heat flow rate of the pastes b. Heat flow rate of the pastes with fly ash

Figure 5. Heat flow rate of the pastes with or without NC.

reaction rate greatly. Fig. 5(b) shows the heat flow rate of the pastes containing high volume fly ash with or without NC. The early hydration reaction rate of the paste with NC moves to the left, suggesting that NC can accelerate the early hydration process.

3.6 *Hydration products*

The XRD patterns of the cement pastes with 0% and 2% NC at 7d and 28d are shown in Figure 6 and Figure 7 respectively. The hydration products in the harden pastes are $Ca(OH)_2$, $CaCO_3$ and unhydrated Ca_3SiO_5, while SiO_2 diffraction peak is from quartz sand. The addition of NC does not result in any new hydration product. However, the peak intensity of $Ca(OH)_2$ in samples with NC is lower than that of the reference plain paste at 3d and 28d, suggesting the enhanced reaction of $Ca(OH)_2$ due to NC addition.

3.7 *SEM images*

SEM images of the pastes at 7d and 28d curing age are shown in Figure 8. A large amount of well-developed hydroxide plates can be observed in the plain specimen in Figure 8a, while more calcium hydroxide appears to be consumed to the sample with NC addition, in

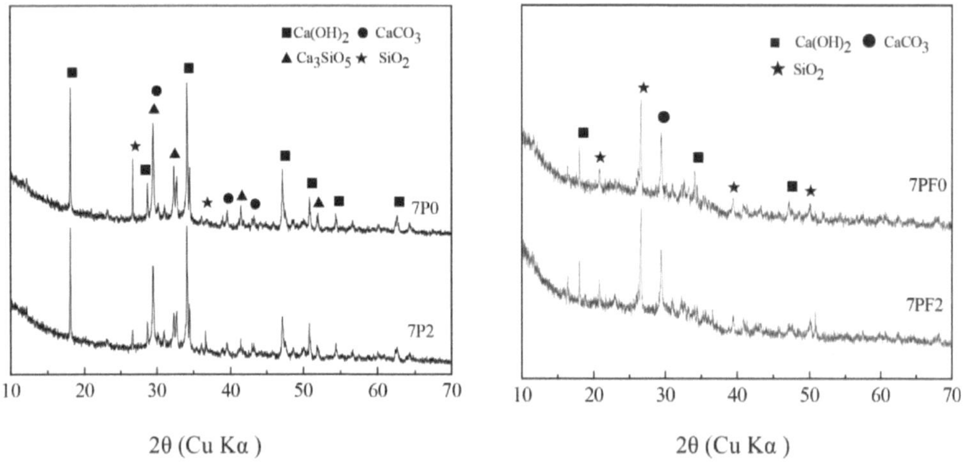

Figure 6. XRD patterns of the cement pastes at 7d with 0, 2.0% NC.

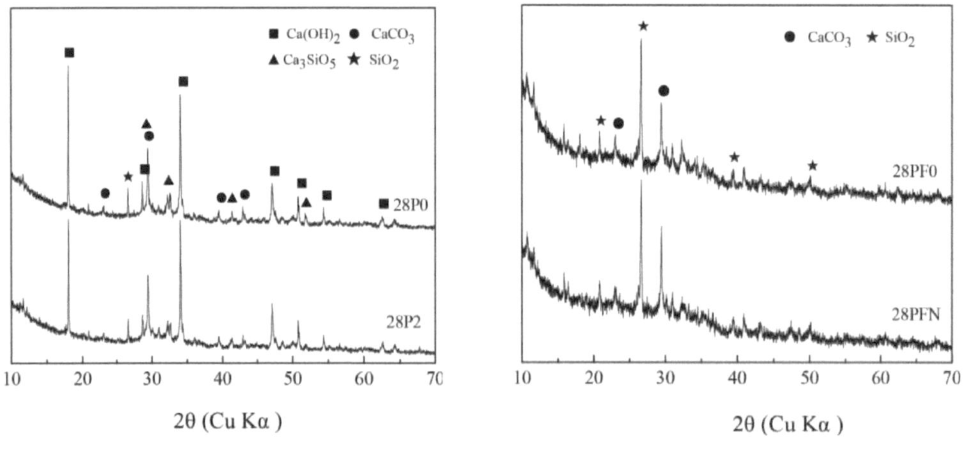

Figure 7. XRD pattern of the cement pastes at 28d.

(a) Plain paste at 7d

(b) Paste with NC at 7d

(c) Plain paste at 28d

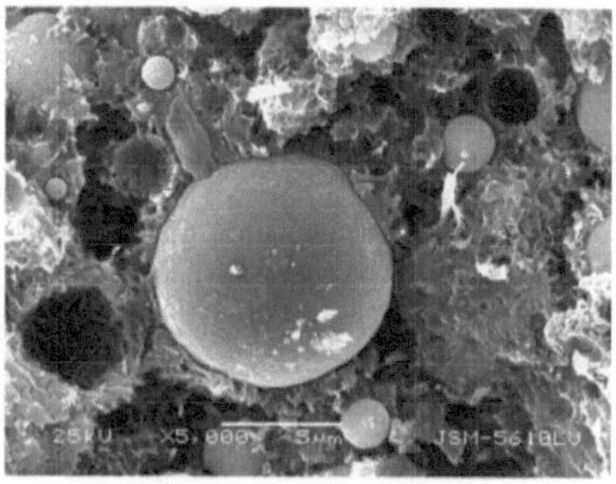

(d) Paste with NC at 28d

Figure 8. SEM images of paste at 3d and 28d.

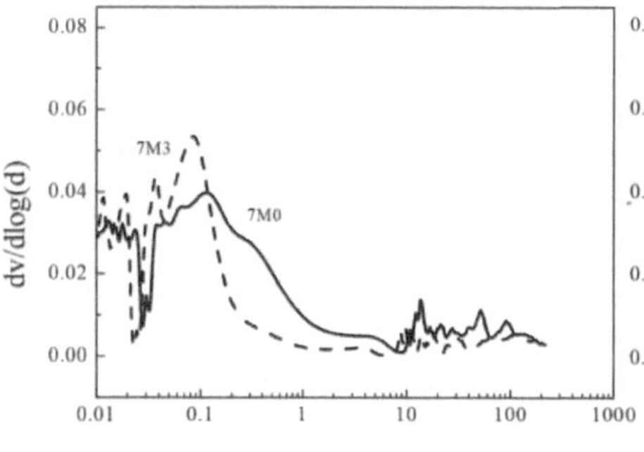

a

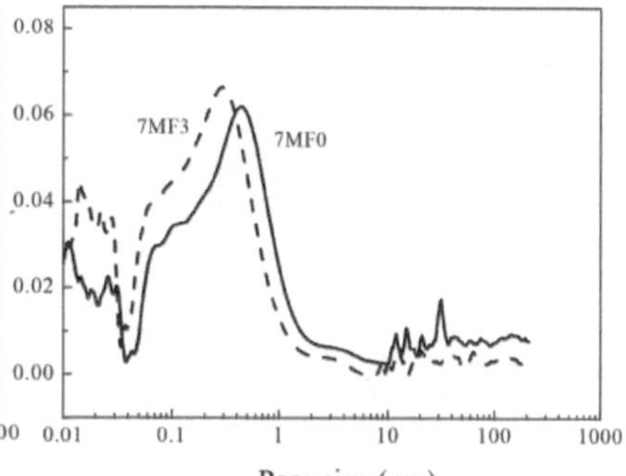

b

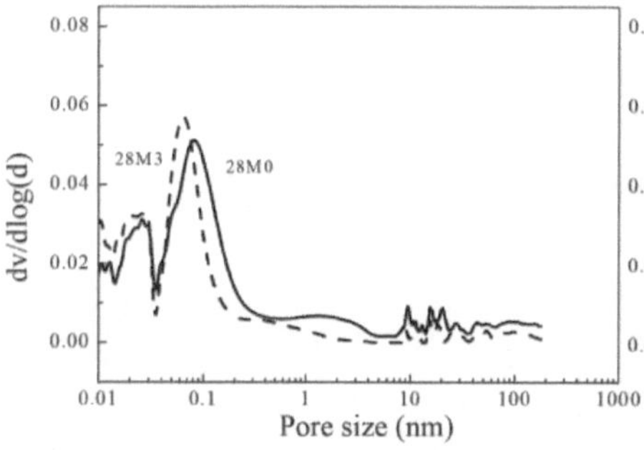

c

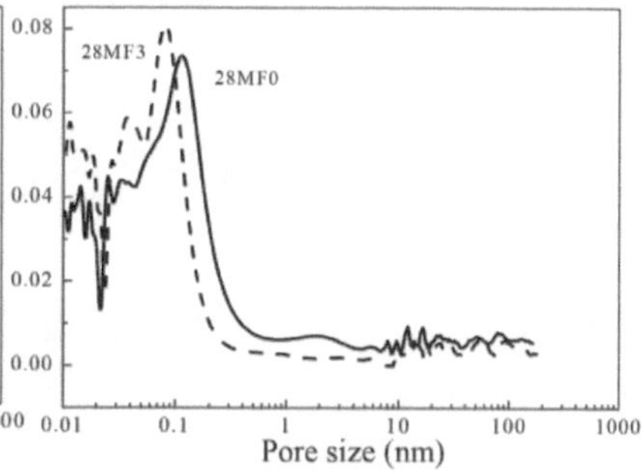

d

Figure 9. MIP analysis of the mortar for 7d and 28d.

accordance with the XRD results. Compare Figure 8c with d, it can be seen that with the addition of NC, the enhanced pozzolanic reaction can generate more C-S-H gels, so it is denser than that without NC.

3.8 *MIP*

The results of pore structure tests with MIP are shown in Figure 9. Due to the nano-filler effect more C-S-H was generated, consequently the large pores changing to fine ones though the total porosity is similar.

4 CONCLUSIONS

Based on the experimental results, the following conclusions can be drawn:

1. The NC addition into cement and concrete can increase their compressive strength both at early and later stages.
2. The introduction of NC into concrete decreases the carbonation depth and diffusion chloride coefficient substantially.
3. The addition of NC can accelerate the hydration process of the cement and decrease the content of calcium hydroxide at early age.
4. The addition of NC reduces the pore size of cement paste.

ACKNOWLEDGEMENTS

This research is financially supported by the Shenzhen Municipal Commission of Science and Technology Innovation.

REFERENCES

Bentz, D.P. et al. 2012. Fine limestone additions to regulate setting in high volume fly ash mixtures. Cem. Concr. Compos. 34(1): 11–17.

Bjornstrom, J. et al. 2004. Accelerating effects of colloidal nano-silica for beneficial calcium–silicate–hydrate formation in cement. Chem. Phys. Lett. 392(1–3): 242–248.

Camileti, J. et al. 2013. Effects of nano- and micro-limestone addition on early-age properties of ultra-high-performance concrete. Mater. Struct. 46: 881–898.

Hou, P.K. et al. 2012. Effects of colloidal nano SiO_2 on fly ash hydration. Cem. Concr. Compos. 34: 1095–1103.

Kakali, G. et al. 2000. Hydration products of C_3A, C_3S and Portland cement in the presence of $CaCO_3$. Cem. Concr. Res. 30(7): 1073–1077.

Pera, J. et al. 1999. Influence of finely ground limestone on cement hydration. Cem. Concr. Compos. 21: 99–105.

Sanchez, F. & Sobolev, K. 2010. Nanotechnology in concrete—A review. Build. Mater. 24: 2060–2071.

Shaikh, F. & Supit, S. 2015. Chloride induced corrosion durability of high volume fly ash concretes containing nano-particles. Constr. Build. Mater. 99: 208–225.

Sobolev, K. & Gutierrez, M. 2005. How nanotechnology can change the concrete world: Part 1. Am. Ceram. Soc. Bull. 84(10): 14–7.

Applied Mechanics and Civil Engineering VI – Xie (Ed.)

Study of rock loosening deformation and security characteristics of columnar jointed basalt based on monitoring information fusion

Long Jiang, Xiaogang Wang, Xiuwen Li, Jianhui Sun, Yufei Zhao & Lipeng Liu
China Institute of Water Resources and Hydropower Research, Beijing, China

Huijuan Zhang & Aijing Zhang
Institute of Remote Sensing and Digital Earth Chinese Academy of Sciences, Beijing, China

ABSTRACT: On the basis of the monitoring results of columnar jointed basalt rock stress and strain deformation, this study investigated the evolution of rock loosening deformation and deformation mechanism and established a surrounding rock safety evaluation system by using energy rate and unit deformation force. This study shows that the stability of columnar jointed basalt is poor and the surrounding rock deformation decreases with the increase in depth. Excavation with support construction methods helps control the deformation of the convergence range, pre-splitting blasting helps control the deformation of surrounding rock, and excavation without support and pre-split blasting is not conducive to the control of the deformation convergence range. Invading surrounding rock formed an arch after the columnar jointed basalt rock was supported; it is conducive to the absorption of surrounding rock deformation energy released. Pre-splitting blasting can effectively prevent the deformation of surrounding rock in shallow joints, but not conducive to the control of the late surrounding rock deformation associated with hidden joints deformed. Excavation is conducive to the release of the energy generated by shallow surrounding rock joint deformation; timely support can help control the energy generated by hidden joint deformation produced by surrounding rock deformation. Energy rate and unit deformation force can effectively evaluate the security characteristics of columnar jointed basalt wall rock.

1 INTRODUCTION

Basalt is a type of volcanic rock, and because of diagenesis and geological structure, it will develop an irregular or regular columnar morphological primary tensile structure—columnar joint. The rock joints and fissures of columnar jointed basalt are more developed and the poor self-stability of rock mass leads to its relaxation and partial collapse of the rock mass during the construction process.

In recent years, the influence and restriction of columnar jointed basalt in engineering construction has attracted significant attention of scholars with the development of large-scale hydropower station. Chen Xu et al. [Chen, 2008] conducted a study on the causes and rock quality evaluation of columnar jointed basalt. Xu Weiya et al. [Di, 2008] studied columnar jointed rock mass transversely isotropic constitutive relation, macroscopic equivalent elastic modulus, the macro equivalent strength parameters, and the size effect. Shi Anchi et al. [Shi, 2008] studied the deformation characteristics of rock mass. Wang Sijing et al. [Liu, 2010] studied the true triaxial model test and the size effect of shear strength parameters. Zhu Daojian et al. [Zhu, 2009] conducted a model test on the deformation and strength anisotropy of columnar jointed rock mass. Feng Xiating et al. [Hao, 2015] studied the stability of the tunnel, the failure mode, and the mechanical mechanism of columnar jointed basalt. In this paper, the relaxation deformation evolution law and the excavation supporting effect of columnar jointed basalt are studied.

Combined with the field monitoring results during the construction of the right diversion tunnel of the surrounding rock deformation of Baihetan Hydropower Station and anchoring structure stress, field monitoring and numerical calculation methods are adopted, with the structure characteristics, deformation characteristics, and mechanical characteristics of column jointed basalt are provided, and the relaxation-excavation supporting the collaboration safety evaluation system of column jointed basalt surrounding rock are established initially. The results have great practical significance and application value.

2 GENERAL SITUATION OF THE ENGINEERING

Baihetan Hydropower Station is located in the lower reaches of the Jinsha River in Ningnan County, Sichuan Province, and Qiaojia County, Yunnan Province. Cofferdam damming and river diversion tunnel are used as the way of construction diversion. Diversion tunnel adopted the gateway-like structure, and the cross-section dimension of the tunnel body is 19 m × 24 m (w × h). The excavation of tunnel is divided into three strata: upper layer, middle layer, and lower layer. The typical section and the monitoring arrangement are shown in Figure 1.

3 CONSTRUCTION CHARACTERISTICS OF COLUMNAR JOINTED BASALT SURROUNDING ROCK

After tunnel excavation, the surrounding rock mainly in quadrangular prism or pentagonal prism (block) shows an irregular columnar jointed three-dimensional grid intersected by several joint surfaces; generally, the prism has a length of 10–25 cm. With the cutting of DIP joint and intra-layer (inter) fracture zone, it formed columnar elements with length of tens of centimeters to tens of meters, as shown in Figure 2. In the process of geological drilling of the surrounding rock, it is unable to drill and collect complete core and all in clip formation, as shown in Figure 2.

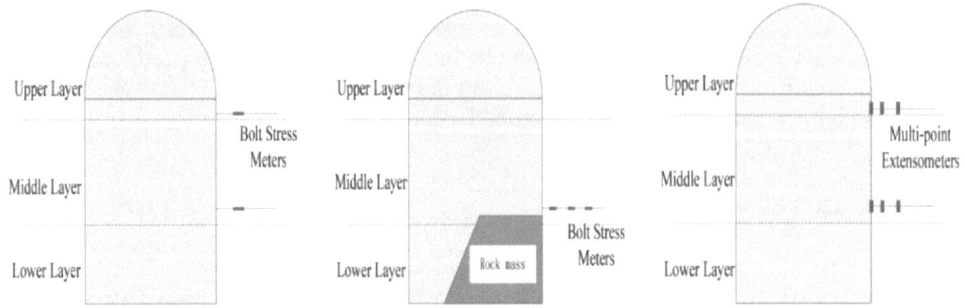

Figure 1. Sidewall monitoring layout of the typical section of the diversion tunnel.

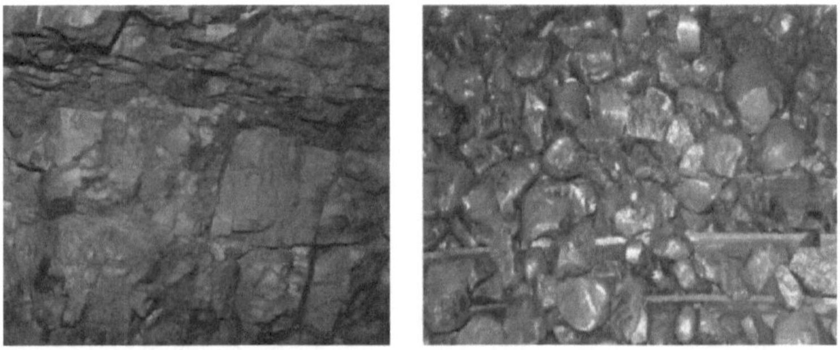

Figure 2. Typical columnar jointed basalt surrounding rock structure.

The joint cylinder of columnar jointed basalt surrounding rock is cut into irregularly arranged rock mass by joints and fractures, and in closed or micro-tension state; furthermore, no fillings are present in joint seam and it is filled with micro-bit cuttings. Under the influence of normal tension after unloading, the hidden joints in the jointed strata open to spread to the joint surface and crack surface, then the joint strata breaks into smaller and unstable pieces. The geological drill core is smaller than the surface of the surrounding rock mass, and it is also evidence of this situation.

4 DEFORMATION CHARACTERISTICS OF COLUMNAR JOINTED BASALT SURROUNDING ROCK

4.1 *The construction situation of surrounding rock excavation and supporting area A*

In February 2013, we arrange multi-point extensometers in the middle layer of the columnar jointed basalt surrounding rock excavation and supporting area A, and the typical curves of the sidewalls of the cavern are as shown in Figure 3.

As shown in Figure 3, there was a sudden change, about 19 mm, in the maximum deformation of columnar jointed basalt after the excavation of the lower layer, and after the bolt support, it grew by about 3 mm. After the concrete lining in August 2013, the increased amplitude of deformation slowed down, by about 8 mm. In this stage, the deformation remained stable.

4.2 *The construction situation of surrounding rock excavation without supporting area B*

In January 2013, we arrange multi-point extensometers in the upper layer of the columnar jointed basalt surrounding rock excavation without supporting area B, and the typical curves of the sidewalls of the cavern are as shown in Figure 4.

Figure 3. Variation of the curves of columnar jointed basalt tunnel section typical excavation and supporting area of surrounding rock with time.

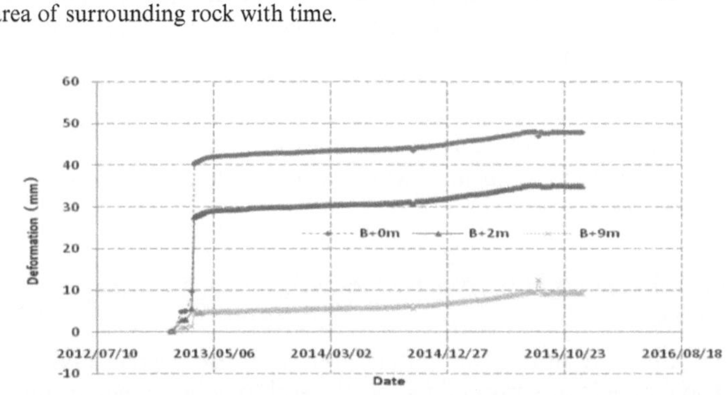

Figure 4. Variation of the curves of columnar jointed basalt tunnel section typical excavation without supporting area of surrounding rock with time.

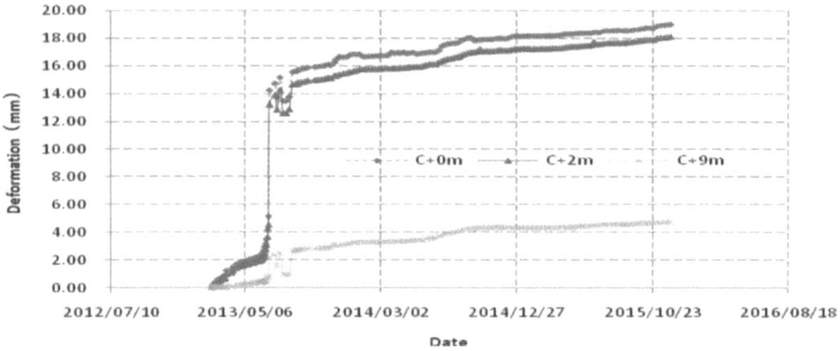

Figure 5. Variation of the curves of columnar jointed basalt tunnel section typical pre-splitting blasting and supporting area C of surrounding rock with time.

As is shown in Figure 4, there was a sudden change, about 9 mm, in the maximum deformation of columnar jointed basalt after the excavation of the middle layer, 30 mm in the excavation of the lower layer, and after the bolt support, it grew by about 2 mm. After the concrete lining in August 2013, the deformation trend slowed down, and the increased amplitude was about 5 mm. In this stage, the deformation remained stable.

4.3 *The construction situation of surrounding rock pre-splitting blasting and supporting area C*

In February 2013, we arranged multi-point extensometers in the middle layer of the columnar jointed basalt surrounding rock pre-splitting blasting and supporting area C, and the typical curves of the sidewalls of the cavern are as shown in Figure 5.

As is shown in Figure 5, there was a sudden change, about 3 mm, in the maximum deformation of columnar jointed basalt after the pre-splitting blasting, and after the excavation of the lower layer, it grew by about 11 mm, and the growth was about 3 mm after the bolt support. After the concrete lining, the amplitude of deformation grew by about 3 mm. In this stage, the deformation remained stable.

5 THE ROCK MECHANICAL PROPERTIES OF COLUMNAR JOINTED BASALT SURROUNDING ROCK

5.1 *The construction situation of surrounding rock excavation and supporting area A*

In January 2013, after excavation of the upper layer, we buried the bolt stress meters in top sidewall and bottom sidewall of surrounding rock excavation and supporting area A. The typical curves of the top sidewall and bottom sidewall are as shown in Figure 6.

As is shown in Figure 6, there was a sudden change, about 65 and 55 MPa, in the maximum anchor bar stress of columnar jointed basalt tunnel after the excavation of the middle layer, and it was about 90 and 180 MPa after excavation of the lower layer. After the concrete lining, the increased amplitude of anchor bar stress slowed down. In this stage, the anchor bar stress remained stable.

5.2 *The construction situation of surrounding rock pre-splitting blasting and supporting area C*

In February 2013, after excavation of the middle layer, we buried the bolt stress meters in 2 m hole, 4 m hole, and 6 m hole of surrounding rock pre-splitting blasting and supporting area C. The typical curves of the sidewall are as shown in Figure 7.

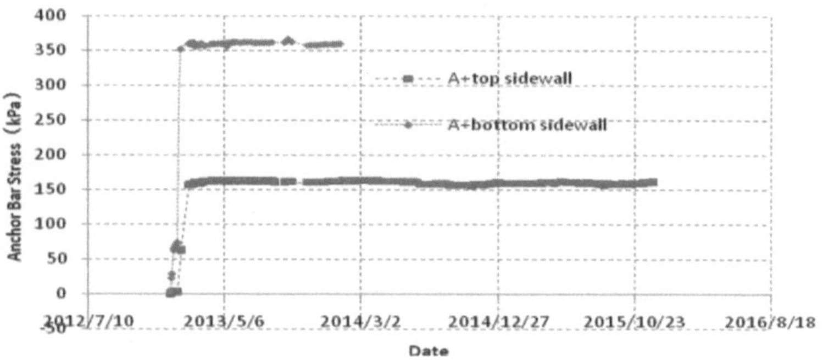

Figure 6. Variation of the curves of columnar jointed basalt tunnel section typical excavation and supporting area's surrounding rock structure stress with time.

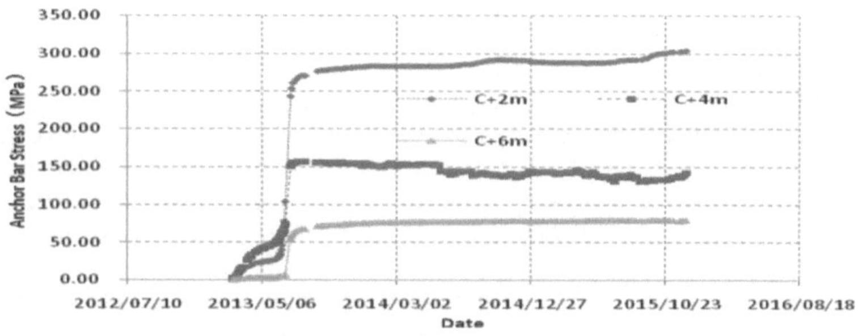

Figure 7. Variation of the curves of columnar jointed basalt tunnel section typical pre-splitting blasting and supporting area's surrounding rock structure stress with time.

As is shown in Figure 7, there was a sudden change, about 100, 70, and 5 MPa, in the anchor bar stress of columnar jointed basalt tunnel after pre-splitting blasting, and it was about 1400, 70, and 50 MPa after excavation of the lower layer. After the concrete lining, the increased amplitude of anchor bar stress was about 30, −10, and 5 MPa. In this stage, there is an increased amplitude of anchor bar stress in 2 m hole, 4 m hole, and 6 m hole, and the anchor bar stress remained stable.

Baihetan Hydropower Station diversion tunnel adopts the gateway-like structure, and the cross-section dimension of the tunnel body is 19 m × 24 m (w × h). We arrange monitoring facilities and select three sections of columnar jointed basalt tunnel section to set up the excavation with supporting area A, the excavation without supporting area B, and pre-splitting blasting and supporting area C with possible construction scheme. After the construction of excavation, supporting, and the concrete lining of diversion tunnel, variations of the curves of typical surrounding rock with depth are shown in Figure 8.

As is shown in Figure 8, we can see that the deformation of diversion tunnel columnar jointed basalt surrounding rock decreases with the increase in depth. The deformation of the surrounding rock in the excavation and supporting area has smaller convergence range, followed by pre-splitting blasting, supporting, and excavation without supporting. The results indicate that, first, the stability of columnar jointed basalt is poor. Second, the excavation supporting is conductive to control the deformation convergence range. Third, pre-splitting has advantages to control the deformation of the whole surrounding rock. Fourth, the excavation without supporting and pre-splitting are detrimental to control the deformation convergence range.

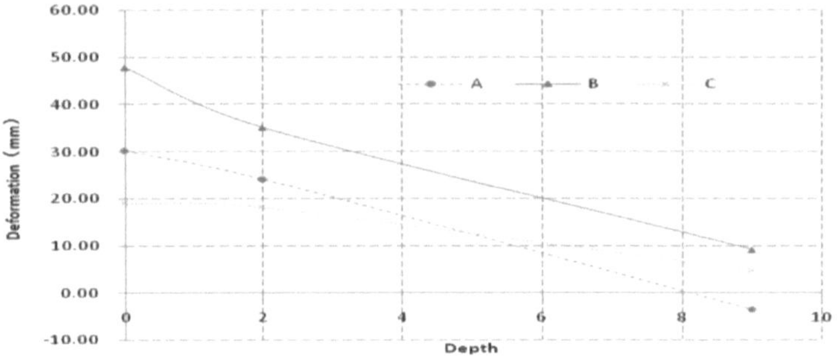

Figure 8. Variation of the curves of the diversion tunnel columnar jointed basalt surrounding rock deformation with depth (07/12/2015).

6 THE RELAXATION EVOLUTION LAW AND DEFORMATION MECHANISM OF COLUMNAR JOINTED BASALT SURROUNDING ROCK

Before tunnel construction, the natural mountain is in the process of natural stress dynamic adjustment with the influence of regional geological structure and groundwater, and once affected by the tunnel construction, its stress state will be changed, and there is no exception to columnar jointed basalt. With the influence of excavation, bolt support, and concrete lining, the surrounding rock of tunnels adjusts its stress dynamically to make it reach a new stress balance.

Baihetan Hydropower Station diversion tunnel adopts the gateway-like structure, and the cross-section dimension of the tunnel body is 19 m × 24 m (w × h). We arrange monitoring facilities and select three sections of columnar jointed basalt tunnel section to set up the excavation with supporting area A, the excavation without supporting area B, and pre-splitting blasting and supporting area C with possible construction scheme. After the construction of excavation, supporting, and the concrete lining of diversion tunnel, variations of the curves of the typical surrounding rock with depth are shown in Figure 8.

As shown in Figure 8, the deformation of diversion tunnel columnar jointed basalt surrounding rock decreases with the increase in depth. The deformation of the surrounding rock in the excavation and supporting area has smaller convergence range, followed by pre-splitting blasting and supporting and excavation without supporting. The results indicate that, first, the stability of columnar jointed basalt is poor. Second, the excavation supporting is conductive to control the deformation convergence range. Third, pre-splitting has advantages to control the deformation of the whole surrounding rock. Fourth, the excavation without supporting and pre-splitting are detrimental to control the deformation convergence range.

7 EFFECTIVE EVALUATION OF THE SECURITY CHARACTERISTICS OF COLUMNAR JOINTED BASALT WALL ROCK

Tunnel construction is an open stress rebalance system, and the excavation–deformation–supporting structure stress of surrounding rock is synergies. According to the earlier monitoring results of columnar jointed basalt stress and strain and surrounding rock deformation, we assessed the safety of the columnar jointed basalt surrounding rock based on the rate of energy and unit deformation force. Energy rate, the changing rate of coupling between surrounding rock deformation and structure by force at different stages (excavation, support, standing), shows the speed of energy release at different stages of the surrounding rock deformation. Unit deformation force, the change of coupling between surrounding rock deformation and structure by force at different stages, shows the energy release at different stages of the surrounding rock deformation.

The energy curves of diversion tunnel excavation and supporting area A sidewalls surrounding rock are shown in Figure 9, and the energy curves of diversion tunnel pre-splitting blasting area C sidewalls surrounding rock are shown in Figure 10.

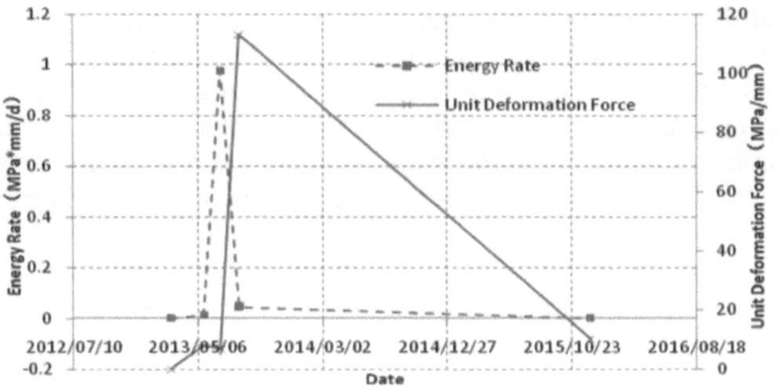

Figure 9. Energy curves of diversion tunnel excavation and supporting area A sidewalls surrounding rock.

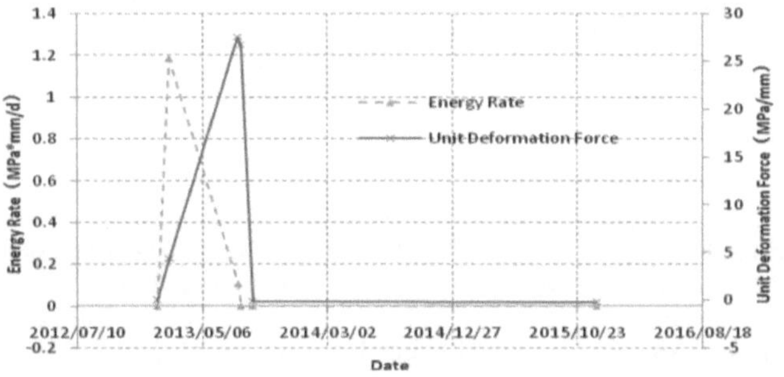

Figure 10. Energy curves of diversion tunnel pre-splitting blasting area C sidewalls surrounding rock.

As is shown in Figures 9 and 10, we can see that the energy rate of the diversion tunnel surrounding rock excavation and supporting area A significantly varied during the tunnel cutting and construction period, unit deformation force greatly varied in the stage of concrete lining, and energy rate and unit deformation force remained steady after concrete lining. First, it shows that the self-stability of columnar jointed basalt surrounding rock is poor. Second, support in the rear part of the free surrounding rock forms an arch ring, which is helpful to absorb energy released from the deformation of surrounding rock. Third, pre-splitting blasting can prevent surrounding rock deformation effectively, but it goes ill with the control of implied joint deformation, which is created by the late deformation of surrounding rock. Fourth, the excavation can prevent surrounding shallow joint deformation, and the supporting has advantages of controlling the implied joint deformation, which is created by the late deformation of surrounding rock. Fifth, the excavation and supporting are conductive to concrete lining structure stress, and pre-splitting blasting is not conductive to concrete lining structure.

8 CONCLUSIONS

This paper started with the characteristics of columnar jointed basalt structure, stress, strain, deformation, and so on, and has thoroughly and systematically studied the evolution law and the deformation mechanism of columnar jointed basalt surrounding rock relaxation. Furthermore, it preliminarily discussed the evaluation of surrounding rock security characteristics by means of energy rate and unit deformation force. The following conclusions are drawn:

1. The self-stability of columnar jointed basalt is poor; deformation of surrounding rock decreases with the increase in depth. Excavation and supporting is helpful to control the deformation convergence range, pre-splitting blasting is helpful to control the deformation of the whole surrounding rock, and excavation without supporting as well as pre-splitting blasting is harmful to control the deformation convergence range.

2. Support the rear part of the free surrounding rock to form an arch ring, which is helpful to absorb energy released from the deformation of surrounding rock. Pre-splitting blasting can prevent surrounding rock deformation effectively, but it is harmful to control implied joint deformation created by the late deformation of surrounding rock. The excavation is helpful to release the energy of surrounding rock shallow joint deformation, and timely supporting is helpful to control the energy of implied joint deformation created by the late deformation of surrounding rock.
3. Energy rate and unit deformation force can evaluate the columnar jointed basalt surrounding rock security characteristics effectively.

ACKNOWLEDGMENTS

This study was sponsored by the China Institute of Water Resources and Hydropower Research Project (GE0145B452016 and 2014-93-08-01) and the National Key Research and Development Project (2016YFC0401804).

REFERENCES

Chen Xu, Xu Mo, Kang Xiaobing, et al. 2008. The Research on Genetic Habit of Columnar Cleavage in Basalt and the Influence on the Rock Mass [J]. Geological Prospecting Review 23(3): 260–263 (in Chinese).

Di Shengjie, Xu Weiya, Wang Wei, et al. 2011. Transversely Isotropic Constitutive Properties of a Columnar Jointed Rock Mass [J]. Journal of China University of Mining & Technology 40(6): 881–887 (in Chinese).

Hao Jianjie, Feng Xiating, Li Shaojun, et al. 2015. Failure Mode of Columnar Jointed Basalt Tunnel and its Mechanism Simulation [J]. Rock and Soil Mechanics 36(3): 837–846 (in Chinese).

Jiang Quan, Feng Xiating, Fan Yilin, et al. 2013. Survey and Laboratory Study of Anisotropic Properties for Columnar Jointed Basaltic Rock Mass [J]. Chinese Journal of Rock Mechanics and Engineering 32(12): 2527–2535 (in Chinese).

Liu Haining, Wang Junmei, Wang Sijing. 2010. Experimental Research of Columnar Jointed Basalt with True Triaxial Apparatus at Baihetan Hydropower Station [J]. Rock and Soil Mechanics, 31(Supp. 1): 163–171 (in Chinese).

Liu Shungui, Chi Yongxiang, Wang Sijing, et al. 2009. Size Effect on Shear Strength of Basalt rock Mass with Columnar Joints [J]. Journal of Engineering Geology 17(3): 367–370 (in Chinese).

Shi Anchi, Tang Mingfa, Zhou Qijian. 2008. Research of Deformation Characteristics of Columnar Jointed Basalt at Baihetan Hydropower Station on Jinsha river [J]. Chinese Journal of Rock Mechanics and Engineering 27(10): 2079–2086 (in Chinese).

Wang Hongbin, Hao Xianjie, Li Shaojun, et al. 2015. Properties of Columnar Jointed Rock Mass in Rock Structure and Unloading Loose and its Engineering Prevention [J]. Civil foundation 29(3): 84–88 (in Chinese).

Wu Wenping, Feng Xiating, Zhang Chuanqing, et al. 2011. Failure Mode Dynamic Recognition and Control for Surrounding Rock of Large-scale Cavern Group [J]. Chinese Journal of Rock Mechanics and Engineering 30(9): 1782–1802 (in Chinese).

Xiang Tianbing, Feng Xiating, Jiang Quan, et al. 2011. Failure Mode Dynamic Recognition and Control for Surrounding Rock of Large-scale Cavern Group [J]. Chinese Journal of Rock Mechanics and Engineering 30(5): 871–883 (in Chinese).

Yan Dongxu, Xu Weiya, Wang Wei, et al. 2012. Research of Size Effect on Equivalent Elastic Modulus of Columnar Jointed Rock Mass [J]. Chinese Journal of Geotechnical Engineering 34(2): 243–250 (in Chinese).

Zhu Daojian, Yang Linde, Cai Yongchang. 2009. Research on Anisotropic Characteristics and Size Effect of Columnar Jointed Rock Mass [J]. Chinese Journal of Rock Mechanics and Engineering 28(7): 1405–1414 (in Chinese).

Zhu Daojian, Yang Linde, Cai Yongchang. 2009. Simulation of Compressive Failure Process of Columnar Jointed Rock Mass and its Failure Mechanism Analysis [J]. Chinese Journal of Rock Mechanics and Engineering 28(4): 716–724 (in Chinese).

Applied Mechanics and Civil Engineering VI – Xie (Ed.)
 ISBN 978-1-138-62631-7

Low-energy environmental improvement for glass curtain wall buildings

M. Cheng
Department of Structural Engineering and Disaster Reduction, Tongji University, Shanghai, China

X.Z. Zheng & W. Cao
Department of Civil Engineering, Tongji University, Shanghai, China

ABSTRACT: This paper starts with a description of the energy-extensive consumption problems in China and analyzes some methods to reduce the energy consumption. It suggests an integration of natural ventilation, sunshade, and building intellectualization technology to present a new project in order to promote building environment with low energy consumption. Moreover, this paper analyzes 1-year temperature data of two models, compares the different effects in different situations, concludes that the energy conservation efficiency is 27.5%, and proves the value of the project.

1 INTRODUCTION

According to the literature (Allouhi, Fouih, Kousksou, Jamil, Zeraouli, Mourad. 2015), in 2013, the whole building energy consumption in China increased to 800 million TEC, comprising 21% of the total energy consumption. Large public buildings (with area higher than 20,000 m^2) mostly use glass curtain wall, and their windows are hardly opened, making it hard to use natural ventilation (Zhu. & Li. 2015). Most of these buildings use air conditioner to keep ventilating environment and comfortable temperature, which leads to high power consumption. Large public buildings in China, including stations and supermarkets, have a total area of about 50 million m^2, only one hundredth of the whole building proportion. However, they comprised 10% of the total energy consumption. According to the literature (Zheng, Li, Yu, Li, Shang. 2016), power consumption of large public buildings is about 200 kWh/(m^2.a), while that of normal buildings is 29 kWh/(m^2.a).

At present, electricity consumption of large public buildings in China has neared the level in Europe; however, this situation could be improved, because these buildings are associated with a serious problem of power wastage. Considering the high power consumption per unit area and the large proportion of using building electricity, an energy conservation of 30% is anticipated. Hence, the research and analysis on current energy using situation and the proposal of an energy conservation plan are crucial. In addition, too less ventilation and overdosed sunshine radiation lead to inferior comfort level. This problem is urgent to be solved as well.

This study puts forward a new energy conservation plan creatively, which integrated the technology of natural ventilation, sunshade, and promotion of building environment. The effect of this plan on environment improvement and energy conservation is verified by small-scale experiments.

2 EXISTING ENERGY CONSERVATION TECHNOLOGIES

2.1 *Material improvement*

New types of glass material can be helpful to reduce building power consumption. These materials can be divided into radiation reducing, conduction reducing, and combination types based on their principles (Lin, Li. 2015).

Low-E glass is the commonly applied new material. In addition to a high reflective rate on far-infrared region, Low-E glass possesses fine light-admitting quality. It can also reduce the impact of outdoor environment, in order to function as an energy conservator (Giovannetti, Föste, Ehrmann, Rockendorf. 2014). Buildings with monolayer Low-E glass can save about 50% energy while that figure for those with double-layer Low-E glass is 70% (Kong, Liu, Yang, Zhong, Qi. 2016). However, this method has the problem of too high cost and too fewer manufacturers in China, and it is not suitable with building renovation.

2.2 *Structural improvement*

Glass curtain walls with new structural design such as energy-saving breathing-type double-layer skin façade can readjust the ventilation situation in the building (Wang, Zhang, Tan, Li. 2011). Comparing with monolayer glass curtain wall buildings, buildings with structural improvement have competitive edges on energy saving, and better heat and noise insulation (Francesca, Fabrizio, Rosa, Filippo, Maria. 2015).

However, this method has high costs of cleaning and operation, and will be less effective with incorrect operation.

2.3 *Natural ventilation*

Nowadays, most buildings in China use air conditioners to adjust indoor temperature. However, using air conditioner would bring the problem of high energy consumption as well as high financial and labor expenditure. This problem could be amplified in large public buildings.

The use of natural ventilation technology is the best way to save the energy used by air conditioners. In traditional Chinese buildings, the courtyard in Hui buildings is the example of natural ventilation. In modern buildings, natural ventilation is achieved by means of roof ventilation, façade ventilation, and cowl ventilation (Sanchez, Rolando, Sant, Ayuso. 2016).

Considering the high building density in cities, the effect of natural ventilation is not very high.

2.4 *Sunshade technology*

Sunshade technology means preventing too much absorption of sunshine with improvements in material or structural properties. In this way, solar radiation absorbed by buildings would be reduced, indoor temperature would be controlled, and the goal of energy saving would be achieved. This method would be categorized into outside layer, inside layer, and middle layer. In the long term, this technology would bring a huge profit on energy conservation.

3 INTELLIGENT ENERGY CONSERVATION PLAN FOR GLASS CURTAIN WALL BUILDINGS

3.1 *Key points*

In view of the glass curtain wall building in existence, this work raises a renovation plan combining tradition and modern technology. Its key points are:

a. Avoiding the high cost and difficulty of changing glass material; this plan chooses to use ventilation and sunshade technology rather than saving energy by the new materials.

b. To make up the problem of insufficient natural ventilation; this plan combines wind pressure ventilation and thermal pressure ventilation. In order to form the channel of wind, there will be a serious problem of renovation on structure, such as an opening on the top and side ventilating sashes.
c. Sunshade baffles are expecting to be programmed to increase the energy-saving efficiency. Energy used by rotatable baffles is provided by the solar energy absorbed by solar panel on baffles.
d. Under the opening on the top, there is an ecological pond, which improves indoor humidity and environmental situation at the same time with recycling rainfall. The plan is shown in Figure 1.

3.2 *Model experiment*

For investigating the energy conservation effect of this plan, reduced-scale model tests are conducted in this study.

3.2.1 *Model design*

The reduced-scale model parameters are shown in Table 1. The model is designed to be flattening just like most public buildings. The real models are shown in Figure 2.

3.2.2 *Natural ventilation experiment*

In this study, the natural ventilation experiment was conducted for 271 days. To prove the results of natural ventilation, we collected and compared the temperature data of opening and closed model. The experimental data are shown in Table 2.

Moreover, we conducted 12-h experiments in fine and cloudy days without sunshade baffle.

3.2.3 *Integrated plan experiment*

The scheme integrates the baffle sunshade technology, in order to verify the integrated energy-saving effect of baffle sunshade and the natural ventilation, from 8 September 2015 to 10 September 2015, experiment with baffles on a sunny day, no baffle on cloudy day, and no baffle on sunny

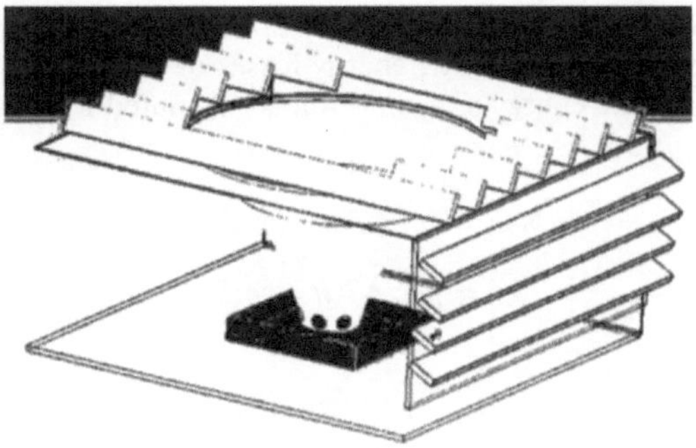

Figure 1. Diagrammatic sketch on renovation plan.

Table 1. Model parameters.

Model size	700 mm × 700 mm × 400 mm
Model material	Monolayer armored glass
Model quantity	2
Closed glass box	Airtight
Opening glass box	Opening on the top (diameter: 15 cm) Ventilation sashes on two sides (diameter: 5 cm) Ecological pond (external diameter: 20 cm internal diameter: 15 cm)

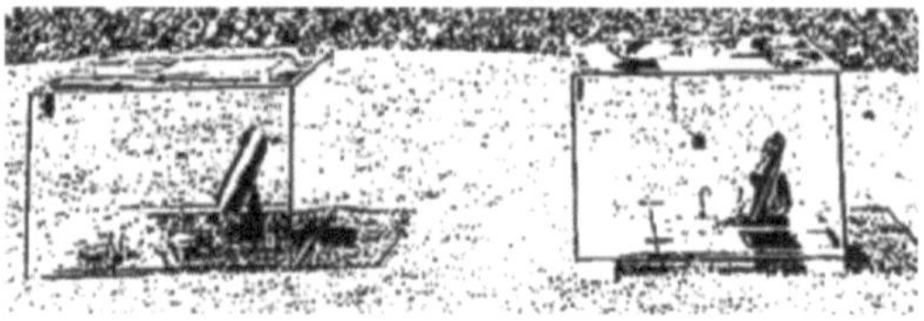

Figure 2. Real models.

Table 2. Natural ventilation experiments.

Experiment content	Record the temperature in two boxes
Experiment equipment	Opening box, closed box, thermometer
Experiment time	21 March 2015 to 16 December 2015
	Each day 7:30–8:00; 11:30–12:30; 16:30–17:30

day tests are conducted for 12 consecutive hours. We recorded hourly temperatures of opening glass box and closed glass box from 6:00 to 18:00, and conducted a comparative analysis.

3.3 *Data processing and analysis*

Before analyzing, the data were corrected by removing singular points and triangle interpolation processing, to improve its reliability. In this paper, the way to find singular points is comparison of the temperature at the same time in the vicinity of a few days.

3.3.1 *Energy consumption calculation*

In general, recognized and applied building energy analysis method is divided into two major categories: (1) static energy consumption analysis method and (2) dynamic energy analysis method established on the unsteady heat transfer theory. The two methods have different application ranges due to their respective characteristics.

For higher precision, the energy consumption of models is calculated as an entirety. The whole energy consumption is assumed equal to the difference between energy states of models and appropriate situation. Because the energy losses of different power supply methods are different, in this paper, the energy consumption is calculated without losses. The calculation idea is shown in Figure 3.

$$W_i = cm\,\Delta T dt = cmS_i \tag{1}$$

$$W_s = \int_i cm\,\Delta T dt = cmS \tag{2}$$

where c = specific heat; m = quality; S = the area difference between two curves; and W_i = the energy to change the temperature inside to appropriate temperature in the micro period.

3.3.2 *Data processing*

In the ventilation experiments, the analysis period is divided by seasons. The experimental data are shown in Figure 4. After triangle interpolation processing, the collected data are connected to broken lines. Using the integral, we accumulate the area of continuous small micro regions, and this area can reflect the energy consumption of glass houses, and the calculation results of energy consumption are shown in Table 3. The total annual energy is 2056.34 J, and the energy saving ratio is 18.27%:

$$R^2 = 1 - \frac{SSE}{SST} = \frac{\sum_1^n (\hat{y}_i - \bar{y})^2}{\sum_1^n (y_i - \bar{y})^2}$$

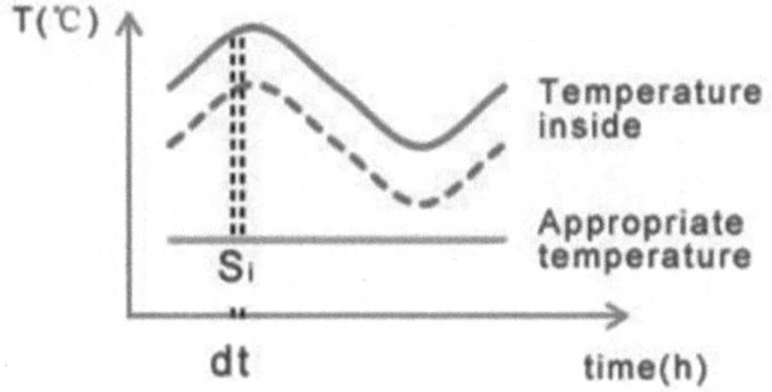

Figure 3. Calculation idea.

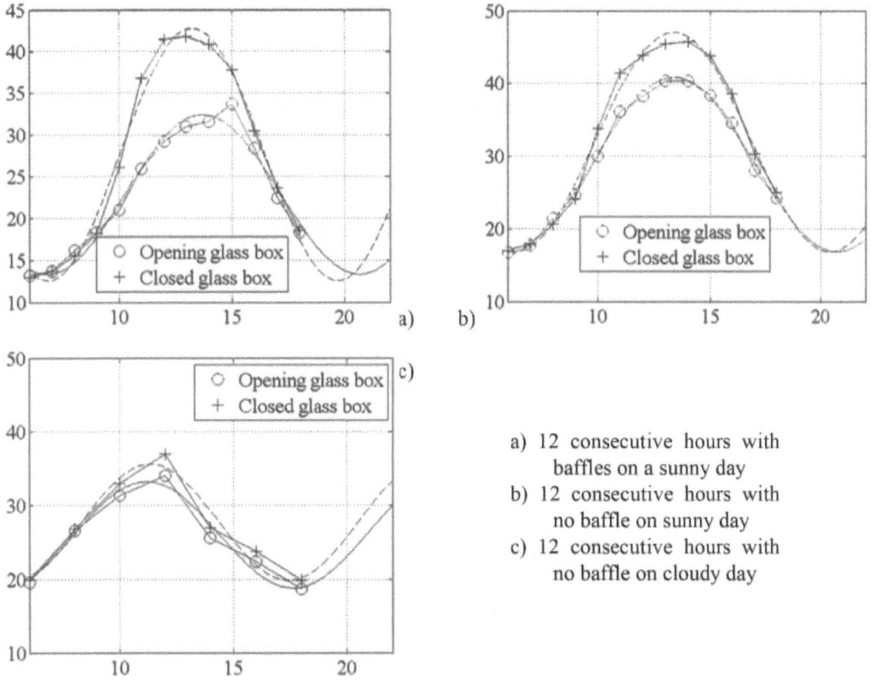

Figure 4. Temperature curves simulated by MATLAB.

Table 3. Energy consumption of two models during thenatural ventilation experiments (J).

	Total	Spring	Summer	Autumn
Opening glass box	4600.89	1173.36	2768.58	658.98
Closed glass box	5629.08	1581.19	3144.38	903.51
Difference of energy consumption	1028.18	407.83	375.80	244.53
Energy-saving ratio (%)	18.27	25.79	11.95	27.06

where n = samples; y_i = temperature recorded at one point; $\bar{y}$ = average temperature; and $\hat{y}_i$ = temperature simulated at one point.

In integrated experiments, collected data were fitted by several ways, as shown in Figure 5. This study also compared the fitting results between different methods according to R square test and Equations 2–3, and the trigonometrically fitting method is chosen. The results of three fitting methods are shown in Table 4.

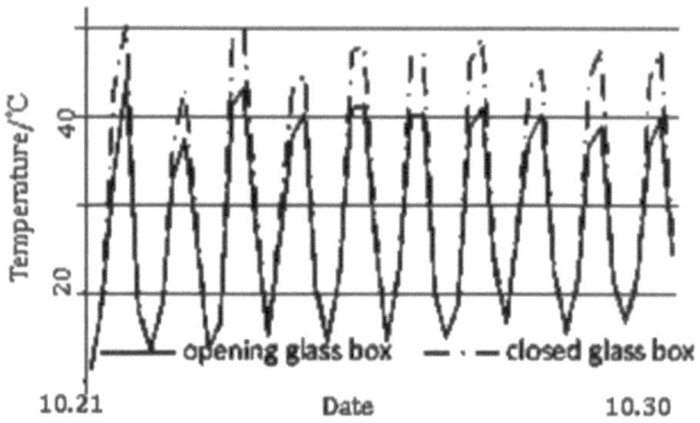

Figure 5. Temperature data from 21 to 30 October 2015.

Table 4. Simulation of total energy-saving effect.

Method	Fitting formula	R^2
Trigonometrically fitting	$T = a \times \sin(b \times t + c) + d$	0.97
Gaussian fitting	$T = a \times exp\left(\left(-\frac{t-b}{c}\right)^\wedge 2\right)$	0.88
Cubic fitting	$T = a \times t^3 + b \times t^2 + c \times t + d$	0.92

Table 5. Energy consumption of two models during the integrated plan experiment (J).

Experiment	Subject	Energy consumption (J)	Energy-saving ratio (%)
12 consecutive hours with no baffle on sunny day	Opening glass box	318.34	21.5
	Closed glass box	405.45	
12 consecutive hours with no baffle on cloudy day	Opening glass box	170.78	10.9
	Closed glass box	191.61	
12 consecutive hours with baffles on a sunny day	Opening glass box	290.24	27.5
	Closed glass box	400.35	

The fitting equation is $T = a \times \sin(b \times t + c) + d\ (2-4)$, a, b, c, and d are fitting coefficients, and t denotes time. The fitting process was finished by MATLAB. Using the calculation methods mentioned above, the energy consumptions of two boxes are calculated, and the results are presented in Table 5.

3.3.3 *Data analysis*

a. Reliability of trigonometrically fitting

It can be seen that the fitted values are close to 1 though R square test, indicating the goodness of curve fitting. In the figure, the goodness of fit is presented as the observed values of curve fitting results and the original data are approximately the same.

The natural ventilation energy saving ratio is 18.27%, which is between the 10.9% of no baffle cloudy experiment and 21.5% of no baffle fine day experiment. This phenomenon can also indicate the rationality of trigonometrically fitting.

b. Difference between ventilation energy-saving effect in different seasons.

During the experiment, the energy conservation effect would change with the season. This effect is more obvious during spring and autumn. For example, from 21 to 30 October 2015, as a period in autumn, the maximum temperature difference between opening glass box and

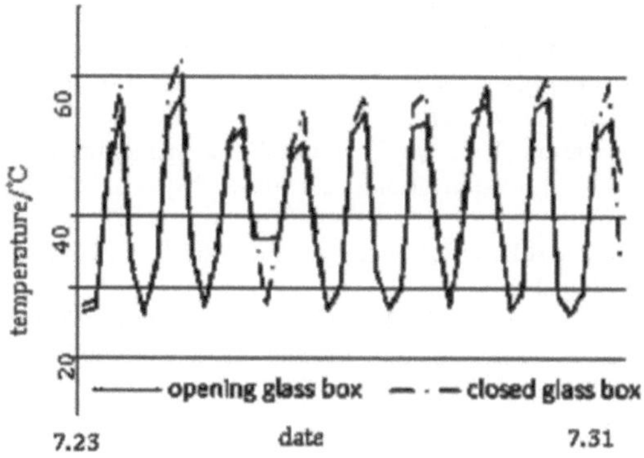

Figure 6. Temperature data from 23 to 31 July 2015.

closed box reached 9.4°C, while the average temperature difference is 3.13°C. However, in summer, for instance, from 23 to 31 July, the average temperature difference is only 0.94°C. The temperature data are shown in Figures 5 and 6.

This phenomenon can be explained by the different climate situation during different seasons. In autumn, the outdoor temperature is about 25°C, lower than that in summer, and the wind is stronger. It can be considered that at this time the volume of the box of ventilation is larger, and the temperature difference is more obvious, so that the energy saving effect is more remarkable.

c. Comparison between various plans.

The comparison between closed box with and without baffle in fine days could be concluded to the result that the energy-saving ratio of sunshade baffle is 15.2%. Three integrated experiments are conducted in summer, and the best energy conservation ratio reached 27.5%, which is much higher than the ratio 12% of plans only with ventilation. Moreover, the use of sunshade baffle could increase the volume of ventilation as it reduces indoor temperature. Overall, the combination of sunshade baffle and natural ventilation would reach the best energy-saving effect among different plans.

Model tests show the effectiveness of the scheme under current conditions. However, reduced-scale model experiments have boundary condition difference with full-scale buildings. The energy consumption situation is distinguishing in different scales. Therefore, energy-saving effect of this scheme in the large-scale case needs further research.

4 CONCLUSION

The experimental data prove that the energy conservation rate of natural ventilation is 18.27% in a relatively long time, as well as the influence of this plan. Different types of subdivision experiments illustrate the efficiency of this plan in different situations. Energy conservation rate of 27.5% in the fine day case with sunshade baffle is the best result, which would prove the superposition effect of natural ventilation and sunshade baffle.

Building energy conservation, especially that of large buildings with glass curtain wall, is crucial. Traditional methods cannot meet the expectation albeit they have advantages. However, the new intelligent plan proposed in this study has been proved by related experiments. Furthermore, it is an excellent plan for energy conservation of glass curtain wall buildings. Even though large-scale experiments are lacking, data and laws concluded from this study are beneficial to further research.

REFERENCES

Allouhi A., Fouih Y.El., Kousksou T., Jamil A., Zeraouli Y., Mourad Y. 2015. Energy consumption and efficiency in buildings: current status and future trends. *Journal of Cleaner Production* 109: 118–130.

Francesca C., Fabrizio A., Rosa F.D.M., Filippo D., Maria R.P. 2015. Multidisciplinary approach to structural/energy diagnosis of historical buildings: a case study. *Procedia Engineering* 104: 52–59.

Giovannetti F., Föste S., Ehrmann N., Rockendorf G. 2014. High transmittance, low emissivity glass covers for flat plate collectors. *Solar Energy* 104: 52–59.

Kong X.F., Liu S.N., Yang H., Zhong Y.L., Qi C.Y. 2016. An experimental study of all-season operation strategy for a respiration-type double-layer glass curtain wall system in cold zone of China. *Building and Environment* 97: 166–176.

Lin S.J. & Li H.H. 2015. Reducing solar effect on the glazing material based on using non-linear patterns. *Energy and Buildings* 109: 369–376.

Sanchez E., Rolando A., Sant R., Ayuso L. 2016. Influence of natural ventilation due to buoyancy and heat transfer in the energy efficiency of a double skin facade building. *Energy for Sustainable Development* 33: 139–148.

Wang F., Zhang X.S., Tan J.J., Li X.W. 2011. The thermal performance of double skin façade with Tillandsia usneoides plant curtain. *Energy Procedia* 75: 1325–1334.

Zheng X., Li H.Q., Yu M., Li G., Shang Q.M. 2016. Benefit analysis of air conditioning systems using multiple energy sources in public buildings. *Applied Thermal Engineering* 107: 709–718.

Zhu J.H. & Li D.Y. 2015. Current situation of energy consumption and energy saving analysis of large public building. *Procedia Engineering* 104: 52–59.

Applied Mechanics and Civil Engineering VI – Xie (Ed.)
© 2017 Taylor & Francis Group, London, ISBN 978-1-138-62631-7

Study on the mechanical properties of a damaged steel beam

Xibing Hu, Yuxuan Xiang, Yafang Chen, Zhishen Yuan & Xiaodong Xiang
College of Civil Engineering and Mechanics, Central South University of Forestry and Technology, Changsha, Hunan, China

ABSTRACT: By using the finite element programming, a finite element model of an H-section steel beam (Q345) was established. Then, the influence patterns of the key parameters on the mechanical properties of the damaged steel beam were analyzed, which include the varying damage degree of the cross section, height-to-thickness ratio of the web plate, and the width-to-thickness ratio of flange. The results of the analysis indicate that the damage of the cross-section reduces the stiffness of the steel beam, but it has little influence on the ultimate bearing capacity of the steel beam. Furthermore, the web height-to-thickness ratio and the width-to-thickness ratio of the flange can increase the stiffness and the ultimate bearing resistance of the damaged steel beam.

1 INTRODUCTION

The steel structure engineering in China increased with years, and it has many advantages, but the corresponding security problems are gradually reflected, and the existing steel structure engineering, for instance, natural calamities and man-made misfortunes will cause a different degree of damage as the service life of the steel buildings increases, which could lead to different levels of security risks. Even some steel structure engineering has entered a frequency period of accidents, and the safety issues of some old steel structure engineering service time have attracted much attention. To the existing steel structure engineering, predicting and judging its ability to for continuous service and detecting damage to the original structure have become the key for the safe use of steel structure engineering and protecting people's property. Research on the damage degree of the steel structure can not only ensure the safe use of the existing steel structure, but also guide secured construction of steel structure, improve the ability to predict damage, reduce the probability of steel structure engineering accident, and give full play to the superiority of steel structure in building.

In this paper, the finite element model of H-section steel beam (Q345) was established, and the influence patterns of the key parameters on the mechanical properties of the damaged steel beam were analyzed, such as the varying damage degree of the cross-section, the web height-to–thickness ratio, and the width-to-thickness ratio of the flange.

2 ESTABLISHMENT OF FINITE ELEMENT MODEL

In order to simulate the stress process of the steel beam accurately, a general finite element program was used to analyze the mechanical properties of H-section steel beam. In this paper, the finite element model is established by using entity unit, and the simulation of constraint conditions can be divided into two types: (1) the vertical constraint condition of the simulation test beam, namely the simulation of loading effect and supporting effect of bearing and (2) the simulation of out-of-plane constraint conditions, which prevent the deformation of the test beam, namely the simulation of the role of the lateral restraint device. First, eigenvalue buckling analysis is performed and the initial imperfection is imposed by the zoom factor 0.1%. Then, the calculation results of carrying capacity show that the finite element method can effectively simulate the local buckling of the flange and web of the steel beam.

3 PARAMETER ANALYSIS OF DAMAGED STEEL BEAM

According to the distribution characteristics of stress and strain of the cross-section of the bending member, under bending condition, the flange of H-shaped cross section first yields, after all the flange damage, the web from both sides to the middle of the web to bending, then. From the slight to full bending of the web, the yield of the flange becomes more serious. In this paper, we defined general damage as the damage that causes full flange of H-shaped cross-section buckling while the Web is not buckling. We defined the serious damage as the damage that causes the 1/2 h_w of the web of H-shaped cross-section already buckling while the flange buckling is more serious. In order to study the effect of the general damage and serious damage on the stress of the H-section steel beam and analyze the influence rules of flange width-to-thickness ratio, web thickness ratio, and other parameters on the flexural behavior of steel beam after damage, we established five sets of H-section steel beam models, and the geometric dimensions of the model are shown in Figures 1 and 2 and Table 1. Each model can be divided into three cases: the pure bending section of steel beam under zero damage, general damage, and serious damage.

The experimental results of damaged steel by domestic and foreign scholars found that the elastic modulus of the steel is satisfied with a certain rule, and it can be expressed as follows:

$$0 \le \varepsilon \le 1.6,\ \check{E} = E_0 \tag{1}$$

$$1.6 \le \varepsilon \le 4,\ \check{E} = \frac{(120 - 12.5\varepsilon)}{100} E_0 \tag{2}$$

$$4 < \varepsilon \le 18,\ \check{E} = \frac{(74.64 - 1.16\varepsilon)}{100} E_0 \tag{3}$$

where ε_a = strain; E_0 = initial elasticity modulus; and $\check{E}$ = elasticity modulus in damage.

On the basis of the results of the above research, stress–strain curves of the steel model used in this paper are shown in Figure 3. The stress–strain curve of steel in the model of the undamaged steel beam is curve *A*; The stress–strain curve of the flange of H-shaped pure bending cross section in the model of general damage steel beam is curve *B*, and other place is curve *A*; The stress–strain curve of the flange of H-shaped pure bending cross section in the model of serious damage steel beam is curve *C*, the web edge of pure bending section to the 1/4 h_w segment of the middle line is curve *B*, and other place is curve *A*. After the calculation of the finite element programming, we can draw five load deflection curves, shown in Figures 4–8; thus, the ultimate load F_u can be obtained for model beams under each different damage levels, the Mid span deflection *S* can also be obtained when the model beam reaches the ultimate load, and the ratio of the *S* of model beam at different damage levels to the *S* of model beam at no damage can be calculated. The ratio is defined as *D*, shown in Table 1.

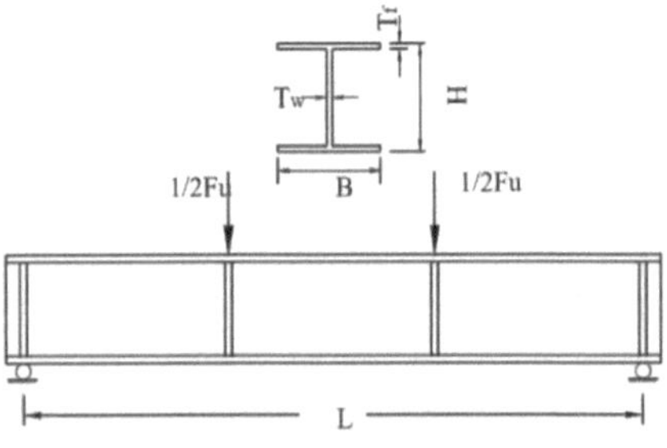

Figure 1. Layout of beam model.

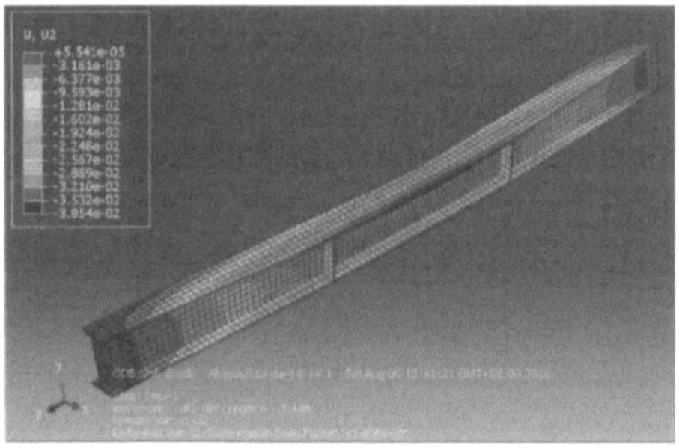

Figure 2. Cloud image of vertical displacement.

Table 1. Beam model size and results of parametric analysis.

Number	L*	B*	H*	T_w*	T_f*	Damage degree	F_u(kN)	*S**	*D*(%)
Beam-1	2100	100	150	6	6	Undamaged	93.30	79.41	
						General	94.88	54.56	68.71
						Serious	94.79	35.57	44.79
Beam-2	2100	150	150	6	6	Undamaged	122.86	80.84	
						General	122.91	67.11	83.02
						Serious	122.44	38.28	47.36
Beam-3	2100	75	150	6	6	Undamaged	77.81	85.03	
						General	78.01	38.31	61.69
						Serious	79.01	35.27	64.73
Beam-4	2100	100	100	6	6	Undamaged	50.71	82.77	
						General	50.69	46.96	53.04
						Serious	51.31	36.18	43.71
Beam-5	2100	100	200	6	6	Undamaged	141.37	80.47	
						General	141.86	63.14	78.46
						Serious	141.42	51.12	63.53

*Unit in millimeters.

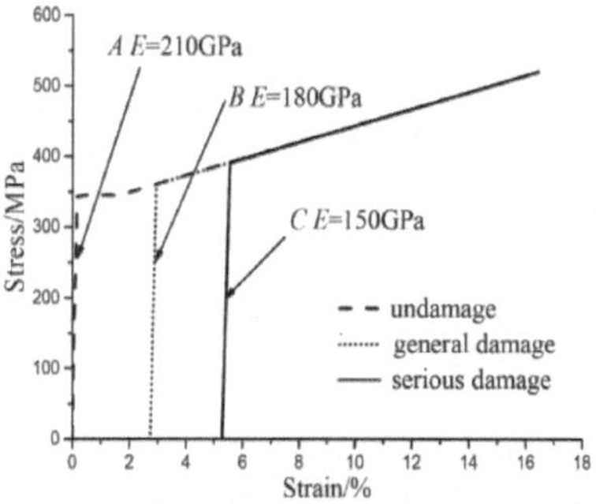

Figure 3. Stress–strain curve of beam model.

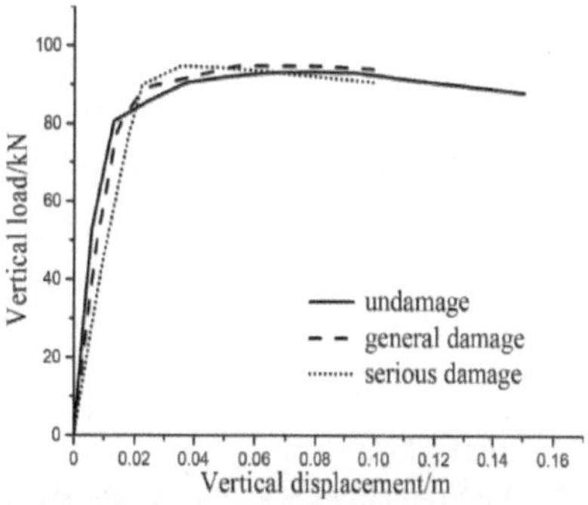

Figure 4. Load–deflection curve of beam-1.

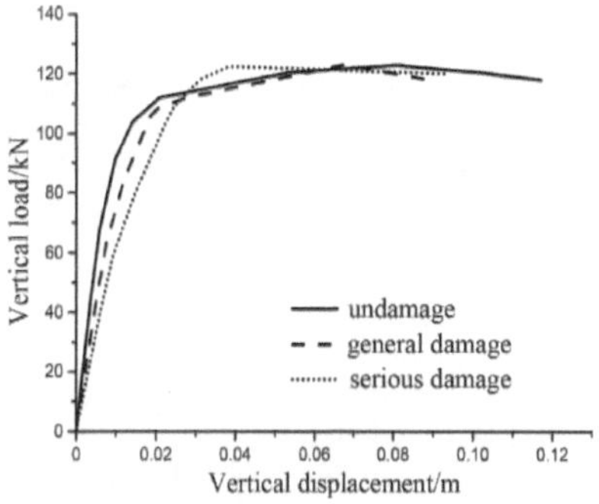

Figure 5. Load–deflection curve of beam-2.

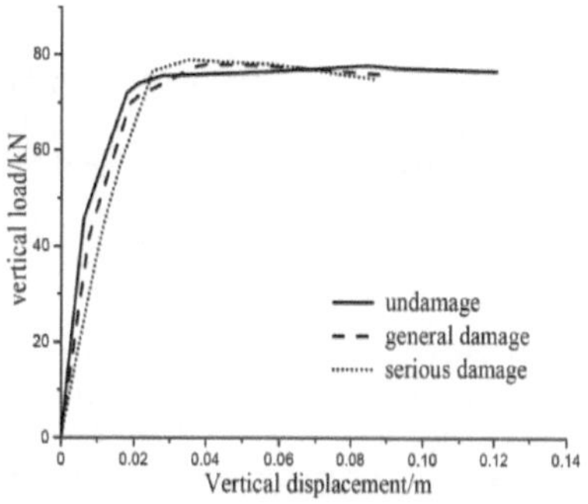

Figure 6. Load–deflection curve of beam-3.

By analyzing the data in Table 1, we show that:

1. For each beam model, F_u is the same under different damage degrees, but S decreases with the increase in the degree of damage.
2. By comparing Beam-1, Beam-2, and Beam-3, D increases with the increase in the width-to-thickness ratio of the flange in the limit range. This is because the damaged flange still

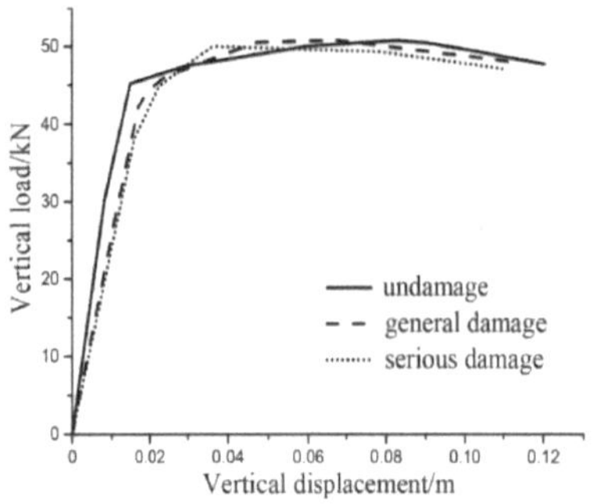

Figure 7. Load–deflection curve of beam-4.

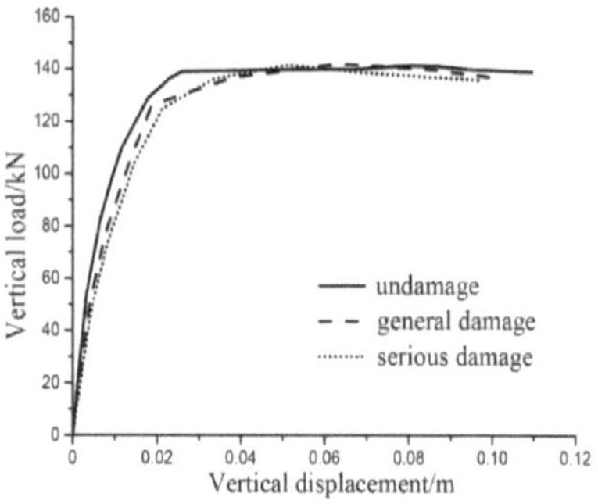

Figure 8. Load–deflection curve of beam-5.

has a certain bearing capacity; the larger the width-to-thickness ratio of the flange, the higher the bearing capacity and ductility of the steel beam.

3. By comparing Beam-1, Beam-2, and Beam-3, D increases with the increase in the height-to-thickness ratio of the web; this is because, web still has a certain bearing capacity while it is damaged. The higher the height-to-thickness ratio of the web, the higher is the bearing capacity and ductility of the steel beam.

4 CONCLUSIONS

We first performed the finite element analysis of damaged steel beam and then studied the damaged steel structure. The following conclusions were drawn by analyzing the different parameters:

1. The section damage reduces the stiffness of the steel beam; the more serious the damage, the greater the decline, but it has no significant effect on the ultimate bearing capacity of steel beam.
2. The increase in the width-to-thickness ratio of the flange will increase the ductility and ultimate bearing capacity of the steel beam in the limit range of width-to-thickness ratio and under the same degree of damage.
3. The increase in the height-to-thickness ratio of the web plate will increase the ductility and ultimate bearing capacity of the steel beam in the limit range of height-to-thickness ratio and under the same degree of damage.

ACKNOWLEDGMENTS

This work was financially supported by Hunan Provincial Ministry of Housing and Urban-Rural Development Science Project (2016-K5-054), Youth project of Hunan Provincial Department of Education (14B187), and the National Natural Science Foundation of China (51608544).

REFERENCES

Fan Qinshan, & Yin Yajun. *Material Mechanics* [M]. Bejing: Tsinghua University Press, 2008.

Hu Xingbing, & Yang Yuwei. Study on mechanical model for analysis of damaged steel structure [J]. *Chinese Journal of Computational Mechanics,* 2016, 33(1): 128–134. (in Chinese).

Lan Shengning & Zhong Xingu. Damage diagnoses and lessons learnt from the failure of light steel structure by heavy snow in Xiangtan [J]. *China Civil Engineering Journal,* 2009, 42(3): 71–75. (in Chinese).

Lemaitre, J. *Damage Mechanics Tutorial* [M]. Ni Jingang, & Tao Chunhu, Translated. Beijing: Science Press, 1996. (in Chinese).

Singhal, A. & Kiremidjian, A.S. Method for probabilistic evaluation of seismic structural damage [J]. *Journal of Structural Engineering*, 1996, 122(12): 1459–1467.

Wang Yuanqing, Hu Zongwen & Shi Yongjiu. Analysis and reflection on snow disaster accidents of steel structures of light-weight buildings with portal frames [J]. *China Civil Engineering Journal*, 2009, 42(3): 65–70. (in Chinese).

Wald, F., Simoes da Silva, L. & Moore D.B., et al. Experimental behavior of a steel structure under natural fire [J]. *Fire Safety Journal*, 2006, 41(7): 509–522.

Applied Mechanics and Civil Engineering VI – Xie (Ed.)

Structural behavior analysis of the steel frame of a damaged beam end

Xibing Hu, Yafang Chen, Yuxuan Xiang, Zhishen Yuan & Zepu Jiang
College of Civil Engineering and Mechanics, Central South University of Forestry and Technology, Changsha, Hunan, China

ABSTRACT: By using the finite element analysis software, the structural behavior analysis of the steel frame of a damaged beam end was carried out to further study the damaged steel frame structures by comparing undamaged and damaged beams (damaged in one end or two ends of the beam). The result of the analysis indicates that damage has a great impact on the horizontal displacement of steel frame structures. Compared with the theoretical calculation method, the results indicate that the linear phase is closer to each other, which means that the finite element model can simulate the structural behavior of damaged steel frames very well. Considering the effect of damage on the layer, a three-storey steel frame is established to analyze the relationship between different layers and the damage index D.

1 INTRODUCTION

During processing and transportation of steel structures, occurrence of some extent of damage in structures is inevitable. The same can be said of natural disasters. The damage of steel frame structures was evaluated based on the arrangement of joints and nodes using the guidelines of the *Code for acceptance of construction quality of steel structures* in China. While this method is not based on the reliability theory for calculation, it is difficult to determine the actual mechanical properties of structures.

In this paper, we mainly analyzed the mechanical properties of damaged steel frame structures based on their arrangement. If the beam end is damaged, the stiffness of the beam end is considerably reduced. By using the index *D* to determine the degree of damage of the steel frame, it uses the finite element software to analyze the mechanical properties, including bearing capacity and horizontal displacement, of the steel frame structure under different degrees of damage.

2 THEORETICAL ANALYSIS

2.1 *Index D*

Steel is one of the stable and continuous materials. The stress–strain relationship is:

$$\sigma = E\varepsilon \tag{1}$$

When the beam end of steel frame structures is damaged, the stiffness of damage section decreases and the elasticity modulus varies. Therefore, the elasticity modulus is modified by damage index D:

$$E^{*} = (1-D)E \tag{2}$$

where E^* = the average elasticity modulus of damage section; E = the elasticity modulus of undamaged section; and D = damage index, when $D = 0$, $E^* = E$, steel frame is undamaged and when $D = 1$, the section is damaged completely.

2.2 *Theoretical calculation*

The principle of structural mechanics can be used for steel frame mechanics analysis.

Horizontal displacement of steel frame can be calculated by the equation of static equilibrium:

$$M_{AB} + M_{AC} = 0 \tag{3}$$

$$M_{BA} + M_{BD} = 0 \tag{4}$$

$$F_H + F_A + F_B = 0 \tag{5}$$

According to formula (3), formula (4), and formula (5), it could evolve the horizontal displacement of steel frame.

3 FINITE ELEMENT ANALYSIS

Finite element software is used for modeling analysis, and in the model, the elasticity modulus of the beam end is used as a variable to simulate its damage.

3.1 *A single-storey planar steel frame structure*

3.1.1 *Calculation model and material properties*

In order to simulate the actual situation, the stress–strain constitutive relation refers to the material properties from experiment. Assuming steel yield strength as 360 *MPa*, elasticity modulus as 1.80×10^5 *MPa*, ultimate strength as 520 *MPa*, the ultimate damage area length is 200 mm.

3.1.2 *Meshing and loading system*

Model is mainly mesh by system intelligently. Element shape is hexahedron to avoid convergence problem as far as possible. Restrain all degrees of freedom of frame column base in the calculation process. During the load process, the vertical load is 40 kN constant load in trisection and horizontal load is increased by 0, 20 kN increase per level. Record horizontal displacement values and sum up the table.

3.1.3 *Finite element analysis*

According to the calculation results of finite element model as in Figure 3, the change of horizontal displacement is showed in Tables 1 and 2.

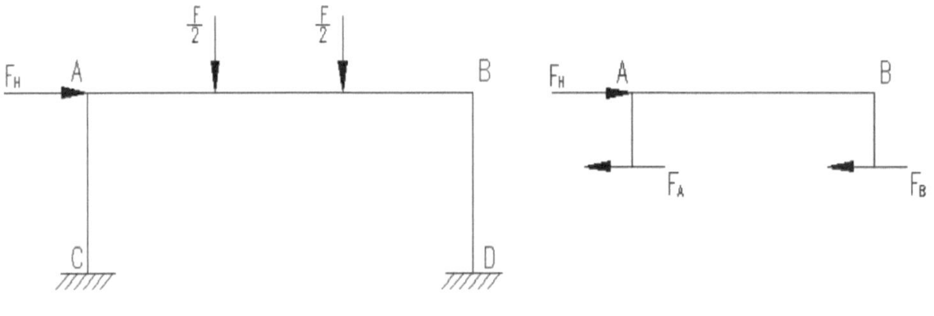

Figure 1. Diagram of calculation.

As the horizontal displacement under different extent of damage in Tables 2 and 3 can be seen:

1. Compared to the undamaged steel frame structure, horizontal displacement of the damaged frame is increased. When the horizontal load is 100 kN, the maximum displacement value increases by 2.52% if D = 0.3; 13.73% if D = 0.7; and 44.16% if D = 0.9. It is obvious that the damage degree has a great impact on the horizontal displacement.
2. When D = 0.2, the ultimate bearing capacity of steel frame structure is approximately 175 kN. When D = 0.9, the ultimate bearing capacity of steel frame structure with one end of beam damaged is approximately 150 kN, the ultimate bearing capacity with two ends of beam damaged is approximately 120 kN. This shows that the ultimate bearing capacity is

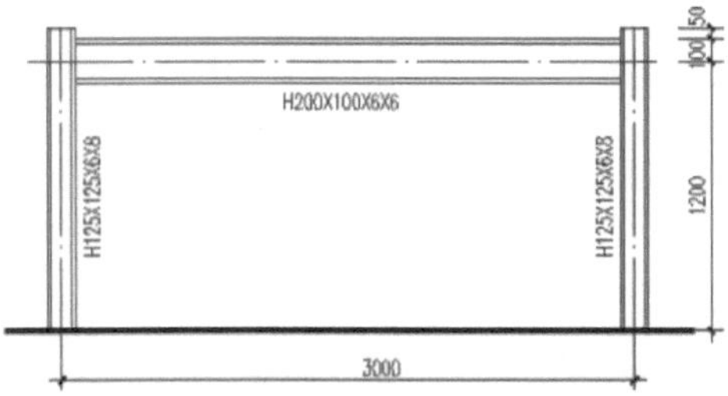

Figure 2. Layout of steel frame structure model.

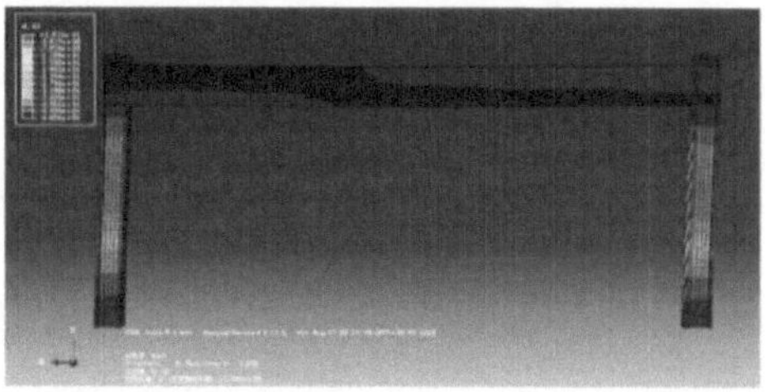

Figure 3. Cloud image of horizontal displacement.

Table 1. Horizontal displacement under different extent of damage (one end of the beam is damaged).

	Horizontal displacement (mm)								
D	20 kN	40 kN	60 kN	80 kN	100 kN	120 kN	140 kN	160 kN	170 kN
0	2.696	4.63	6.573	8.537	10.71	13.22	16.65	27.22	48.06
0.1	2.730	4.672	6.623	8.594	10.75	13.26	16.67	27.29	48.26
0.2	2.771	4.723	6.682	8.664	10.81	13.30	16.70	27.38	48.49
0.3	2.822	4.787	6.761	8.751	10.88	13.35	16.73	27.55	48.78
0.4	2.887	4.869	6.860	8.864	10.98	13.43	16.78	27.58	49.12
0.5	2.974	4.979	6.993	9.019	11.12	13.53	16.85	27.77	49.56
0.6	3.096	5.133	7.179	9.237	11.32	13.70	16.97	28.12	50.20
0.7	3.283	5.368	7.463	9.570	11.69	13.98	17.20	28.93	51.26
0.8	3.606	5.774	7.953	10.14	12.34	14.60	17.76	29.32	53.10
0.9	4.315	6.663	9.024	11.40	13.78	16.28	20.75	–	–

Table 2. Horizontal displacement under different extent of damage (two ends of the beam are damaged).

	Horizontal displacement (mm)								
D	20 kN	40 kN	60 kN	80 kN	100 kN	120 kN	140 kN	160 kN	170 kN
0	2.696	4.63	6.573	8.537	10.71	13.22	16.65	27.22	48.06
0.1	2.713	4.665	6.626	8.608	10.78	13.30	16.73	27.57	48.66
0.2	2.733	4.708	6.692	8.695	10.87	13.39	16.83	28.05	49.37
0.3	2.758	4.762	6.776	8.805	10.98	13.51	16.99	28.75	50.27
0.4	2.79	4.832	6.884	8.949	11.13	13.67	17.19	29.49	51.4
0.5	2.834	4.927	7.03	9.145	11.34	13.88	17.48	30.6	52.92
0.6	2.897	5.062	7.239	9.427	11.65	14.22	17.94	32.28	55.13
0.7	2.994	5.273	7.563	9.865	12.18	14.76	18.76	34.87	58.07
0.8	3.169	5.649	8.141	10.65	13.16	15.85	20.68	39.83	–
0.9	3.586	6.529	9.486	12.46	15.44	18.53	–	–	–

Table 3. Horizontal displacement of different layers.

D	Horizontal displacement (mm)		
	Third layer	Second layer	First layer
0	27.48	20.74	9.64
0.1	27.91	21.02	9.74
0.2	28.44	21.37	9.87
0.3	29.12	21.82	10.02
0.4	30.02	22.41	10.23
0.5	31.30	23.32	10.63
0.6	33.81	24.39	10.91
0.7	36.01	26.26	11.54
0.8	41.54	29.71	12.67
0.9	56.17	38.42	15.47

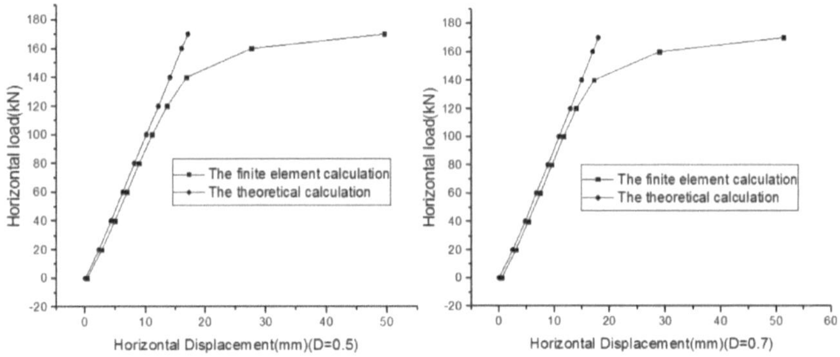

Figure 4. Contrast of finite element calculation and theoretical calculation (one end of the beam is damaged).

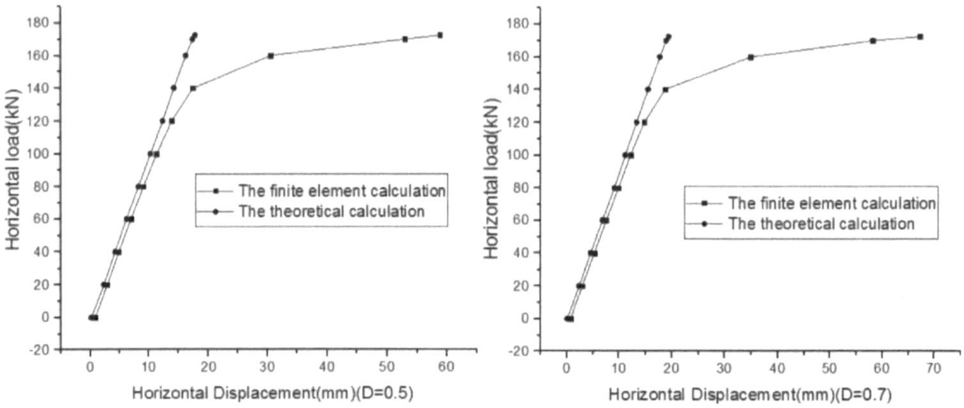

Figure 5. Contrast of finite element calculation and theoretical calculation (two ends of the beam are damaged).

related to the degree of damage. With a slightly damaged beam end, the effect of index D is inconspicuous for the ultimate bearing capacity. With severely damaged beam end, the effect of index D is conspicuous for the ultimate bearing capacity.

3. The horizontal displacement of steel frame structure with two ends of beam damaged is greater than that with one end of beam damaged under the same vertical load and horizontal load. When severe damage occurs on the beam end (D = 0.9), the ultimate bearing

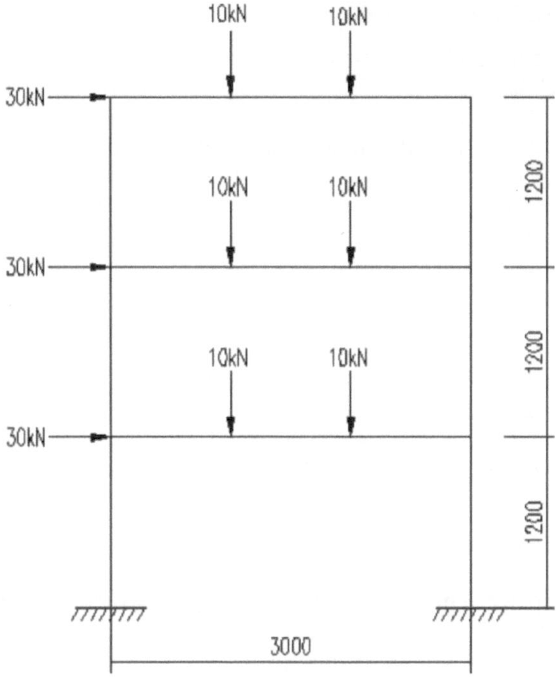

Figure 6. The model of a three-storey planar steel frame structure.

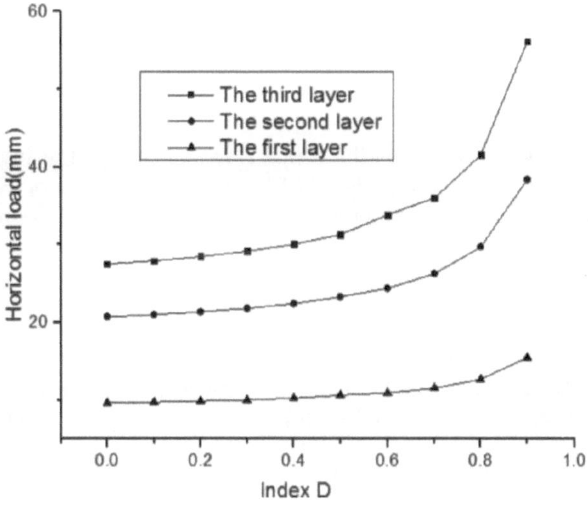

Figure 7. D-Δ curve.

capacity of frame structure with the two ends of beam damaged compared with the frame with one end of beam damaged has been reduced by 20%.

By comparing finite element calculation and theoretical calculation in Figures 4 and 5, the two group values are very close to each other in linear stage. Since the theoretical calculation is more ideal, the horizontal displacement will be slightly less than the calculated value of finite element. As the structure enters the nonlinear stage, the nonlinear finite element becomes closer to the actual situation.

3.2 *Three-storey planar steel frame structure*

According to the above method, a three-storey planar steel frame structure with two ends of beam damaged on each layer was established to analyze the regularity of horizontal displacement on each layer.

Table 3 and Figure 7 show that horizontal displacement on each layer of the frame increased with the increase in the damage index D. Between the two adjacent index D, the horizontal displacement increases by 22.10% in the first layer, 29.32% in the second layer, and 35.22% in the third layer. Obviously, the displacement of the third layer significantly varies with D.

4 CONCLUSION

In this paper, a finite element model was established to analyze the factors affecting different degrees of damage and the quantity of the damaged area or the layer, and compare with the results of the theoretical calculation, which proves the mechanical properties of the steel frame structure of the damaged beam end. The main conclusions drawn are as follows:

1. The finite element calculation is close to the theoretical calculation in the linear stage. However, the finite element simulation is more close to the actual situation in the nonlinear stage.
2. The horizontal displacement value of the steel frame structure is increased with the increase in the damage index D or quantity of the damaged area. The ultimate bearing capacity of the damaged steel frame is decreased with the increase in the index D.
3. For a multistorey steel frame structure, the horizontal displacement on each layer of the frame is increased with the increase in the damage index D. The influence of the index D is not obvious when the layer is low. The higher the layer, the greater the effect of the damage.

ACKNOWLEDGMENTS

This work was financially supported by the China Department of Housing and Urban and Rural Construction Science Project (2016-K5-054), the Youth project of Hunan Provincial Department of Education (14B187), and the National Natural Science Foundation of China (51608544).

REFERENCES

Gu M, Huang Y, Zhao M. 2011. Dynamic instability of light-weight steel structures with portal frames under wind and snow loads [J]. Journal of Tong ji University. Natural Science, 39(9):1266–1272.
Li Liankun. 2010. Structural Mechanics [M]. Higher Education Press.
Yang Yuwei. 2014. Study on structural behavior of Damaged steel frame [D]. Central South University of Forestry and Technology.
2001 G B. 2002. Code for acceptance Of construction quality Of steel structures [S].
2001 G B. 2001. Unified standard for reliability design of building structures [S].

Applied Mechanics and Civil Engineering VI – Xie (Ed.)
© 2017 Taylor & Francis Group, London, ISBN 978-1-138-62631-7

Application of stable isotopes in the study of lake and reservoir leakage

M.J. Wu, Y.J. Lin, T.P. Lu & J. Niu
Key Laboratory of Plateau Lake Ecology and Global Change, Yunnan Normal University, Yunnan, China

W.X. Zhang
Key Laboratory of Plateau Surface Process and Environmental Change, Yunnan Province, China

ABSTRACT: Leakage of lakes and reservoirs is a common geological problem, especially in karst areas, which directly affects the security of regional water. On the basis of the systematic summary of research on water seepage models, the advantages of various models and their applicable conditions are analyzed and compared. The results indicate that the calculation of the leakage amount of lakes and reservoirs, by the stable isotope mass balance and water balance model, is more accurate, and there is a wide scope of application. This study will provide a scientific basis for the quantitative and accurate calculation of the leakage of lakes and reservoirs, the evaluation of water seepage prevention measures and effects, and the rational allocation of regional water resources and deal with global change.

1 INTRODUCTION

The water cycle is linked with core and bond for various spheres of terrestrial surface. Isotope as the main component of natural water bodies is very sensitive to environmental changes, and can record the information of the water cycle. Hydrogen and oxygen stable isotopes are often used as the tracer in the study of water cycle, and it allows us to further understand the geochemical process (Craig H, 1961; Friedman I, 1953; Kohn M J., 2005). By the comprehensive analysis of the temporal and spatial variation of precipitation and reservoir water stable isotopes, the amount of reservoir leakage can be evaluated, and the effects of regional water cycle on ecosystems are quantitatively estimated by water stable isotopes. It also can provide a scientific basis for reasonable allocation of water resources and environmental change.

On the basis of the linear Darcy law, Guangxi mining areas of karst water system were simulated by using a triple-ring medium model (Cheng, 1998). The water balance of the river basin was calculated by the WetSpass model, which evaluated the amount of the utilizable water resources in the northern Ethiopia Geba Basin (Tesfamichael G., 2013). Pang (2005) studied the atmospheric precipitation with stable isotopes. Zhang et al. simulated the stable isotope fractionation of evaporation water (Zhang, 2003) and analyzed the difference between the characteristic of stable isotopes and the remaining water under the balanced mode and the dynamic model. Although the current research on the use of stable isotopes to explore the leakage of water is large, especially in the karst area, there is little research on the safety of the lives of the people of Yunnan in terms of the reservoir water seepage. Therefore, how to make use of hydrogen and oxygen stable isotopes to study the leakage of reservoir has an important theoretical and practical significance.

On the basis of the study of the relevant water leakage model, comparison of the water leakage models and the applicable conditions of the models have been discussed. Furthermore, the research will provide a scientific basis for the quantitative and accurate calculation of the leakage of lakes and reservoirs, the evaluation of water seepage prevention measures and effects, and the rational allocation of regional water resources, and deal with global change.

2 DEVELOPMENT OF LAKE AND RESERVOIR SEEPAGE MODEL

2.1 *Darcy's law*

In the 1850s, Henry Philibert Gaspard Darcy obtained reflect water seepage law in rock pores by a large number of scientific experiments, that is, Darcy's law. This law is widely applied in the research of sandy soil and finely fissured rock. A large number of experimental results show that the penetration is proportional to the first power of friction head loss during the small seepage velocity. Its expression is:

$$Q = K\omega\frac{H_1 - H_2}{L} = K\omega I \quad \text{or} \quad V = K \cdot I. \tag{1}$$

where Q: permeate flow; ω: cross section of water; L: penetrating path; V: seepage velocity; K: permeability coefficient; I: hydraulic gradient; H_1 and H_2: upstream and downstream cross section.

2.2 *Water balance*

Water balance is the area or water of any choice in any period of time; the difference between the amount of water income and the amount of water consumption will be equal to the amount of change impoundment period in the area or in the water (Zhang, 2014). Since the 1960s, because of the development of the need of water resources, many scholars have gradually shifted to the small- and medium-scale area, including the study of water balance in the basin. The formula for reservoir leakage is:

$$W = (W_Q + W_P) - (W_E + W_f) \pm W_N \tag{2}$$

where W: the amount of leakage; W_E: reservoir surface evaporation; W_P: reservoir water directly to accept precipitation recharge; W_Q: upstream inflow; W_f: reservoir water discharge; and W_N: reservoirs storage variable.

2.3 *Bayesian inverse analysis method*

After the Bayesian inverse analysis model of the maximum entropy principle was established, identification of the random parameters is transformed into the optimization problem. After the inversion of dam and dam foundation materials, the direct method has certain limitations because of the huge work, but the analytic method has high convergence speed and good convergence. According to the maximum entropy principle, the model is:

$$\int P(x)\ln\left[P(x/u)\right]dx = f_{\max} \tag{3}$$

In the inverse analysis of the parameters of the dam and the foundation, Bayesian method is used to measure the error distribution and the parameter estimation under the known parameter distribution condition. The prerequisite conditions are:

$$\begin{cases} u^* = u(x) + \varepsilon_u \\ \varepsilon_u \sim N\left[0, \text{cov}(\varepsilon_u)\right] \\ x \sim N\left[u_x, \text{cov}(x)\right] \\ \text{cov}(x, \varepsilon_u) = 0 \end{cases} \tag{4}$$

where u^*, u: the observation data arrays, calculation array; $cov(\cdot)$: covariance array; ε_u: measurement error array; u_x: mean array for the parameter x to be obtained; and $cov(\cdot)$, u_x: known matrix. Therefore, according to formula (3) and the above known condition, the criterion function of the Bayesian inverse analysis method can be introduced:

$$J = (u^* - u)^T \text{cov}^{-1}(\varepsilon_u)(u^* - u) + (x - u_x)^A \text{cov}^{-1}(x - u_x) \tag{5}$$

2.4 *Numerical difference method*

The high-density electrical method is the artificial field source; when the underground is a homogeneous medium, the obtained information is stable. When the weak local anomaly field is superimposed on the strong regional field, the contour line is only shown as slight bending in the local area, and it is difficult to achieve satisfactory geological results. If second-order differential treatment is carried out, then the regional field is suppressed, and the local field is prominent. The useful local anomalies will be separated out, which is helpful to distinguish the abnormal body (Li, 2007). Because the high-density electrical measured data are discrete and the adjacent distance difference is 1 m, the second-order differential operator can be approximated with the second-order differential operation. Limit difference quotient to define the derivative:

$$f'(x_i) = \lim_{\Delta x \to 0} \frac{f(x_i + \Delta x) + f(x_i)}{\Delta x} \tag{6}$$

Taylor series expansion can be obtained:

$$f'(x_i) = \frac{\Delta f(x_i)}{\Delta x} + \sigma(\Delta x) \tag{7}$$

where $\Delta f(x_i)$ is a forward difference.

When the step size Δx is sufficiently small, formula (7) $\sigma(\Delta x)$ tends to 0:

$$f'(x_i) \approx \frac{\Delta f(x_i)}{\Delta x} \tag{8}$$

$$f''(x) \approx \frac{\Delta^2 f(x_2)}{\Delta x^2} \tag{9}$$

When the step size was equidistant change:

$$f''(x_i) \approx k \cdot \Delta^2 f(x_i) \tag{10}$$

where $k = 1/\Delta x^2$.

By the definition of derivative, when $f(x)$ is discrete data, the difference will replace the derivative:

$$f'(x_i) = \frac{f(x_{i+1}) - f(x_i)}{x_{i+1} - x_i} \tag{11}$$

$$f'(x_i) = \frac{f(x_i) - f(x_{i-1})}{x_i - x_{i-1}} \tag{12}$$

$$f'(x_i) = \frac{f(x_{i+1}) - f(x_{i-1})}{x_{i+1} - x_{i-1}} \tag{13}$$

Similarly, the second-order derivative is approximated as:

$$f''(x_i) \approx \frac{f(x_{i+h}) - 2f(x_i) + f(x_{i-h})}{h^2} = \frac{f(x_{i+1}) - 2f(x_i) + f(x_{i-1})}{h^2} \tag{14}$$

where $h = x_{i+1} - x_i = x_i - x_{i-1}$.

When the step size was equidistant change, the second-order differential operation can be approximated by the second-order difference.

In high-density electrical probing, the AB/2 electrode series is 2, 3, 4, 5, 6, 7, 8, 9, ..., 20, according equidistant changes, and the step size is constant. Set ρ_z to $f(x)$, AB/2 to x, and $\rho_z^m = \Delta^2 f(x)$. By the second-order difference:

$$\rho_z^m = \rho_{z(i+1)} - 2\rho_{zi} + \rho_{z(i-1)} \tag{15}$$

where $\rho_{z(i+1)}, \rho_{zi}, \rho_{z(i-1)}$ respectively for $(i+1), i, (i-1)$, the resistivity corresponding to the electrode distance.

In the contour map of the electrical resistivity value, which is processed by the numerical calculation method, the negative abnormal sealing ring appeared, which corresponds to the range of the dam leakage body.

2.5 *Introduction of water stable isotopes*

The lake water isotope reflects the basin precipitation isotope, lake evaporation intensity, hydrological conditions, water residence time, and other factors. The stable isotope method is a useful tool for estimating lake evaporation. It is an important method for the study of lake water quantity, water quality, and water environment. For inland lakes, there is a dynamic equilibrium between precipitation, evaporation, and runoff (Gibson J J, 1998; Gibson J J, 2000); furthermore, the oxygen stable isotopes in the three water vapor samples exist. The leakage of inland lakes can be estimated by the relationship between precipitation and oxygen isotope in the lake water. For an inland lake, water quantity change can be expressed as:

$$\frac{dV}{dt} = I - Q - E \tag{16}$$

where *V*: Lake water volume; *t*: time; *E*: evaporation; *Q*: out of the lake water amount, including surface runoff and leakage of water *L*; and *I*: the total amount of water into the lake, including total precipitation, surface runoff, and groundwater recharge.

Similarly, formula (16) shows as:

$$V\frac{d\delta}{dt} + \delta\frac{dV}{dt} = I\delta_I - Q\delta_Q - E\delta_E \tag{17}$$

where δ, δ_I, δ_Q, and δ_E are the isotopic composition of the lake water, the water in the lake, out of the lake water, and the evaporating water, respectively.

3 COMPARISON AND EXPLORATION OF THE LAKE AND RESERVOIR LEAKAGE MODEL

3.1 *Comparison of the lake and reservoir leakage model*

In general, the penetration rate of sand and clay is very low, which is called laminar flow, the water flow is a flow line parallel to each other. The seepage law of motion is Darcy's law, and the relationship between the penetration rate (v) and the hydraulic gradient (i) can be in the v-i coordinates expressed as a straight line, as shown in Figure 1(a). Very fine particles of high compression of soil, free expansion of clay, and a few clay permeability test showed that: an initial hydraulic gradient (i_b) penetrates, and the soil only after reaching the initial hydraulic gradient can occur in the penetration. This type of soil occurs after the infiltration, and the infiltration rate can be approximated with a straight line, as shown by curve (2) in Figure 1(a). Gravel and other coarse-grained soil test results, as shown in Figure 1(b), show that, because of its large pores, when the hydraulic gradient is small, seepage can be considered as laminar flow. v-i relations undergo a linear change, but Darcy's law is still applicable. When the hydraulic gradient is larger, the flow rate increases, and the seepage flow will be turbulent flow, which is a type of irregular mutual mixed flow form. When the v-i relationship is nonlinear change, Darcy's law is no longer applicable.

Water balance calculation method is simple and adaptable, which in many cases can use and recognize regional hydrogeology conditions that do not require relatively harsh data. Water balance method is applicable to the broader range of water leakage evaluation, and the drawback is the microscale evaluation capacity in the determination of the water balance of each item. Water balance method applies only to a unified complete hydrogeological unit, recharge and discharge can easily determine the amount of water and there are estimates of long-term observational data mining case. When groundwater recharge and discharge condi-

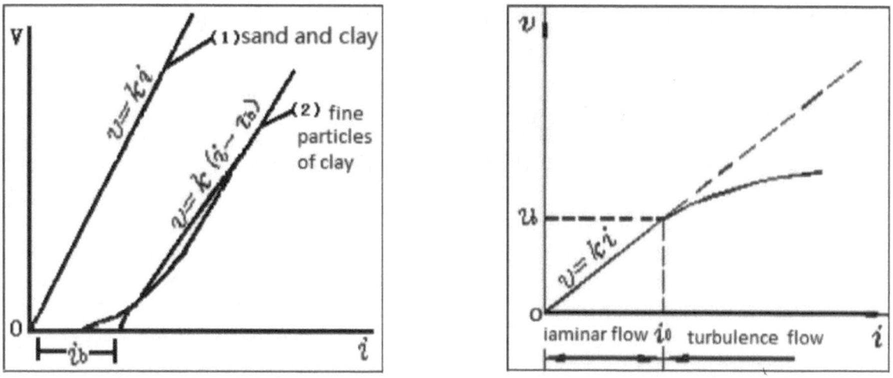

Figure 1(a). v-i relation of fine-grained soil. Figure 1(b). v-i relation of coarse-grained soil. Figure based on Baidu Encyclopedia.

tions are simple, the water balance elements are easy to determine, or a little change occurs in the mining area, the effect of this method is better evaluated for groundwater resources. It is also often used as an evaluation method to verify the results of other methods.

Various uncertain factors exist in dam seepage monitoring; by using the maximum entropy principle, Bayesian model of reservoir water level and leakage was established, while the introduction of matching coefficient realizes the prediction model of optimization inverse analysis and predicts the future condition of water level under the quantity of seepage influence trend (Gibson J J, 1998). The process of the theory of information entropy and Bayesian inverse analysis method combine to make the uncertainty inversion analysis problem into Bayesian criterion function optimization problem, and by setting the matching coefficient to harmonize the relationship between the subjective and objective information. A proper optimization solving method can realize the organic combination of on-site measurement information and a priori information, which fully reflects the random uncertainty inversion feasibility and superiority (Yang, 2006).

The advantage of the numerical difference method is that macro region can be evaluated simultaneously with the regional water resources, but the treatment of heterogeneous and complex boundary conditions pose a problem. It can also be close to mine hydrogeological conditions and drainage condition for the realization of predicting water influx tasks. The disadvantage is the need to establish clear conceptual model of hydrology and geology, the calculation time is more and the cost is high. Numerical difference method requires high-quality relevant data and a higher degree of control seepage field. Numerical difference method is generally used for more complex mining conditions.

3.2 *Application and exploration of water stable isotopes*

Stable isotope method does not require the direct determination of certain parameters (such as the amount of groundwater into the lake and evaporation), but simply measuring its content of stable isotopes. Stable isotope content may be considered by water balance and stable isotope material balance equations. When the lake is difficult to determine the revenue and expenditure items, isotope method to estimate is undoubtedly a viable new method. It should be noted that because the isotope sampling is more difficult, the isotopic content of various water bodies is not a constant, and is greatly affected by temperature, humidity, and other factors. Therefore, the stable isotope method is used to make estimates, and it should be compared with other measuring methods, but further improved isotope sampling techniques are required (Hu, 2007). As a more effective tool, stable isotope techniques reveal the hydrological ecosystem water cycle mechanism and provide a new method to observe the relationship between water use efficiency and water sources. At present, the international community has launched a study of hydrogen and oxygen isotopes in "soil–plant–atmosphere" interface during water circulation. In future research, hydrogen and oxygen stable isotope technique is worth exploring new areas, it will be applied to the entire study water circulation system, and the ultimate solution to the water cycle process has some key problems and practice problem.

4 CONCLUSION AND PROSPECT

1. Darcy's law applies to the area in which a proportional relationship exists between the seepage velocity and the hydraulic gradient. The principle of water balance is most applicable to the area with simple hydrogeological conditions, and it lacks the capacity for microscale evaluation. The inverse analysis method based on the maximum entropy is mainly used for the estimation of structural engineering parameters and load source of uncertainty inverse analysis problems (such as dam seepage field analysis). The numerical difference method can be used to solve the water resource problem of a clear regional complex boundary of the hydrogeological conceptual model, but it requires higher correlation data and seepage field, and the calculation is time-consuming and expensive. The numerical difference method is mainly used to solve the problem of urban geophysical and other engineering geophysical explorations due to various serious interferences between shallow geological information and seepage of the dam.
2. The leakage of lakes and reservoirs can be studied by using the stable isotope method and the lake fractionation model. The research shows that this method has a good prospect in the study of water leakage. Meanwhile, when using the stable isotope of water body to study the water leakage, it is necessary to discuss the effect of different factors on the estimation of leakage, to ensure the accuracy of research results.

ACKNOWLEDGMENT

This study was supported by the Natural Science Foundation of Yunnan Province (Grant No. 2015FB141).

REFERENCES

Cheng Jianmei. Chen Chongxi. A numerical simulation on groundwater in Beishan of Guangxi karst pipeline fracture pore [J]. *Hydro geology and engineering geology*, 1998, (4): 50–54 (in Chinese with English abstract).

Craig H. Isotopic variations in meteoric waters. *Science*, 1961, 133: 1702–1703.

Friedman I. Deuterium content of natural waters and other substances. *Geochimica et Cosmochimica Acta*, 1953, 4(1): 89–103.

Gibson J.J, Edwards TWD. Global Boigeochemical Cycles, 2002, 16(2): 1–13.

Gibson J.J, Reid R, Spence C. Hydrol Processes, 1998, 12: 1779–1792.

Hu Haiying. BAO Wei min ed. Fractionation mechanism of stable hydrogen and oxygen isotope in water body evaporating [J]. *Journal of China Hydrology*, 2007, 3(27): 1–5 (in Chinese with English abstract).

Kohn M.J., Welker J.M. On the temperature correlation of $\delta^{18}O$ in modern precipitation. Earth and Planetary Science Letters, 2005, 231(1): 87–96.

Li Guozhong. Sun Jin long. The application of numerical differential method in dam leakage-prevention engineering [J]. *Water Conservancy & Electric Power Machinery*, 2007, 29(3): 77–79 (in Chinese with English abstract).

Pang Hong xi. He Yuan qing ed. The source of δ^{18}O and monsoon water vapor in monsoon Precipitation [J]. *Chinese Science Bulletin*, 2005, 50(20): 2263–2266 (in Chinese with English abstract).

Tesfamichael G. Smedt F D. Application of a spatially distributed water balance model for assessing surface water and groundwater resources in the Geba basin. *Tigray, Ethiopia. J Hydrol*, 2013, 499: 110–123.

Yang Jie. Hu Dexiu ed. Bayesian uncertainty inverse analysis method based on pome [J]. Journal of Zhejiang University (Engineering Science), 2006, 40(5): 810–815 (in Chinese with English abstract).

Zhang Xin ping. Yao Tan dong ed. Study on the Fractionation Mechanism of Stable Isotope in Evaporating Water Body [J]. *Journal of Glaciology and Geocryology*, 2003, 25: 65–70 (in Chinese with English abstract).

Zhang Yin qin. Study on the Impact of base flow on watershed water balance [D]. *Northwest Agriculture and Forestry University*, 2014, 12: 9–15 (in Chinese with English abstract).

Applied Mechanics and Civil Engineering VI – Xie (Ed.)
 ISBN 978-1-138-62631-7

Root cause analysis of the collapse accident at the Tsinghua university high school construction site using "24 model"

Hong Zhang & Gui Fu
School of Resources and Safety Engineering, China University of Mining and Technology, Beijing, China

ABSTRACT: Collapse is a critical accident type in construction industry. Different from other accidents (e.g. falls and mechanical damages), collapses are normally due to combinative causes involving people from various hierarchies in many organizations besides front-line operators. A systematic approach is necessary to analyze the causes of collapses. The purpose of this paper is to apply a new accident analysis approach (24 Model) to analyze the root cause of a collapse accident at the Tsinghua University High School construction site. The results showed that 24 Model was feasible for accident analysis in this situation. Also, prevention measures were deduced according to the accident analysis.

1 INTRODUCTION

Construction industry is of high risk in China. In particular, collapse occurs with the highest frequency among other accident types in construction industry. But the worst part is, there is a high probability of fatality in the case of collapse. Learning from the past is an effective way to prevent accidents and identify hazards. Therefore, the adoption of the accident analysis approach determines the information or lessons we could find out from an accident. In China, the causes of accidents are classified as direct and indirect causes in accident inquiry reports issued by the State Administration of Work Safety. Direct causes involve unsafe acts, unsafe conditions of equipment and inappropriate working environments; the other causes are attributed to indirect causes. There are no specific analysis models adopted. Also, the relations among causes are not analyzed. However, the contributory factors of collapse normally involve many people from several individual organizations. In this case, a systematic approach could improve the understanding of the causes of the accident. According to Hollnagel (2004), the choice of the model to analyze an accident is crucial because it will determine the analyst's perspective and, therefore, guide the conclusions of the investigations and preventive measures resulting from these conclusions. Accident analysis is guided by the well-known "What-You-Look-For-is-What-You-Find" principle. 24 Model provides a new way to clarify what leads to an accident. The aim of this paper is to apply 24 Model to analyze contributing factors of a reinforced craft collapse accident at the Tsinghua University High School construction site and obtain a systematic analysis result of this accident. This study can also verify the feasibility of 24 Model in accident analysis in construction industry.

2 METHODS

2.1 *Approaches to accident analysis*

There are several accident causation models that dominate the accident analysis domain. These include Swiss Cheese Model, HFACS, AcciMap and STAMP.

The Swiss Cheese Model is undoubtedly the most popular of all accident causation models. This model describes the interaction between system wide 'latent conditions' (e.g. inadequate designs and equipment, supervisory and maintenance failures, inadequate training and procedures) and unsafe acts made by human operators and their role in accidents (Reason, 1997). However, it does not classify the types of the "whole" on the "cheese". So, it needs sufficient experience and knowledge if it is used in conducting accident analysis.

Because of the absence of taxonomies of latent failures and unsafe acts within Reason's Swiss Cheese model, Wiegmann and Shappell (2003) provided a framework for classifying and analyzing human factors associated with accidents and incidents, which is called the Human Factors Analysis and Classification System (HFACS). However, they limited its utility as an aviation accident analysis method. When it is applied into analyzing accidents in other industries, modifications are necessary.

The AcciMap describes the failures involved in an accident and places them across six levels. It is very useful to outline all the contributions to an accident. However, without taxonomies of specific failure modes, AcciMap is highly comprehensive in terms of its ability to identify all of the contributory factors involved in a particular accident. It needs sufficient data and highly skilled analysts to describe the entire accident trajectory in terms of failures across the system and the relationships between them (Salmon, Cornelissen & Trotter, 2012).

STAMP (Leveson, 2004) focuses on the interactions between system components and the control mechanisms used throughout the work system. STAMP views accidents as resulting from the inadequate control of safety-related constraints (Leveson, 2004). When used for accident analysis purposes, STAMP produces a description of the system's control structure and then identifies failures in this control structure that contributed to the accident. STAMP is more suitable for identifying and classifying technical control failures related to complex human decision making and organizational failures.

2.2 *The approach adopted (24 Model)*

24 Model was developed from the view of behavior safety by Fu et al. (2016). The causes are analyzed at both the organizational level and individual level. These causes are divided into four types: root cause, radical cause, indirect cause and direct cause (see Figure 1).

In 24 Model, direct causes include unsafe acts and unsafe conditions. Front-line operator's unsafe acts are one part of unsafe acts here. People from various hierarchies of an organization could have behaviors that may cause accidents. The developers of 24 Model argue that the reasons for individuals committing unsafe acts or the existing of unsafe conditions are lack of knowledge, lack of safety awareness, bad habits relating to safety and adverse psychological or physiological states. These indirect causes related to individuals are consequences of defects of organizations, which are inadequate safety culture and faults in the

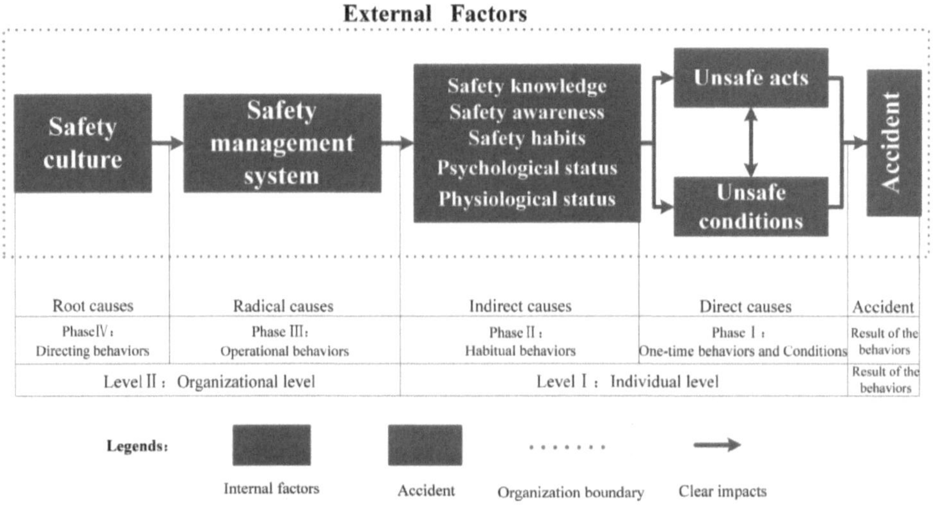

Figure 1. Framework of 24 Model.

safety management system. A safety management system includes necessary documents and safety management human resources.

24 Model is adopted in this paper because of the following two reasons. Firstly, 24 Model is suitable for analyzing accidents related to systems that are not very complicated. Normally, the mechanical systems in a construction site are not very complex. Also, without many decision-making occasions, people will usually comply with fixed procedures. Secondly, accidents in construction industry are predominantly caused by human behaviors. Therefore, 24 Model analyzes accidents based on behavior safety theories.

3 CASE PRESENTATION

At 8:20 am on 29th, December 2014, a reinforced craft collapse occurred at the Tsinghua University High School construction site, which led to the death of 10 people and left four injured. There was a deep foundation pit with 13 meters' height, and a high building was to be constructed. This accident occurred in the foundation pit, and the whole collapsed area reached 2000 m^2 (see Figure 2).

There were two layers of reinforced crafts built. The upper layer was supported by "horse stools" (see Figure 3). The horse stools were designed to be welt by steel of Φ25 mm or Φ28 mm, and the space between each other was maintained from 0.9 m to 2.1 m. The upper layer reinforced craft collapsed in the accident, and there were 31 workers working under it. The workers were fastening steels. At the same time, a tower crane was swinging and piling steels on the reinforced craft. With increasing steel piled on the reinforced craft, it was overloaded and collapsed.

4 RESULTS

Table 1 outlines the descriptions of accident causes using 24 Model.

4.1 *Direct causes*

Both unsafe acts and unsafe conditions led to this accident. Unsafe acts were committed by people from different organizations (construction organization and construction control organization) and various hierarchies (front-line workers and managers). Firstly, in a construction site, safety guidelines should be instructed to the front-line workers before the work is started. However, there was no safety technology guideline communication with front-line workers before they built the steel reinforced craft. Secondly, there was insufficient safety training for front-line workers. Some of the workers even did not attend any kind of safety training before their work. Therefore, there were some low quality steel horse stools built on the site. Thirdly, neither the unsafe acts of workers nor unsafe acts of managers who were responsible for training were noticed or corrected by supervisors.

Figure 2. The construction site after collapse.

Figure 3. The horse stools.

Table 1. Accident analysis using 24 Model.

Cause type		Causes in this accident
Direct causes	Unsafe acts	1) Managers did not communicate safety technology guideline with front-line operators. 2) Safety training for front-line operators was inadequate. 3) Supervisors on site did not correct the unsafe acts of operators and managers.
	Unsafe conditions	1) Steels were not piled up in accordance with the requirements of the procedure. 2) The horse stools were not produced in accordance with the requirements of the production scheme and arrangement of horse stool. 3) There was no effective support among horse stools.
Indirect causes	Safety knowledge	1) Front-line operators lacked training and necessary safety knowledge. 2) The project manager lacked knowledge on how to accomplish the project.
	Safety awareness	1) Front-line operators had less awareness on safety. 2) Managers had less awareness on safety.
	Safety habits	Nobody raised objections during work in case of hazard existence.
	Psychological status	Not identified.
	Physiological status	Not identified.
Radical causes	Safety management system	1) Lack of project manager. 2) There were not enough full time safety management personnel. 3) There were defects in the project design.
Root causes	Safety culture	People in the organization did not take safety seriously and reduced the project duration time.

There were three unsafe conditions contributing to this accident directly. The first one was piling up materials not in accordance with the requirements of the program. Piling up steel onto a focus point led to overload of the reinforced craft. To make the matter worse, the horse stools were not made and put in accordance with the requirements of the production plan, which led to horse stools not providing enough support. Finally, there was no effective support among horse stools. So, the horse stools did not form firm supporting systems to support the load on the reinforced craft.

4.2 *Indirect causes*

24 Model argues that the most unsafe conditions were implemented by people. In this accident, all the unsafe conditions were caused by people without enough experience in their previous work. So, both the unsafe acts and unsafe conditions were caused by the safety knowledge, safety awareness or unsafe habits of front-line operators and managers. Psychological and physiological states were not mentioned in any reports on this accident. Workers lacked knowledge on how to work safely because they did not take enough training on safety, and managers in this project were not qualified and without enough knowledge on how to work safely. Also, both workers and managers worked in the same way on other construction sites, which means that "work unsafely" has become their normal habits.

4.3 *Radical causes (safety management system)*

Documents related to safety and safety human resources constitute the safety management system in 24 Model. People have unsafe acts because they are not regulated to do or there is nobody to do the right thing. In other words, documents related to safety and safety human resources are the foundation to maintain safety. In this accident, lack of project managers and safety management resources contributed much to it. The project manager appointed to this project does not have prior experience on a construction site. Also, two full time safety managers were required on site, which was not the case. The worst part is, there were defects in the project design. On the design drawing, some labels were incorrect, which did not give sufficient instructions to the workers to do the work correctly. The other safety documents had no obvious mistakes.

4.4 *Root causes*

The root cause refers to safety culture of organizations. The main defect of safety culture in the construction organization is the safety value, which is giving first importance to safety. The time period planned for this project is 27.6% less than required time period. The time constraint led to the safety procedures being ignored during the work.

4.5 *External factors*

Organizations' behavior is also affected by the government administration and supervision, because organizational safety culture normally falls into strictly supervised levels and they are not active in improving safety. In this accident, the government authority did not inspect the construction site according to the regulations. Both the frequency and items of inspection were not sufficient.

5 DISCUSSION

Applying 24 Model into collapse accident analysis obtained causes in the organization domain. However, analysis of an individual accident represents for the process and causes of a single accident. Collecting a series of accidents' causes has more guidance on collapse prevention.

Accident analysis results depend on analyst's opinions, knowledge, experience, etc. It is suggested that 24 Model is used by a group of people to get a unified opinion.

In construction industry, normally, several organizations are involved, which include project developers, construction companies, labor dispatch entities and construction supervision authority. There might be errors committed by people from all these organizations. So, it might be better to analyze their errors in each organization.

In analysis results, there were some contributions to the accident which were not identified. However, they did exist. Our analysis is according to the accident inquiry report. Therefore, we can't deduce any information that was not mentioned in it. From another point of view, the accident causing model, which was utilized when investigating, determined what we would find in the accident analysis.

24 Model focuses on developing training as preventive measures. Therefore, the key to prevent this accident is to make the related people know how to work safely and promote their safety awareness so as to work according to the safety regulations. Necessary training is an effective way to prevent collapse in construction industry by reducing unsafe acts of people.

6 CONCLUSION

The case discussed here outlines a series of causes that led to a catastrophic accident. The results showed that people' unsafe acts and defects in organizational management played a great role in the genesis of this accident. In addition, the feasibility of applying 24 Model in a collapse accident is indisputable. Therefore, it is recommended to include 24 Model to accident investigation in collapse accidents at construction sites.

ACKNOWLEDGMENT

This research was funded by the Natural Science Foundation of China Project (51534008).

REFERENCES

Fu, G., Fan, Y., Tong, R., etc., 2016. A Universal Methodology for the Causation Analysis of Accidents (1st Edition). Journal of Accident Prevention 2(1): 7–12.

Hollnagel, E., 2004. Barriers and Accident Prevention. Ashgate Publishing Ltd, Aldershot.

Leveson, N., 2004. A new accident model for engineering safer systems. Safety Science 42 (4): 237–270.

Reason, J., 1997. Managing the Risks of Organizational Accidents. Ashgate Publishing Ltd, Aldershot.

Salmon, P.M., Cornelissen, M., Trotter, M.J., 2012. Systems-based accident analysis methods: a comparison of AcciMap, HFACS, and STAMP. Safety Science 50 (4): 1158–1170.

Wiegmann, D.A., Shappell, S.A., 2003. A Human Error Approach to Aviation Accident Analysis: The Human Factors Analysis and Classification System. Ashgate Publishing Ltd, Burlington, Vermont, USA.

Applied Mechanics and Civil Engineering VI – Xie (Ed.)
© 2017 Taylor & Francis Group, London, ISBN 978-1-138-62631-7

Research on the development law of pore pressure during shearing of the saturated dense silty soil by the rake tooth

Gongxun Liu, Guojun Hong, Qingbo Zhang, Dayi Shen & Zhongwei Zhou
Key Lab of Dredging Technology of CCCC, CCCC National Engineer Research Center of Dredging Technology and Equipment Co. Ltd., Shanghai, China

Liquan Xie
Department of Hydraulic Engineering, Tongji University, Shanghai, China

ABSTRACT: Through a series of cutting tests on the saturated dense silty soil in the Huanghua port area with prototype rake tooth, the influence of cutting parameters (including cutting angle, moving velocity of the rake tooth and cutting thickness) and the type of rake tooth on the pore pressure distribution along the surface of rake tooth was investigated. The relationship between the peak value of negative pore pressure and the cutting parameters was set up, respectively. The test results show that during the cutting process, the soil shows an obvious dilation characteristic and a high negative pore pressure is created. For the chisel rake tooth, the negative pore pressure along the rake tooth surface increases first, decreases gradually after it reaches the maximum value, and ultimately achieves a relatively stable value. With the increase of cutting angle, the maximum negative pore pressure increases. The relationship between the maximum negative pore pressure and the sine of the cutting angle is basically linear. With the increase of tooth cutting speed, the maximum negative pore pressure increases linearly. With the increase of cutting thickness, the increase of the maximum negative pore pressure becomes larger and larger, and the relationship between the maximum negative pore pressure and the cutting thickness conforms to the quadric curve. For the chisel rake tooth, plough rake tooth and semi-plough rake tooth, the relationship between the maximum negative pore pressure and the sine of the cutting angle conforms to different curves. At the same cutting angle, speed and thickness, the value of the maximum negative pore pressure caused by cutting the soil is the lowest for the semi-plough rake tooth, the next for the chisel rake tooth, and the highest for the plough rake tooth. The above research results have very important guiding significance to reduce tooth cutting resistance and improve dredging efficiency by selecting the construction parameters and the type of rake tooth.

Keywords: saturated dense silty soil; shear dilation; negative pore pressure; cutting parameter; the type of rake tooth; cutting resistance; dredging efficiency

1 INTRODUCTION

In the process of dredging, dense silt, "iron sands" and other such soils are often encountered, which are difficult to cut and raked, mainly because of the obvious dilation characteristic. In the process of cutting, a certain dilatancy zone takes shape in the interior of soil, where the pore pressure is negative, so that the soil effective stress and the shear strength increase significantly (Miedema, 2010; Zhao, 2001).

At present, variation laws of pore pressure in the triaxial shearing process about sand and silt have been obtained a lot. Based on the hypothesis that while the stress levels are equal, the body change of consolidated undrained shear and that of consolidated drained shear are identical, Zhang (1987) explained the occurrence mechanism and its calculation method of pore pressure in the process of static load shearing. In use of monotone shear tests, Liu (2010) systematically researched the influence of the initial large principal stress direction angle, initial middle principal stress coefficients and other factors on the pore pressure of undisturbed saturated silty soil from the Yellow River Delta region in the monotonic shearing process. According to the cyclic triaxial tests, Seed et al. (1976) established a pore pressure model under isotropic pressure consolidation, which confirmed the inverse sine function between the pore pressure ratio and the cycle time ratio. Thereafter, the pore pressure increment models were proposed and improved in succession (Ishibashi et al., 1982; Towhata et al., 1985; Sun et al., 2005, 2006; Polito, 2008). The pore pressure increment model does not need to confirm the liquefying cycle numbers in advance, and when to achieve initial liquefaction is determined by the pore pressure increment of each time interval. Therefore, the pore pressure increment model has certain advantages in the dynamic time history analysis by using the effective stress method.

Compared to research studies on the variation law of pore pressure in the triaxial shearing process, research achievements on pore pressure variation laws in the process of soil cutting by rake tooth are relatively few. Apart from studies [1] and [2], domestic scholars (Xie et al., 2011, 2015) made tooth cutting experiments on Huanghua port silt by using the self-made rake tooth cutting soil monitoring device, and preliminary mastered the dynamic process of tooth cutting soil and the evolution law of pore pressure. In addition, through the finite element calculation, the development law of pore pressure for saturated sand in the cutting process was analyzed. The calculation results indicate that the saturated sand has a strong dilatancy effect in the cutting process with the increase of cutting speed. The negative pore pressure in front of the rake tooth tip becomes larger and larger, due to which the soil effective stress and cutting resistance are improved significantly.

In this paper, through a series of prototype rake tooth cutting tests on Huanghua port saturated dense silty soil, the effects of the occurrence and development of pore pressure on the tooth surface caused by the rake tooth cutting angle, cutting speed, cutting thickness and tooth type are systematically investigated and relationships between cutting parameters and pore pressure are analyzed.

2 TEST CONDITIONS AND TEST CASES

2.1 *Test apparatus and equipment*

The rake tooth cutting test is powered by a rake traction system, as shown in Fig. 1. For the dredging test rig, the speed variation range is 0~2 m/s and the maximum cutting resistance is 20 kN. The starting and braking of the test rig are operated by the control system (shown in Fig. 2) installed on the vehicle-mounted console.

As shown in Fig. 3, the rake tooth attitude control system consists of a test stand, hydraulic double cylinder, hydraulic single cylinder and rake tooth.

The laser transmitter and receiver are mounted onto the edge of the port dredging sled, and the laser reflecting plate is fitted on the distal fixation place of dredge test flume, so as to realize the measurement of tooth cutting speed, as shown in Fig. 4.

2.2 *Test soil*

The silty soil samples are taken from the Huanghua port of Hebei Province, whose basic physical properties are listed in Table 1.

Figure 1. Traction system for the rake tooth.

Figure 2. Control system for the trolley.

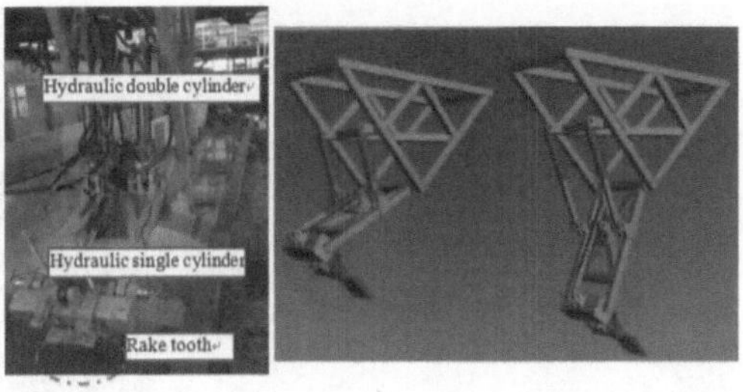

Figure 3. Attitude control system for the rake tooth.

Figure 4. Arrangement of the laser velocimetry sensor.

Table 1. Physical properties of the silty soil for testing.

Median particle diameter d_{50}/(mm)	Relative density G_s	Plastic limit w_P/%	Liquid limit w_L/%	Plasticity index I_P	Permeability coefficient k/(cm/s)
0.038	2.70	16.3	21.6	5.3	1.5×10^{-5}

3 STUDY OF THE INFLUENCE OF CUTTING PARAMETERS ON VARIATION LAWS OF PORE PRESSURE ALONG THE TOOTH FACE IN THE CUTTING PROCESS

In this investigation, the chisel rake tooth was used to study the influence of cutting parameters on variation laws of pore pressure along the tooth face in the cutting process.

In order to monitor the changes of pore pressure, 12 holes are drilled into the rake tooth and a pore pressure gauge is placed in each hole to measure the variation of pore pressure. As shown in Fig. 5(a), hole numbers are indicated in red letters. Fig. 5(b) shows for the pore pressure data acquisition instrument.

As shown in Table 2, a total of 16 cases are carried out in this study. For each one, two sets of parallel tests are made to study the influence of the tooth cutting angle, cutting speed, cutting thickness and tooth type on pore pressure occurrence and development.

3.1 *The influence of cutting angle*

In order to examine the influence of the cutting angle on the pore pressure, cases 2, 8, 9 and 10 are compared.

Fig. 6 shows the variation of pore water pressure with time at different cutting angles. It can be found that the negative pore pressure gradually increases with the passage of time, and when reaching the maximum, it gradually decreases and finally reaches a relatively stable value.

Fig. 6 also shows that at a cutting angle of 45 degrees, the largest negative pore pressure is −35.2 kPa; at 60 degrees, −48.69 kPa; at 75 degrees, −55.72 kPa; at 90 degrees, −58.40 kPa. The above data indicate that with the increase of cutting angle, the maximum negative pore pressure increases. It is noted that the maximum negative pore pressure and the sine value of the cutting angle are linear correlation basically; therefore, a linear fitting is put forward as shown in Fig. 7, and the relationship can be expressed by the following equation:

$$y = 20.282 - 78.874x \tag{1}$$

3.2 *The influence of cutting speed*

In order to study the influence of cutting speed on the pore pressure, cases 1, 2, 3 and 4 are chosen to carry out a comparative analysis.

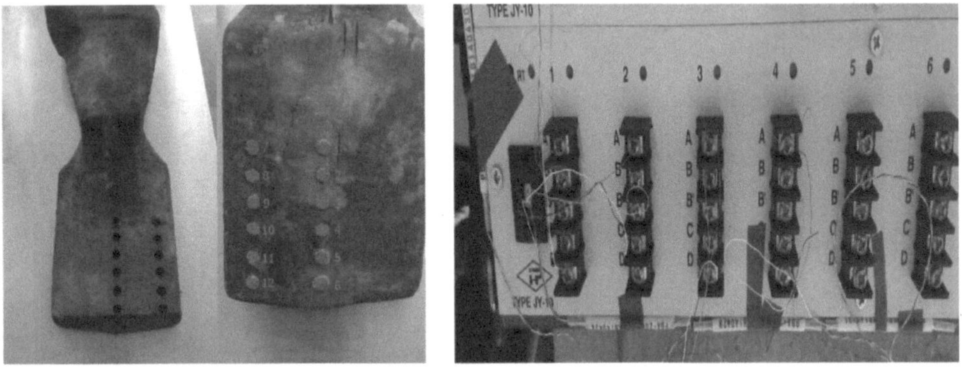

(a) Layout of monitoring points on the rake tooth (b) Data acquisition instrument

Figure 5. Distribution of monitoring points for pore pressure and monitoring instrument.

Table 2. Test cases.

Case	Cutting angle/°	Cutting speed/(m/s)	Cutting thickness/cm	Remark (Analysis factor)
1	60	0.17	5.5	Chisel tooth
2	60	0.29	5.5	Chisel tooth
3	60	0.47	5.5	Chisel tooth
4	60	0.54	5.5	Chisel tooth
5	60	0.30	4.5	Chisel tooth
6	60	0.29	6.5	Chisel tooth
7	60	0.31	7.5	Chisel tooth
8	45	0.31	5.5	Chisel tooth
9	75	0.31	5.5	Chisel tooth
10	90	0.30	5.5	Chisel tooth
11	45	0.17	5.5	plough rake tooth
12	60	0.18	5.5	plough rake tooth
13	75	0.17	5.5	plough rake tooth
14	45	0.17	5.5	semi-plough rake tooth
15	60	0.17	5.5	semi-plough rake tooth
16	75	0.18	5.5	semi-plough rake tooth

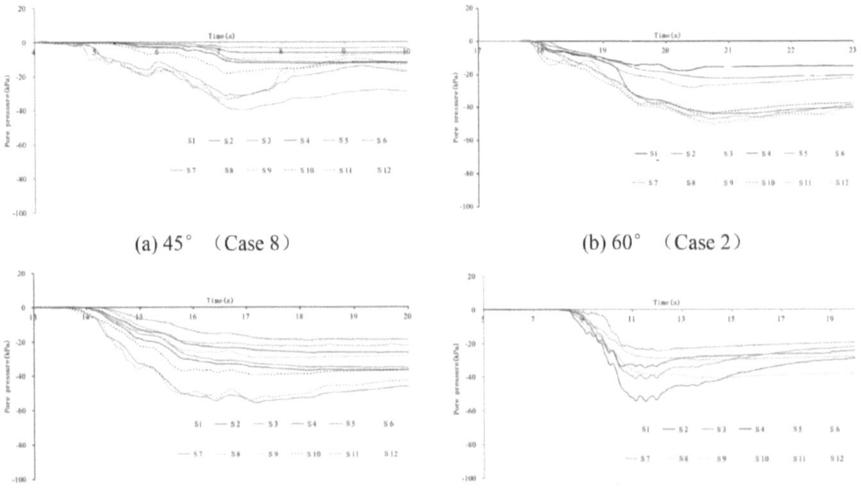

(a) 45° (Case 8) (b) 60° (Case 2)

(c) 75° (Case 9) (d) 90° (Case 10)

Figure 6. Time history curves of pore pressure at different cutting angles.

Fig. 8 shows the variation of pore water pressure with time at different cutting speed values. It can be found that the variation law of pore pressure caused by cutting speed is basically the same as caused by the cutting angle, i.e. the negative pore pressure increases with the passage of time and after reaching the maximum, it gradually decreases and finally reaches a relatively stable value.

Fig. 8 also shows that, at a cutting speed of 0.17 m/s, the largest negative pore pressure is −36.89 kPa; at 0.29 m/s, −48.69 kPa; at 0.47 m/s, −60.51 kPa; and at 0.54 m/s, −74.85 kPa. The above data indicate that with the increase of cutting speed, the maximum negative pore pressure increases. It is noted that the maximum negative pore pressure and cutting speed are linear correlation basically, so a linear fitting is put forward as shown in Fig. 9 and the relationship can be expressed by the following equation:

$$y = -94.455x - 20.523 \tag{2}$$

3.3 *The influence of cutting thickness*

In order to study the influence of cutting thickness on pore pressure, cases 2, 5, 6 and 7 are chosen to carry out a comparative analysis.

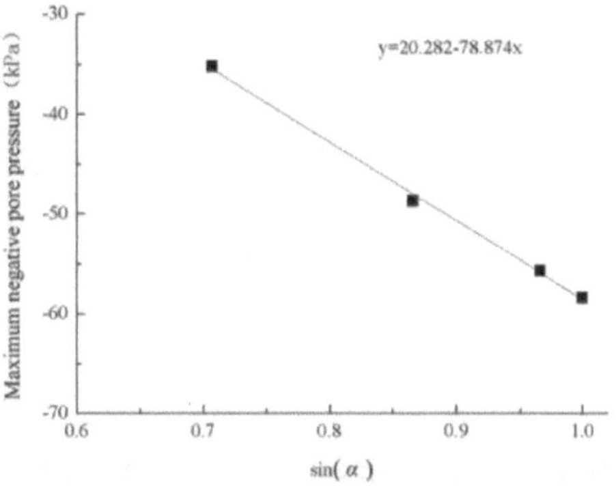

Figure 7. Relation curve between the maximum negative pore pressure and the cutting angle.

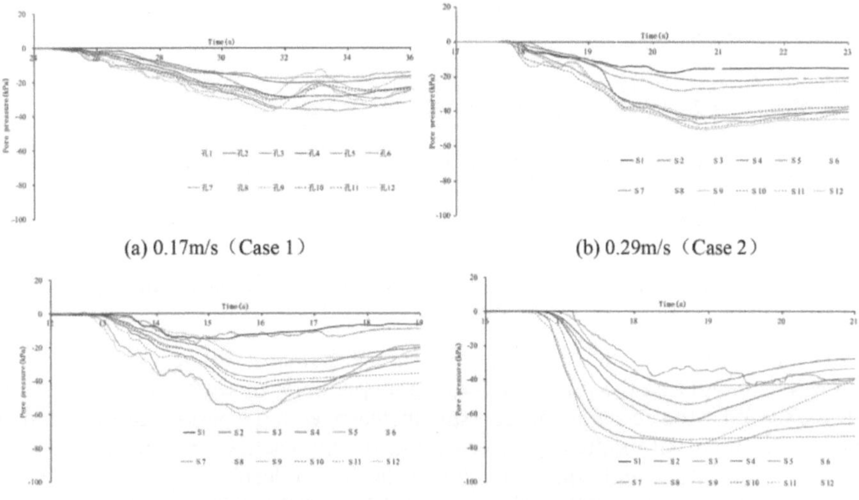

(a) 0.17m/s（Case 1）

(b) 0.29m/s（Case 2）

(c) 0.47m/s（Case 3）

(d) 0.54m/s（Case 4）

Figure 8. Time history curves of pore pressure at different cutting speed.

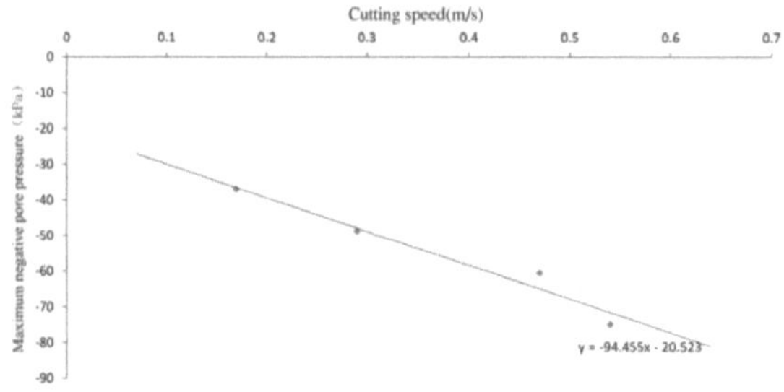

Figure 9. Relation curve between the maximum negative pore pressure and the cutting speed.

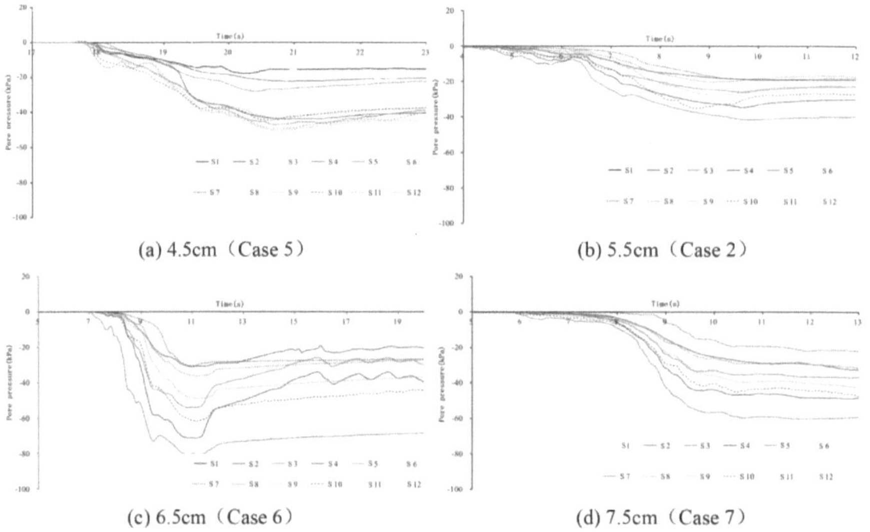

(a) 4.5cm（Case 5）

(b) 5.5cm（Case 2）

(c) 6.5cm（Case 6）

(d) 7.5cm（Case 7）

Figure 10. Time history curves of pore pressure at different cutting thicknesses.

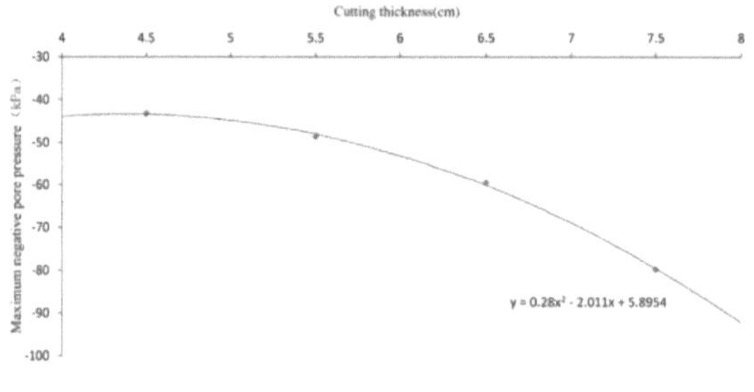

Figure 11. Relation curve between the maximum negative pore pressure and the cutting thickness.

Fig. 10 shows the variation of pore water pressure with time at different cutting thickness values. It can be found that the variation law of pore pressure caused by cutting thickness is basically the same, i.e. the negative pore pressure increases with the passage of time, and after reaching the maximum, it gradually decreases and finally reaches a relatively stable value. Moreover, the greater the depth, the more obvious this law reflects.

Fig. 10 also shows that at a cutting thickness of 4.5 cm, the largest negative pore pressure is −43.26 kPa; at 5.5 cm, −48.69 kPa; at 6.5 cm, −59.48 kPa; and at 7.5 cm, −79.79 kPa. The above data indicate that with the increase of cutting thickness, the maximum negative pore pressure increases. Therefore, it is found that with the increase of cutting thickness, the maximum negative pore pressure increases largely. It is noted that the relationship basically conforms to a quadratic curve. The fitting curve is shown in Fig. 11, and the equation is given as follows:

$$y = -3.72x2 + 32.602x - 114.85 \tag{3}$$

4 STUDY OF THE INFLUENCE OF TOOTH SHAPE ON VARIATION LAWS OF PORE PRESSURE ALONG THE TOOTH FACE IN THE CUTTING PROCESS

In this study, three kinds of tooth types are analyzed, such as the chisel type, plough type and semi-plough type. For the plough rake tooth type and semi-plough rake tooth type, two holes for monitoring pore pressure are drilled at the places close to and away from the tooth tip, respectively, as shown in Fig. 12 and Fig. 13.

Figure 12. Distribution of monitoring points for the pore pressure of the plough rake tooth.

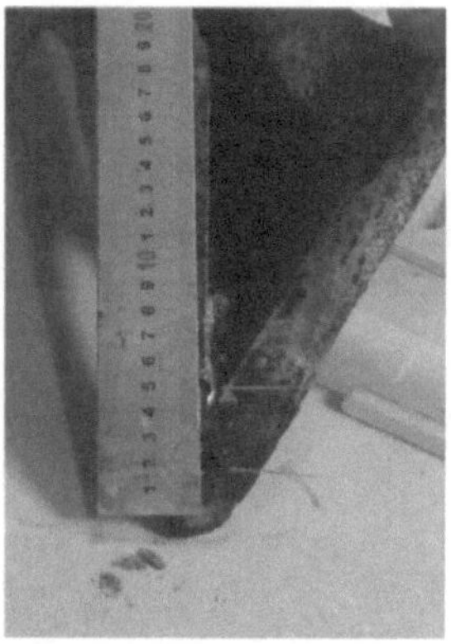

Figure 13. Distribution of monitoring points for the pore pressure of the semi-plough rake tooth.

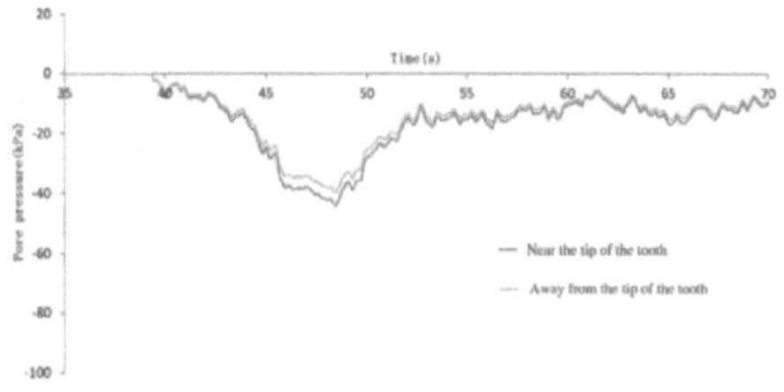

(a) 45° (Case 11)

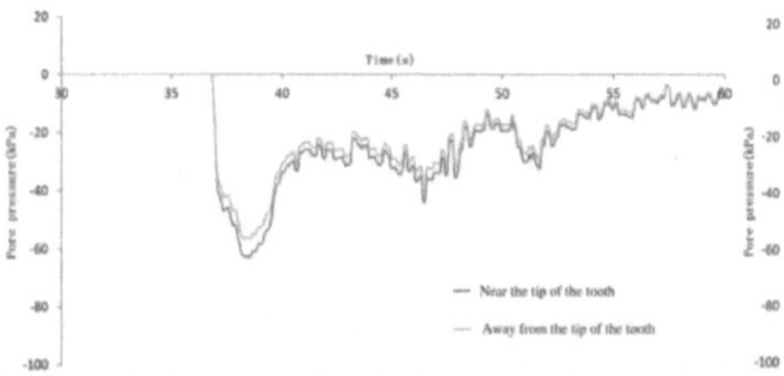

(b) 60° (Case 12)

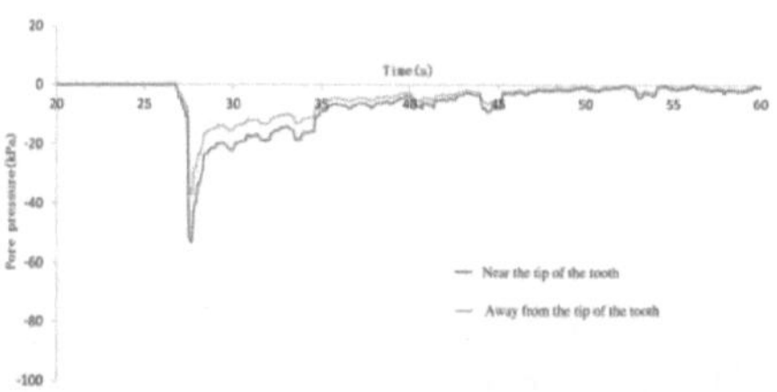

(b) 75° (Case 13)

Figure 14. Time history curves of pore pressure with different cutting angles for the plough rake tooth.

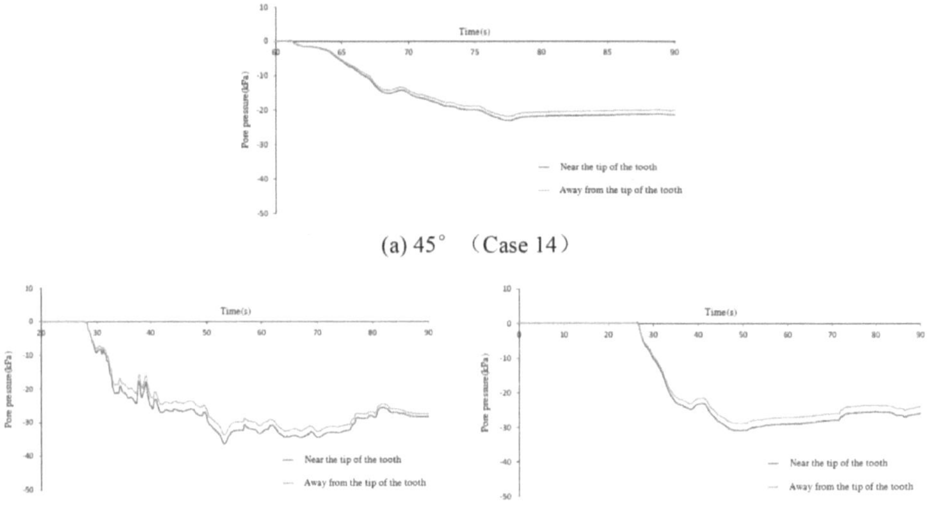

(a) 45° (Case 14)

(b) 60° (Case 15)

(b) 75° (Case 16)

Figure 15. Time history curves of pore pressure with different cutting angles for the semi-plough rake tooth.

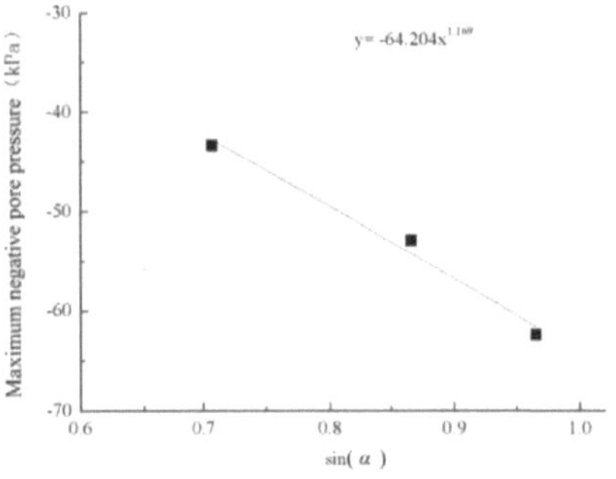

Figure 16. Relation curve between the maximum negative pore pressure and the cutting angle for the plough rake tooth.

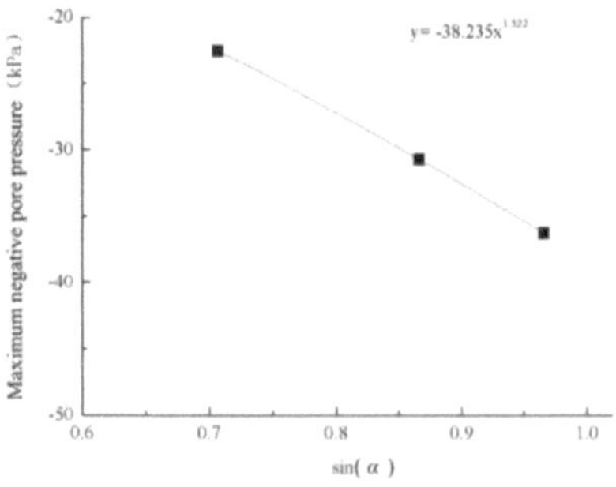

Figure 17. Relation curve between the maximum negative pore pressure and the cutting angle for the semi-plough rake tooth.

The pore pressure time history curves for different cutting angles of plough rake tooth and semi-plough rake tooth are shown in Fig. 14 and Fig. 15. Comparing Fig. 14 and Fig. 15 with Fig. 6, it can be found that three types of tooth have the following common characteristics: (1) the closer to the tooth tip, the greater the maximum negative pore pressure; (2) the negative pore pressure increases first, then decreases and eventually tends to be stable. The distinct characteristics of the three tooth types are the following: (1) for the plough rake tooth, the negative pore pressure appears in a certain position suddenly, then rapidly increases to the maximum value, after which it slowly decreases to a stable value; for the chisel rake tooth and semi-plough rake tooth, negative pressure gradually increases to maximum, then decreases gradually and finally tends to a stable value; (2) when the cutting angle, cutting speed and cutting thickness are corresponding to one another, the maximum negative pore pressure produced by semi-plough rake tooth is minimum, and that by plough rake tooth is maximum; (3) for chisel rake tooth, the fitting curve of the cutting angle and the largest negative pore pressure is shown in Fig. 7, and the relationship can be expressed by Eq. (1); for plough rake tooth, the fitting curve of the cutting angle and the largest negative pore pressure is shown in Fig. 16, and the relationship can be expressed by Eq. (4); for semi-plough rake tooth, the fitting curve of the cutting angle and the largest negative pore pressure is shown in Fig. 17, and the relationship can be expressed by Eq. (5):

$$y = -64.204x^{1.169} \tag{4}$$

$$y = -38.235x^{1.522} \tag{5}$$

5 CONCLUSION

1. Cutting test results of chisel rake tooth prototype tooth indicate that, firstly, the negative pore pressure on the tooth surface gradually increases with the passage of time, and after reaching the maximum, it gradually decreases and finally reaches a relatively stable value. With the increase of cutting angle, the maximum negative pore pressure increases, and they are basically linearly dependent. With the increase of tooth cutting speed, the maximum negative pore pressure also increases, and they are basically linearly dependent as well. With the increase of cutting thickness, the maximum negative pore pressure increases, and the relationship between them can be expressed by a quadratic equation.
2. For chisel rake tooth, plough rake tooth and semi-plough rake tooth, the pore pressure distribution on the tooth surface has some common and distinct characteristics. For the three tooth types, the relationships between the maximum negative pore pressure and the cutting angle meet different multinomial fittings.
3. In summary, this research study has an important guiding significance for reducing the tooth cutting resistance and enhancing the dredging efficiency by appropriate selection of construction parameters and tooth types in the dredging process.

REFERENCES

Ishibashi I, Sherif M.A, Cheng W.L. The effects of soil parameters on pore-pressure-rise and liquefaction prediction. Soils and Foundations, 1982, 22(1): 39–48.

Liu G. Study on Shear Behaviors of Saturated Marine Soils Under Complex Stress Conditions. Dalian University of Technology, 2010.

Miedema S.A. New developments of cutting theories with respect to offshore applications, the cutting of sand, clay and rock. 20th International Offshore and Polar Engineering Conference, Beijing, China, 2010.

Polito C.P, Green R.A, Lee J. Pore pressure generation models for sands and silty soils subjected to cyclic loading. Journal of Geotechnical and Geoenvironmental Engineering, 2008, 134(10): 1490–1500.

Seed H.B, Martin P.P. and Lysmer J. Pore-water pressure changes during soil liquefaction. Journal of Geotechnical Engineering Division: American Society of Civil Engineering, 1976, 102(4): 323–346.

Sun R, Yuan X. Simplified incremental formula for estimating pore pressure of saturated sands under anisotropic consolidation. Chinese Journal of Geotechnical Engineering, 2005, 27 (9): 1021–1025.

Sun R, Yuan X. A method for calculating time-dependent pore pressure rise of saturated sands under earthquake loading. Earthquake Engineering and Engineering Vibration, 2006, (3): 261–263.

Towhata I, Ishihara K. Shear work and pore pressure in undrained shear. Soils and Foundations, 1985, 25(3): 73–84.

Xie L, Wang X, Liang X et al. Experimental study on pore pressure of saturated soil subjected to cutting motion by draghead teeth. Rock and Soil Mechanics, 2015, 36(supp.1): 1–5.

Xie L, Zhang J, Qin S et al. Laws of pore pressure development in cutting process of densely compacted sand under water. China Harbour Engineering, 2011, (4): 10–12.

Zhang L. Pore-water pressure of sand during triaxial shear process. Journal of Chengdu University of Science and Technology, 1987, (4): 69–77.

Zhao Y, Miedema S.A. Finite element calculations to determine the pore pressures when cutting water saturated sand at large cutting angles. CEDA Dredging Days, Amsterdam, The Netherlands, 2001.

Applied Mechanics and Civil Engineering VI – Xie (Ed.)
 ISBN 978-1-138-62631-7

Study on the punching shear strength and deformation capacity of RC slabs based on a database

C.T. Huang, S. Pu & Q. Ma
Zunyi Normal College, Guizhou, P.R. China

ABSTRACT: The punching shear strength and deformation capacity of Reinforced Concrete (RC) slab-column connection are the basic performances in design, as well as the core factors that control the performance design of a flat plate system. Although various studies and hundreds of laboratory experiments related to the punching shear strength of RC slabs have been published and conducted, respectively, code provisions have been developed based on a rather limited subset of the available test results and those studies have ignored the details of test slabs and experiments. Worse still, the equations presented so far have not been able to calculate the punching shear strength and deformation capacity simultaneously. A database of 678 specimens was categorized and structured depending on the type of test set-up in the present study. In this study, eight equations for the punching shear strength prescribed in specifications were evaluated based on the database. A simplified strength equation and a simplified deformation equation were also proposed from the database.

1 INTRODUCTION

The RC flat plate structure has proven to be the most appropriate ceiling system for structural and industrial engineering, because there are no visible beams, column capitals or drop panels, making the simplicity both for construction (less formwork and simple reinforcement) and use (easy placement of equipment and pipeline installation), meanwhile providing high space utilization (decreasing storey height significantly and arranging partition freely). However, the flat plate structure is susceptible to punching shear failure at slab-column connection under bending-shearing composite forces. Once punching shear failure occurs, it makes gravity redistribution and, worse still, the structure may progressively collapse.

In the late 1940s, flat plates were initially developed in the US. Since then, hundreds of laboratory experiments have been conducted to investigate the punching shear behavior of two-way RC slabs at interior supports. In general, most research studies distinguished the controlling factors of punching shear strength based on the regression analysis of experimental data and elastic theory, and formulated all the controlling factors empirically to calculate the punching shear strength.

Unfortunately, because of the arbitrariness by researchers and code developers in selecting reference data together with ignored details of test slabs and experiments, code provisions have been developed based on a rather limited subset of the available test results and also based on the not detailedly categorized tests. Until now, almost all structural design codes have judged punching shear failure by the nominal stress in the critical section. However, there are divergences with respect to the location of the critical section, the shape of the critical section, the calculation method of the nominal stress, the value of the ultimate stress and other controlling factors, as shown in Figure 1 and Table 1.

At the same time, with the development of a disaster prevention and mitigation strategy, the cost-benefit criterion and performance-based seismic design require deformation and force demands for different seismic hazards, and also require deformation and force capacities for

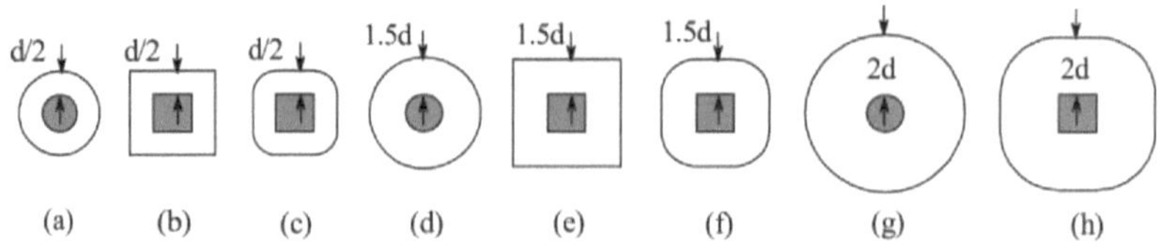

Figure 1. Position and perimeter of the critical section.

Table 1. Specifications of different codes for punching shear strength.*

Design code	Critical section (u) in Fig. 1	Effective height (d)***	Concrete strength***	Size effect (S_e)	Other factors (η)****	Punching shear strength (V)	Inclination of critical section (θ)
GB (2010)	(a)(b)	$(d_x + d_y)/2$	f_t	1.0-(d-800)/12000; $0.9 \le S_e \le 1.0$	$\eta_1 = f_1(c_1/c_2)$; $\eta_2 = f_2(d/u)$; $\eta = \min(\eta_1, \eta_2)$	$0.7S_ef_t\eta ud$	63.4
ACI (2011)	(a)(b)	$(d_x + d_y)/2$	$f_c'^{1/2}$	–	$\eta_1 = f_3(c_1/c_2)$; $\eta_2 = f_4(d/u)$; $\eta = \min(\eta_1, \eta_2, 0.33)$	$f_c'^{1/2}\eta ud$	63.4
CEB-FIP (2010)	(a)(c)	$(d_x + d_y)/2$	$f_{ck}^{1/2}$	–	$\eta = f_5(E_s, f_y, d_g, r_s)$	$f_{ck}^{1/2}\eta ud$	63.4
BS (1997)**	(d)(e)	$(d_x + d_y)/2$	$f_{cu}^{1/3}$	$(400/d)^{1/4}$; $S_e \ge 1.0$	$\eta = f_6(\rho)$; $\rho \le 0.03$	$0.79S_e(f_{cu}/25)^{1/3}\eta ud^{*2}$	33.7
CAN (2010)	(a)(b)	$(d_x + d_y)/2$	$f_c'^{1/2}$	$1300/(1000 + d)$; $d \ge 300$	$\eta_1 = f_7(c_1/c_2)$; $\eta_2 = f_8(d/u)$; $\eta = \min(\eta_1, \eta_2, 0.38)$	$S_ef_c'^{1/2}\eta ud$	63.4
EN (2004)	(g)(h)	$(d_x + d_y)/2$	$f_{ck}^{1/3}$	$1 + (200/d)^{1/2}$; $S_e \le 2.0$	$\eta = f_6(\rho)$; $\rho \le 0.02$	$0.18S_ef_{ck}^{1/3}\eta ud$	26.6
DIN (2008)	(d)(f)	$(d_x + d_y)/2$	$f_{ck}^{1/3}$	$1 + (200/d)^{1/2}$; $S_e \le 2.0$	$\eta = f_6(\rho)$; $\rho \le 0.02$	$0.21S_ef_{ck}^{1/3}\eta ud$	33.7
JSCE (2002)	(a)(b)	$(d_x + d_y)/2$	$f_c'^{1/2}$	$(1000/d)^{1/4}$; $S_e \le 1.5$	$\eta_1 = f_6(\rho)$; $\eta_2 = f_9(c_1, c_2, d)$; $\eta_1 \le 1.5$, $\eta = \eta_1 \times \eta_2$	$0.2S_ef_c'^{1/2}\eta ud$	63.4

Note

* Units are expressed in MPa and mm;

** When $f_{cu} < 25$MPa, $f_{cu} = 25$; and $f_{cu} \le 40$MPa;

*** d_x and d_y are the effective height in two orthogonal directions; f_t, f_c', f_{ck}, f_{cu} are the axial tension strength, cylinder compressive strength, eigenvalue of compressive strength, cubic (150mm × 150mm × 150mm) compressive strength of concrete, respectively;

**** c_1 and c_2 are the long side and the short side of the rectangular column, respectively, when $c_1/c_2<2$, $c_1/c_2 = 2$, and $c_1/c_2= 2$ for the round column; the functions in Table 1: f_1: $0.4 + 1.2/(c_1/c_2)$; f_2: $0.5 + 0.25\alpha_s d/u$; f_3: $0.166 + 0.332/(c_1/c_2)$; f_4: $0.166 + 0.083\alpha_s d/u$; f_5: $1/[1.5 + 43.2r_sf_y/(16 + dg)/E_s]$; f_6: $(100\rho)1/3$; f_7: $0.19 + 0.38/(c_1/c_2)$; f_8: $0.19 + 0.1\alpha_s d/u$; f_9: $1 + 1/[1 + 0.5(c_1 + c_2)/d]$; where α_s is 40 for interior columns, 30 for edge columns, 20 for corner columns; ρ is the reinforcement ratio; r_s denotes the position where the radial bending moment is zero with respect to the support axis; f_y is the yield strength of steel; d_g is the maximum size of the aggregate; E_s is the elastic modulus of steel.

various performance levels. Thus, it is necessary to analyze and quantify the deformation of slab-column connection. Even in regions of high seismic risk where moment frames or shear walls are provided as the main lateral-force-resisting system, the deformation compatibility is also required for the slab-column connection. However, the models or equations presented so far have not been able to calculate the punching shear strength and deformation capacity simultaneously.

The purposes of the present study are to create a database for the punching shear strength and deformation capacity of RC slabs, to evaluate the strength equations proposed in eight specifications and to propose two rational strength and deformation equations based on the database.

2 DATABASE AND EVALUATION OF DESIGN SPECIFICATIONS

2.1 *Database and categorization*

Ospina et al. (2011) reported the basis of the punching databank containing 512 test results of slabs without shear reinforcement subjected to concentric punching at interior column locations. However, Ospina missed Chinese and Japanese papers written in Chinese and Japanese, respectively. The specifications of eight different codes for punching shear strength are given in Table 1. Huang (2015) collected 166 test results of slabs without shear reinforcement subjected to concentric punching at interior column locations, which was conducted in China and Japan. Along with Ospina, a database of 678 specimens was created. As shown in Table 2, the 678 specimens were categorized into eight main types according to the test set-up and the specimen geometry. Figure 2 shows the plan view of typical test specimens, indicating the different variables and notation used to describe their geometric properties in plan and support conditions. It should be noted that the tests missing the key data will be removed.

2.2 *Evaluation of design specifications*

Figure 3 provides the Average Value (AVG) and Standard Deviation (ST) of V_{test}/V_{Codes}. From the different code results in each group, due to limited test data, the strength equations and

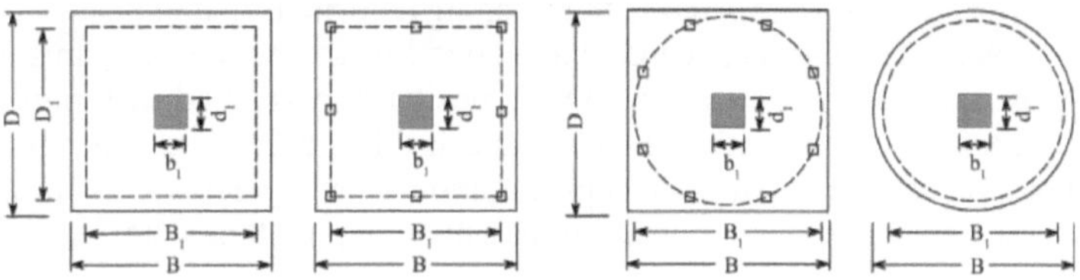

Figure 2. Test specimen descriptors.

Table 2. Classification of test specimens.

Group	Slab supported or loaded pattern—Column shape	Number of specimens	Filter out	Surplus specimen
CC	Circular – Circular	136	22	114
CS	Circular – Square	67	10	57
R(RCS)	Rectangular – (Rectangular, Circular, Square)	4 + 5 + 7	0	16
SC	Square – Circular	30	2	28
(SC)R	(Square, Circular) – Rectangular	21 + 2	0	23
SS	Square – Square	324	6	318
O	Opposite –	35	0	35
U	Slab Uniformly Loaded –	47	0	47
Total		678	40	638

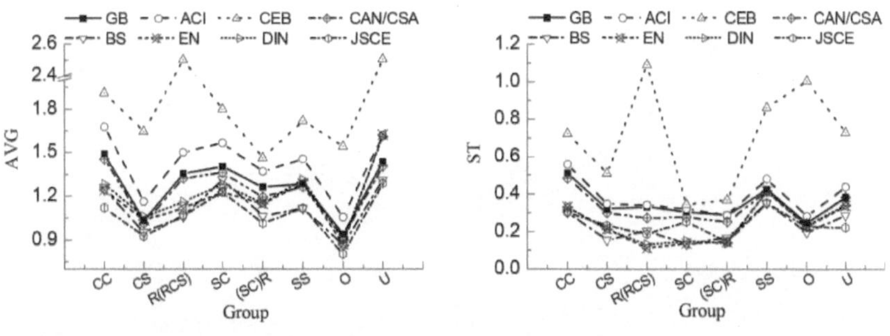

Figure 3. Calculation results within different groups.

computational accuracy vary widely. The strength equations provided in the specifications are mostly based on specific test data, which contain a small-scale specimen geometry and material parameter. From the results of each specification among the eight groups, it is obvious that the test set-up and the specimen geometry can influence the computational accuracy significantly. That is to say, the arbitrariness by researchers and code developers in selecting reference data can lead to poorer applicability of the specifications.

3 PROPOSED RATIONAL EQUATION FOR PUNCHING SHEAR STRENGTH AND DEFORMATION

3.1 *Simplified calculation of punching strength and deformation*

To study the punching behaviors of slab-column connection, based on the experimental phenomena, a relatively comprehensive Bending-Shearing Critical Crack Model (BSCCM), which can be used to calculate the punching load and deformation simultaneously, was established by Huang (2015).

The BSCCM indicated that the factors that affect punching behavior include the reinforcement ratio (ρ), concrete strength (f_c), yield strength of steel (f_y), the punching span ratio (λ) and the perimeter thickness ratio of the critical section (b_0/d). The reinforcement ratio and yield strength of steel are negatively correlated with punching deformation and positively correlated with punching capacity. In contrast, the concrete strength, the punching span ratio and the perimeter thickness ratio of the critical section are positively correlated with punching deformation and negatively correlated with punching capacity. Therefore, this paper constructs the comprehensive parameter K_0 here (as shown in Equation 1), which represents the comprehensive influence of the above factors on punching capacity and punching deformation:

$$K_0 = \frac{\rho f_y}{\sqrt{f_c}\lambda b_0 / d} \tag{1}$$

Figure 4 and Figure 5 show the overall scatter relation, scatter relation in key areas, and the fitting curve of K_0-$\psi\lambda d$ and $\psi\lambda d$-v_0, respectively. Here, ψ is the punching rotation, $\psi \cdot \lambda \cdot d$ represents the deflection deformation when punching occurs on the component; $v_0 = V/(f_c^{1/2} \cdot b_0 \cdot d)$,

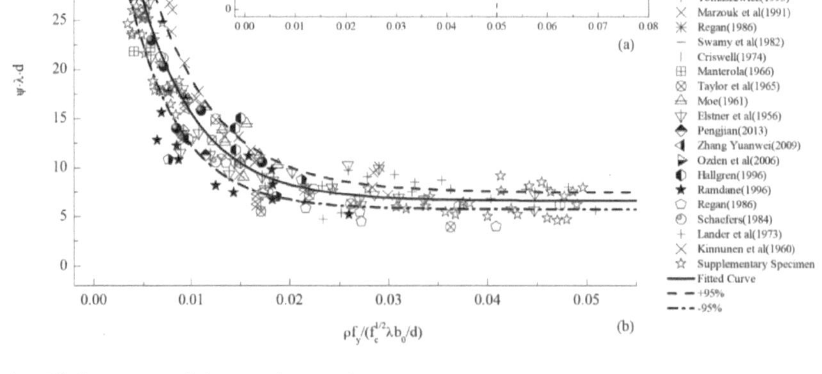

Figure 4. Fitting curve of the rotation angle.

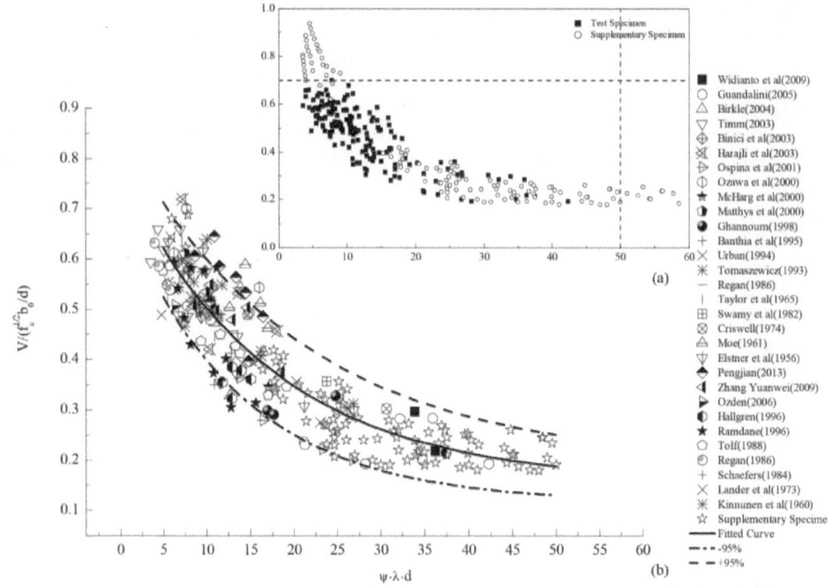

Figure 5. Fitting curve of the punching load-rotation angle.

V is the punching load, $f_c^{1/2}$ is the concrete item, referring to the standards of different countries to select 1/2 time of concrete strength for consideration, b_0 is the perimeter of the critical section (calculated according to the fillet section which is 0.5d far from the column edge), d is the effective thickness of the slab, and v_0 represents the regular nominal punching capacity.

As shown in Figure 4(b), when the K_0-$\psi\lambda d$ correlation shown in Equation 2 is adopted according to the test points and supplementary example points, the fitting curve has the smallest residual sum of squares:

$$\psi\lambda d = 6.613 + 46.469 \cdot e^{-K_0/0.00603} \tag{2}$$

As shown in Figure 5(b), when the $\psi\lambda d$-v_0 correlation shown in Equation 3 is adopted according to the test points and supplementary example points, the fitting curve has the smallest residual sum of squares:

$$100 \cdot v_0 = 15.498 + 62.780 \cdot e^{-\psi\lambda d/16.913} \tag{3}$$

3.2 *Verification*

This paper selects two groups of tests, CC and SS, to make the detailed discussion. There are two reasons for extracting these two test data: first, the component shape and the loading mode are the most regular and the closest to the theoretical method built previously; secondly, most researchers have chosen these two ways of test, being 136 and 324, respectively, accounting for 67.85% of the total test components collected; therefore, the analysis results of these two groups of components are the most representative ones.

The calculation results are summarized in Table 3. The statistical results in the table are the ratios of the test values to the calculated values of punching strength. It can be seen that the calculated method proposed in this paper (Equation 1~3) has the best results. Since it needs to calculate the rotation angle $\psi\lambda d$ according to Equation 2 when calculating the punching capacity according to Equation 3, it indirectly indicates that the method has high accuracy in the calculation of the rotation angle.

Table 3. Calculation results of punching shear strength.

	CC Group		SS Group	
Calculate method	AVG	COV	AVG	COV
Simplified method proposed in this paper	1.00	0.188	1.075	0.173
GB50010-2010	1.491	0.341	1.285	0.330
ACI318-11	1.676	0.332	1.455	0.329
CEB/FIP MC2010	1.911	0.379	1.718	0.501
BS 8110-97	1.244	0.244	1.115	0.316
CAN/CSA A23.3-04 (R2010)	1.456	0.331	1.265	0.328
EN 1992-1-4: 2004	1.240	0.264	1.289	0.321
DIN 1045-1: 2008	1.280	0.256	1.309	0.316
JSCE2002	1.119	0.269	1.117	0.315

4 CONCLUSIONS

A database of 678 specimens was categorized and structured depending on the type of test set-up in the present study. In this study, eight equations for the punching shear strength prescribed in specifications were evaluated based on the database. The calculation results indicated that all eight design codes had great deviations, among which the JSCE code had the best calculation results. The calculation results of China's concrete code were more accurate than those of ACI but poorer than European and JSCE codes. The simplified strength equation and deformation equation considered the influence of size effects, concrete strength, reinforcement ratio, perimeter thickness ratio of the critical section and the punching span ratio. Based on the comparison with the database, it was found that the simplified equations had a high precision.

ACKNOWLEDGMENTS

This study was funded by research grant from the Doctoral Program Foundation of Zunyi Normal College (BS[2015]17#) and Science and Technology Department of Guizhou Province (LH[2016]7005), which are gratefully acknowledged.

REFERENCES

ACI 318-11. *Building Code Requirements for Structural Concrete and Commentary*, 2011. Detroit, Farmington Hills: American Concrete Institute.

BS 8110-97. *Structural use of concrete, Part 1: Code of Practice for Design and Construction*, 1997. British Standards Institution, London, UK.

BS EN 1992-1-1:2004. *Design of Concrete Structures. General rules and rules for buildings*, 2004. Brussels, Belgium.

CEB-FIP. *Mode Code for Concrete Structures (2010)*, 2010. Wilhelm Ernst & Sohn, Berlin, Germany.

CAN/CSA-A23.3-04 (R2010). *Design of Concrete Structures*, 2010. Canadian Standards Association, Rexdale, ON, Canada.

DIN 1045-1. *Plain, Reinforced and Prestressed Concrete Structures-Part 1: Design and Construction*, 2008. Normenausschuss Bauwesen (NABau) im DIN Deutsches Institut für Normung e.V. Beuth Verl. Berlin.

Huang, C.T. 2015. *Study on the Analysis Methods of Hollow Floor and Punching Issues of Slab-column Connection*, PhD thesis, School of Civil Engineering, Chongqing University, Chongqing, China.

JSCE. *Standard specification for concrete structures (2002)*, 2002. Japan Society of Civil Engineering.

Ospina, C.E., Gerd Birkle & Widianto Widianto, 2011. *ACI 445 Punching Shear Collected Data-bank, Network for Earthquake Engineering Simulation (database)*, Dataset, DOI: 10.4231/D3TX35618.

Applied Mechanics and Civil Engineering VI – Xie (Ed.)
© 2017 Taylor & Francis Group, London, ISBN 978-1-138-62631-7

Experimental study on impact localization based on wavelet analysis

Yulin Chen, Huiqin Chen & Liwei Huang
Department of Electronics, Chengxian College, Southeast University, Nanjing, China

ABSTRACT: Composite materials are sensitive to the impact load. Therefore it is necessary to carry out the online structural health monitoring for the impact on composite structures. In this paper, experimental study on impact localization based on fiber Bragg grating measurement and wavelet analysis is presented. A composite plate with the size of 480 mm × 480 mm is adopted in the experiment. Impact responses are measured using the fiber Bragg grating. Wavelet packet transform is used to extract the energy feature vector. Similarity matching algorithm is used to determine the impact localization on the composite laminate. Results show that the positioning errors of eight impact points are small.

1 INTRODUCTION

Composite material has been rapidly-developed and widely used. It has the advantage of high stiffness-to-density ratio. But also it has the disadvantage of low impact resistance ability. Impact is easy to cause the invisible internal damages to composite material, which leads to the degeneration of strength and stability and significantly reduces the carrying capacity. Therefore, it is necessary to monitor the impact on composite material.

There are a varieties of impact localization methods, most of them could be divided into two categories: (1) Acoustic emission analysis method. (2) Inverse analysis method. The first kind of method has better precision in the orientation of isotropic material. However, due to the anisotropy of the composite structure, the reflection and scattering of the stress wave and the influence of the noise, the wave velocity and the wave arrival time are difficult to be accurately determined, which makes it difficult to achieve accurate localization. Therefore, the impact localization methods based on time delay have some limitations. There are many relevant researches on this method: Tribikram Kundu et al. pasted three sensors in the composite plate, then by accepting signal time delay in every direction and the propagation velocity of the three sensors, the impact localization could be determined. Chan Yik park pasted four sensors in composite material plate to measure time delay of the impact signal of each sensing point, then the position of the impact point was determined by the neural network. Yongzhen Su, Yong Teng Zhong and Yingdi Xu made further improvement on that basis. The direction of the impact was estimated by the MUSIC algorithm. The wavelet transform was used to extract specific frequency of Lamb wave, then the arrival time of signal wave was got. Finally, the distance from the impact point was solved by the relationship between the speed, time and distance. In brief, the method is mainly based on the inverse operation of the impulse response. With the rapid development of the sensing technology, this method has been developed and applied widely. But due to the short response time and wide frequency excitation, identifying the impact will be more complex. Although the inverse problem solving method is widely used, the ill-posedness brings great difficulty. Meanwhile, huge number of unknowns makes the localization process to be computation intensive.

Recently, a novel impact localization method has been proposed. Shrestha Pratik et al. collected the impact signal of each point, after the normalization of the database, the correlation coefficient method and the root mean square method were used to determine the impact localization. Jiyun Lu and other researchers adopted wavelet packet to extract key frequency band characteristic value, and proposed the sixth order wavelet packet energy value as the impact localization of the feature vector. Shizeng Lu et al. used Fourier Transform method to extract the frequency domain characteristic of the impact signal amplitude as a feature, and implemented the low speed impact localization after data matching.

In this paper, experimental study on impact localization based on fiber Bragg grating measurement and wavelet analysis is presented. According to the relationship that the closer two impact positions, the higher the degree of similarity signal amplitude-frequency characteristics. Energy feature vector was extracted through the signals from the Fiber Bragg Grating (FBG) sensor using wavelet packet transform. Then the database of energy feature vector was established, and finally the impact localization was determined according to the similarity function.

2 THEORETICAL BACKGROUND

2.1 *The principle of fiber Bragg grating sensor*

The principle of FBG sensor is that when the light passes through the fiber Bragg grating, the light accordant with the fiber grating will be reflected back. The rest light will be transmitted through the fiber grating. The central wavelength of the fiber grating can be determined.

FBG sensor is sensitive to temperature and stress. The pitch and refractive index of sensor will be changed with the change of temperature and strain. On this basis, these laws can be used to measure it.

The central wavelength of each FBG reflected light λ is shown in Formula (1):

$$\lambda = 2n\Lambda \tag{1}$$

where, λ is a reflection wavelength, n is the effective refractive index in fiber grating and Λ is grating period.

This experiment was carried out in a small room. The temperature of room seldom change. Therefore the wavelength of the reflected light wave caused by the temperature is small enough to be negligible. The influence of strain was considered. Therefore the pitch Λ changed after the stress period. The variation is $\Delta\Lambda$. Meanwhile, $\varepsilon = \frac{\Delta\Lambda}{\Lambda}$ was used to represent the corresponding amount of a strain. The change of wavelength can be concluded in Formula (2):

$$\Delta\lambda = (1-P)\lambda\varepsilon = \alpha\varepsilon \tag{2}$$

where, $\Delta\lambda$ is the change of reflection wavelength. ε represents a strain. P is elasto optical coefficient, which $P = \frac{\Delta n}{n}$. α represents sensitivity of FBG sensor. For the FBG sensors, when the center wavelength is 1510 nm–1590 nm, α is about $1.2\,pm/\mu\varepsilon$.

When the fiber grating is subjected to axial stress, the waveguide effect was ignored. The shift of fiber grating induced by wavelength can be simplified in Formula (3):

$$\Delta\lambda_B = 2\Lambda\left(\frac{\partial n_{eff}}{\partial L}\Delta L + \frac{\partial n_{eff}}{\partial a}\Delta a\right) + 2\frac{\partial\Lambda}{\partial L}\Delta L n_{eff} = 2\Lambda\left[-\frac{n_{eff}^3}{2}\Delta\left(\frac{1}{n_{eff}^2}\right)\right] + 2n_{eff}\varepsilon_{zz}L\frac{\partial\Lambda}{\partial L} \tag{3}$$

where, ΔL represents the longitudinal expansion of the fiber. Δa indicates the change of the optical fiber diameter (caused by the longitudinal tension). $\partial n_{eff}/\partial L$ is the elastic optical effect. $\partial n_{eff}/\partial a$ shows the waveguide effect. $\varepsilon_{zz} = \Delta L/L$ expresses the longitudinal expansion strain.

When the experimental composite plate was tapped, the stress is generated. The strain was instantly produced on the composite plate. Then the signal of strain will be received by the FBG sensor. The wavelength of the reflected wave was changed. Eventually, the signal emitted will be demodulated by the SM130 FBG sensor demodulation. The monitoring of impact signal was completed.

2.2 *Extraction of wavelet packet feature vector*

Daubechies (Db) wavelet has the orthogonality and double orthogonality of the tight support, which can realize the fast algorithm and the perfect reconstruction. Similarly, the Db wavelet family can be more flexible to weigh the boundary problem caused by the increase of the length of the branch. Therefore the Db wavelet series of wavelet family was used in this paper.

Wavelet packet decomposition can also be referred to as the wavelet packet, the sub tree or the best sub tree structure. The concept is to use the analysis tree to represent the wavelet packet as shown in Figure 1.

The signal of each frequency band was represented by each wavelet packet of tree. The low frequency component A1 and the high frequency component D1 can be got by the a layer of decomposition of the original signal S. Then the signal components DA2, AD2, AA2, DD2 can be got using the two layers of decomposition of signal S. According to the number of layers of decomposition, each layer of wavelet packets can be successively got. Decomposing averagely the frequency band of the upper signal to the two signals of the next layer is called the decomposition of wavelet packet. If the signal on the upper layer is 250 Hz-500 Hz frequency band, the two wavelet packets of decomposition successively represent the 250 Hz-375 Hz and 375 Hz-500 Hz frequency bands.

2.3 *Similarity algorithm*

The similarity measurement is divided into two main categories: (1) the distance measure method. (2) the similarity function method.

Several methods of distance measure were commonly used at present: (1) Minkowsky distance. (2) Euclidean distance. (3) Manhattan distance. (4) Jffreys & Matusita distance. (5) Mahalanobis distance. (6) Camberra distance. In the above methods, the Minkowsky distance is the general form of the distance. The distance between Euclidean and Manhattan is its special form. Taking the most widely used Euclidean distance as an example, its similarity measure formula is shown in Formula (4). If (x, y) is a two-dimensional vector, the similarity is plane distance. If (x_i, y_i) is a three-dimensional vector, the similarity is spatial distance. So the closer the distance $(d(x, y))$, the more similar the two vectors.

$$d(x,y) = \left(\sum_{i=1}^{n} |x_i - y_i|^2 \right)^{1/2} \tag{4}$$

Similarly, the similarity function method also has the method of centralized maturity. The similarity function of normal vectors has included angle cosine method, generalized Dice coefficient method and correlation coefficient method etc. The similarity function can be divided into the similarity function of two vectors and the similarity function of general vector. This paper is based on the consideration of the similarity function of general vector. In this paper, the correlation coefficient method in the similarity function method is used as the similarity matching algorithm of the wavelet packet energy feature vector. The greater the similarity, the closer the localization of impact.

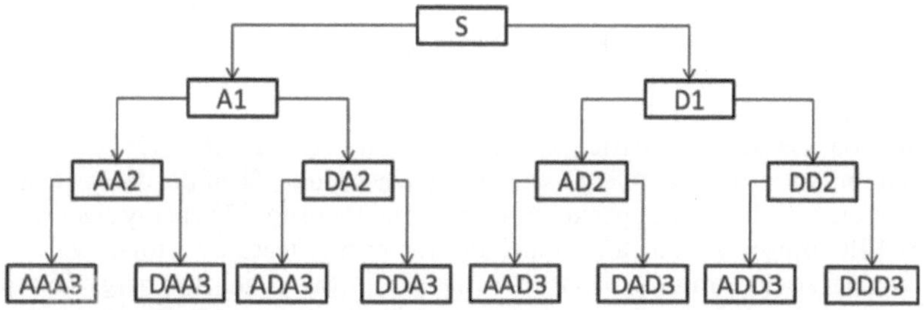

Figure 1. Wavelet packet tree.

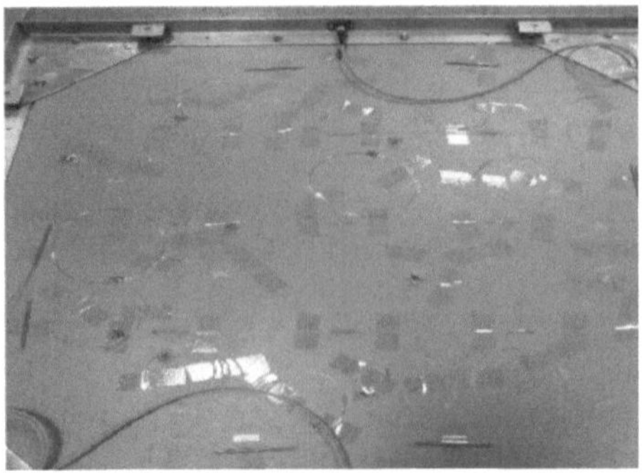

Figure 2. Composite plate pasted sensor.

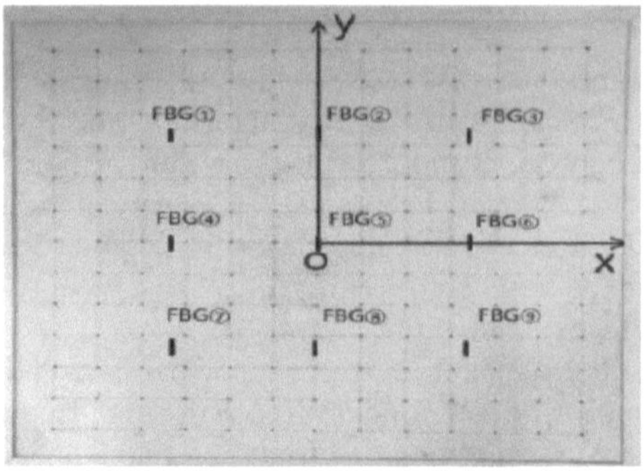

Figure 3. The arrangement of sensor.

3 EXPERIMENTAL SYSTEM

3.1 *Composite laminated plate*

A square glass fiber reinforced composite laminated plate with four edges clamped was adopted in this experiment. A square composite laminates with four sides of 620 mm and the thickness of 2 mm was selected. The laminate is fixed with aluminum alloy frame. The back of the board was pasted with nine FBG sensors as shown in Figure 2. The arrangement of FBG sensor can be shown in Figure 3. The position is represented by the thick black line. The coordinate of FBG sensor numbers ascend respectively: (–137.5 mm, 137.5 mm), (0 mm, 137.5 mm), (137.5 mm, 137.5 mm), (–137.5 mm, 0 mm), (0 mm, 0 mm), (137.5 mm, 0 mm), (–137.5 mm, –137.5 mm), (0 mm, –137.5 mm), (137.5 mm, –137.5 mm). The middle of the plate is divided into 13 × 13 sample points. The distance between every two adjacent sample points is 40 mm. The impact localization experiment is the middle of the 480 mm × 480 mm region. When taking the sample point, the lower left corner is the origin coordinates. Then the data was collected according to the coordinate of the entire value.

3.2 *Data collection of impact localization*

The data collection of impact localization part includes impact hammer, computer and FBG demodulator. The impact hammer is divided into the automatic hammer and the manual hammer. During the experiment, the first step is to pound plate by impact hammer. Then the sensor was usd to extract the signal components. The data will be transmitted to the computer by the demodulator. Finally, the MATLAB is used for analysis and processing to get the impact localization.

4 IMPACT LOCALIZATION EXPERIMENT

4.1 *Establishment of energy characteristic vector database*

Each sample point of the impact composite laminated plate was struck by a manual hammer and an automatic hammer. Meanwhile, the signal received by the FBG sensor at each point was stored by computer software. The time of receiving was roughly same. Two groups of the data were used to establish the database. The energy characteristic vector of each FBG fiber grating sensor was extracted from each test point using MATLAB software: $[\lambda_1, \lambda_2, \lambda_3, \ldots, \lambda_{15}]$.

After each impact, a total of 9 FBG sensors were accepted. Then the energy feature vectors were extracted by the wavelet packet transform and there are 135 energy eigenvalues. The numbers of FBG sensor were arranged into the 135 energy feature vectors: $[\lambda_1, \lambda_2, \ldots, \lambda_{135}]$. The same wavelet packet transform was also carried out to extract the energy feature value in the other set of data. The average value was sought to obtain a vector. The vector, which represents the energy feature vector of the sample point, was used as a database to match the energy feature vector of the needed impact point. There are 169 sample points on the composite laminated plate. Each sample point will need to be impacted and will have a feature vector. Finally, a matrix with 169 feature vectors was got: $[t_1, t_2, \ldots, t_{169}]$, where $t_i = [\lambda_1, \lambda_2, \ldots, \lambda_{135}]_i$.

4.2 *Achieving impact localization by similarity matching*

When any localization were pounded in the composite laminate, the impact signal will be received using a FBG sensor. Then the data was transmitted into the computer by the fiber optic demodulator. Finally, the db10 wavelet is used to decompose the six layers wavelet packet. Then the energy feature vector of each FBG fiber grating sensor was calculated. All the energy feature vectors were extracted by the all FBG sensors. Finally, the energy feature vector of the impact point was got: $t_0 = [\lambda_1, \lambda_2, \ldots, \lambda_{135}]_0$. The similarity was calculated between the feature vectors of the test points and the 169 energy feature vectors in the database by the correlation coefficient. Then the similarity between the test points and each sample point can be got: $[sim_{1-0}, sim_{2-0}, \ldots, sim_{135-0}]$, where sim_{i-0} was calculated by Formula (5):

$$sim_{i-0} = \frac{\operatorname{cov}(t_i, t_0)}{\left(\operatorname{cov}(t_i, t_i) \cdot \operatorname{cov}(t_0, t_0)\right)^{1/2}} \tag{5}$$

Thus, the localization point of the maximum value of the similarity can be obtained, which is the impact localization.

4.3 *Experimental results*

The experiment did not take the actual length to establish the axis in recording data but in accordance with the way of the sample point. The composite laminated plate is 13 × 13 sample points. So the 13 × 13 coordinate point was taken as to carry out the following experiment record and contrast. Firstly, the database was built with the experimental data. Then the random impact point and the database were used to match the similarity to get the localization. Meanwhile, the distance error s and the percentage error p are defined as in Formulas (6) and (7):

$$s = \sqrt{(x - x_0)^2 + (y - y_0)^2} \tag{6}$$

$$p = \frac{\sqrt{(x - x_0)^2 + (y - y_0)^2}}{480\sqrt{2}} \times 100\% \tag{7}$$

where, x, y represents the localization of the set point. x_0, y_0 represents the coordinate localization of actual impact. The percentage of the error is the percentage of the maximum error. The experiment was divided into two groups: the group of the automatic hammer and the group of the manual hammer. Eight points was tested for contrast in each group.

4.3.1 *Impact of the automatic hammer*

The eight points (0, 9), (1, 6), (3, 5), (5, 6), (6, 6), (9, 3), (11, 3) and (12, 11) were pounded by the automatic hammer. The impact energy produced by the automatic hammer was same. The data was shown in Table 1.

The results show that after the impact of the 7 points (0, 9), (1, 6), (5, 6), (6, 6), (9, 3), (11, 3) and (12, 11), energy feature vectors obtained were matched with 169 energy feature vectors in the database. The similarity of these seven points was the highest. Because of the highest similarity, the 7 points were maximum point. The distance error and the percentage error are zero by calculation. After the impact of the (3, 5) point, energy feature vectors obtained were matched with 169 energy feature vectors in the database. The localization was deviated to (4, 5). The distance of deviation is 40 mm. The percentage error is 5.89%. The two-dimensional curve and three-dimensional curve of (3, 5) point were shown in Figures 4 and 5.

Table 1. Localization results of the automatic hammer.

Actual coordinate	Location coordinate	Distance error/mm	Percentage error	Actual coordinate	Location coordinate	Distance error/mm	Percentage error
(0, 9)	(0, 9)	0	0	(6, 6)	(6, 6)	0	0
(1, 6)	(1, 6)	0	0	(9, 3)	(9, 3)	0	0
(3, 5)	(4, 5)	40	5.89%	(11, 3)	(11, 3)	0	0
(5, 6)	(5, 6)	0	0	(12, 11)	(12, 11)	0	0

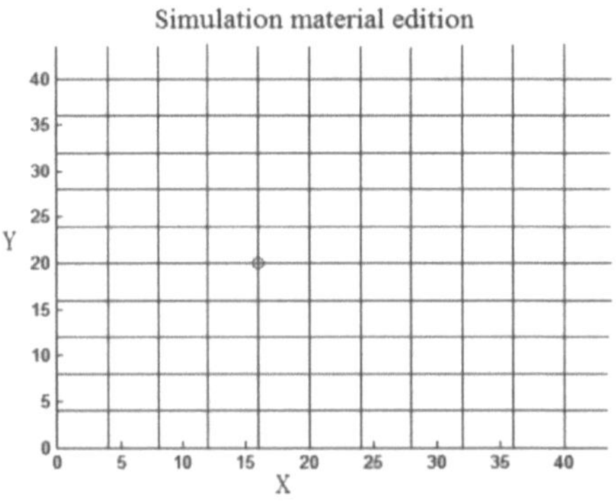

Figure 4. 2D localization of (3, 5).

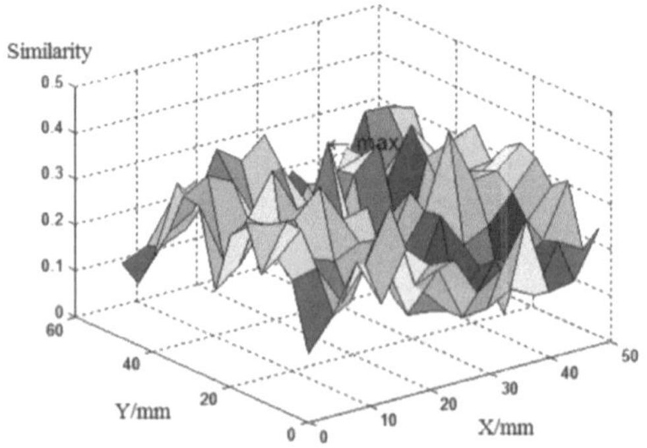

Figure 5. 3D localization of (3, 5).

Table 2. Localization results of the manual hammer.

Actual coordinate	Location coordinate	Distance error/mm	Percentage error	Actual coordinate	Location coordinate	Distance error/mm	Percentage error
(0, 9)	(0, 9)	0	0	(6, 6)	(6, 6)	0	0
(1, 6)	(1, 6)	0	0	(9, 3)	(9, 3)	40	5.89%
(3, 5)	(4, 5)	0	0	(11, 3)	(11, 3)	0	0
(5, 6)	(5, 6)	0	0	(12, 11)	(12, 11)	0	0

4.3.2 *Impact of the manual hammer*

The eight points (0, 9), (1, 6), (3, 5), (5, 6), (6, 6), (9, 3), (11, 3) and (12, 11) were pounded by the manual hammer. The impact energy produced by the manual hammer was different. The data was shown in Table 2.

The results show that after the impact of the seven points (0, 9), (1, 6), (3, 5), (5, 6), (6, 6), (11, 3) and (12, 11), energy feature vectors obtained were matched with 169 energy feature vectors in the database. The similarity of these seven points was the highest. Because of the highest similarity, the seven points were maximum point. The distance error and the percentage error are zero by calculation. After the impact of the (9, 3) point, energy feature vectors obtained were matched with 169 energy feature vectors in the database. The localization was deviated to (10, 3). The distance deviation is 40 mm. The percentage error is 5.89%. The two-dimensional curve and three-dimensional curve of (10, 3) point were shown in Figures 6 and 7.

In summary, it can be found that the impact of the manual hammer and the automatic hammer do not have much influence on the positioning process through the localization of 8 random points as well as the comparison of the manual and automatic hammer. Therefore, only once time is located to the near point in the 8 random impact points. The maximum distance error is 40 mm. The maximum percentage error is 5.89%. Two errors are all in the acceptable range.

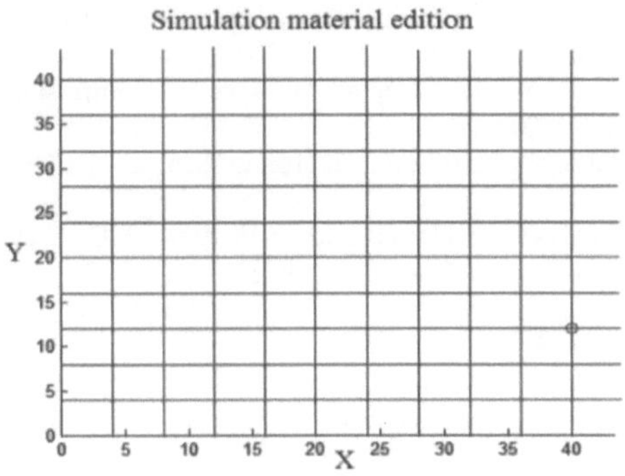

Figure 6. 2D localization of (9, 3).

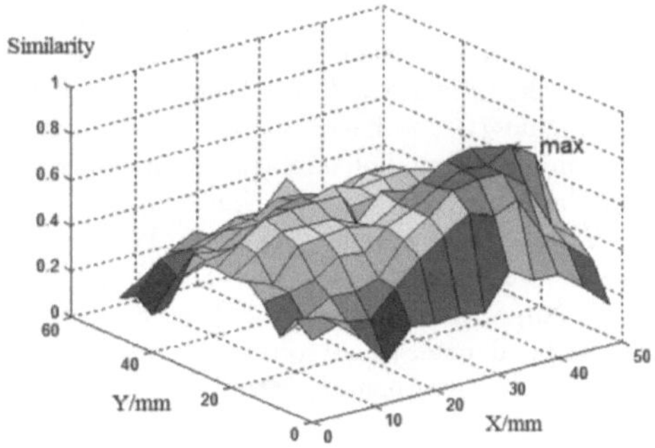

Figure 7. 3D localization of (9, 3).

5 CONCLUSIONS

In this paper, a composite plate with the size of 480 mm × 480 mm was adopted in the experiment. Impact responses are measured using the fiber Bragg grating sensor. Wavelet packet transform is used to extract the energy feature vector. Similarity matching algorithm is used to determine the impact localization on the composite laminate. The 8 impact localization experiment were carried out on 480 mm × 480 mm composite laminate using the automatic hammer and the manual hammer. Impact localization was successfully achieved. The conclusions could be drawn that:

1. Whatever the automatic hammer or the manual hammer, only one impact is located to the near point in the 8 random impact localization points.
2. The maximum distance error is 40 mm. The maximum percentage error is 5.89%. Two errors are all in the acceptable range.

REFERENCES

Baoqi Tao. Smart Material Structure [M]. Beijing: National Defense Industrial Press, 1997: 59–64.

Chambers A R, Mowlem M C, Dokos L. Evaluating Impact Damage in CFRP Using Fiber Optic Sensors [J]. Composites Science and Technology, 2007, 67(6): 1235–1242.

Chan Yik Park, Jong Heon Kim, Seung-Moon Jun, et al. Localizations and force reconstruction of low-velocity impact in a composite panel using optical fiber sensors [J]. Advanced Composite Materials, 2012, 21(5–6): 357–369.

Dan Xu. Research on the Application of Fiber Optic Sensor Strain Monitoring [D]. Dalian: Dalian University of Technology, 2010.

Jiemin Zhou. Application of Wavelet Analysis and ART Neural Network in Impact Localization of Composite Plate [J]. Aeronautical Journal, 2001, 22(1): 94–96.

Jiyun Lu, Bangfeng Wang, Dakai Liang. Fiber Prague Grating Impact Positioning System Based on Wavelet Packet Feature Extraction and Support Vector Regression Machine [J]. Optics and Precision Engineering, 2012, 20(4): 712–718.

Pratik Shrestha, Jin-Hyuk Kim, Yurim Park, et al. Impact localization on composite wing using 1D array FBG sensor and RMS/correlation based reference database algorithm [J]. Composite Structures, 2015, 125: 159–169.

Shizeng Lu, Mingshun Jiang, Qingmei Sui, et al. FBG Impact Location System Based on Data Template Matching Algorithm [J]. Journal of Optoelectronics. Laser, 2014, 25(2): 305–310.

Tonghua Ling, Yancheng Liao, Sheng Zhang. Wavelet Packet Analysis of Acoustic Emission Signal Energy Characteristics of Rock Under Impact Load [J]. Journal of Vibration and Shock, 2010, 29(10): 127–130.

Tribikram Kundu, Samik Das, Steven A. Martin, et al. Locating point of impact in anisotropic fiber reinforced composite plates [J]. Ultrasonics, 2008, 48(13): 193–201.

Tsuda H, Lee J R. Strain and Damage Monitoring of CFRP in Impact Loading Using a Fiber Bragg Grating Sensor System [J]. Composites Science and Technology, 2007, 67(7–8): 1353–1361.
Weiyao Zhu, Xiwu Xu. Finite Element Simulation of Low Velocity Impact Damage of Composite Laminates [J]. Journal of composite materials, 2010, 27(6): 200–206.
Xiaowen Xiao. Study on damage identification by wavelet energy method [D]. Hunan: School of Civil Engineering, Central South University, 2012.
Yan Wang, Dakai Liang, Bing Zhou. Wavelet Packet Analysis for Damage Identification of Optical Fiber Grating Smart Structures [J]. Optics and Precision Engineering, 2007, 15(11): 1731–1737.
Yingdi Xu, Shenfang Yuan, Ge Peng. Research on active Lamb wave localization technique for 2D structure damage [J]. Aeronautical Journal, 2004, 25(5): 476–479.
Yongteng Zhong, Shenfang Yuan, Lei Qiu. Composite Material Omnidirectional Impact Positioning Method Based on Plum Blossom Array [J]. Acta Materiae Compositae Sinica, 2014, 31(5): 1369–1374.
Yongzhen Su, Shenfang Yuan, Yu Wang. Composite Impact Localization Based on Multiple Signal Classification Algorithm [J]. Journal of composite materials, 2010, 27(3): 105–110.
Yu Zhang, Yudong Liu, Zhao Ji. Vector Similarity Measure Method [J]. Acoustic Technique, 2009, 28(4): 532–536.
Zhiyu Lin, Xiwu Xu. Residual Compressive Strength of Composite Laminates after Low Velocity Impact [J]. Journal of composite materials, 2008, 25(01): 140–146.

Applied Mechanics and Civil Engineering VI – Xie (Ed.)
 ISBN 978-1-138-62631-7

Analysis of shield tool wear problems in a variable compound stratum

Jiaqi Guo
Key Laboratory of Road and Traffic Engineering of Ministry of Education, Tongji University, Shanghai, China

Shuguang Qiao
Nanning Rail Transit Co. Ltd., Nanning, Guangxi, China

Runze Xu
China Railway Construction Bridge Engineering Bureau Group Co., Ltd, Tianjin, China

Dajiang Geng
Key Laboratory of Road and Traffic Engineering of Ministry of Education, Tongji University, Shanghai, China

ABSTRACT: The wear of a cutting tool in a variable compound stratum is one of the tough problems encountered during the process of shield construction. Based on the shield tunnels built in the Line 1 phase 1 project of the NanNing Rail Transit, a section from the MinZu University Station to the QingChuan Station, first, the tool configuration and wear condition of the earth pressure balance shield machine in a variable compound stratum is analyzed. Second, the wear condition of tools is discussed through the analysis of data collected from the field. Third, advice is provided to avoid the cutter wear problem. Finally, conclusions are drawn based on the results of the analysis, which indicate that the given tool has satisfied the need of the section under study, while disk cutter wear mainly consists of eccentric wear and uniform wear of the cutter ring.

1 INTRODUCTION

Shield construction technology has an increasing popularity in tunnel construction, which is mainly used in modern construction of tunneling nowadays. However, problems concerning serious tool wear and rupture are encountered during the process of shield tunneling construction. These problems have caused engineering accidents and a serious influence on the timing and pricing of a project. Some of them may even be life-threatening to workers. Thus, to analyze the wear problem of the shield tunneling tool and study the influencing factors and solutions to the problem, it is exclusively important to guarantee the efficiency and safety of the whole shield working process. Many scholars in China have conducted some studies on this problem.

Zhang (2007) conducted studies on the wear problem of shield tools. Based on the Beijing Underground Sector 4 of line 4, the author tried to study this kind of phenomenon by using laboratory tests, in situ monitoring and theoretical analysis. Pei (2009) made some accomplishments through the research on the wear and layout of shield tools. The cut-in point of this study is listed as follows. On the one hand, the author analyzed the influence factor of the wear problem; on the other hand, the author tried to determine how to reduce the wear and then prolong the service life of the shield machine. When Wu et al. (2015) conducted wear monitoring of a shield tunneling tool used in Dalian Metro line 1 section 101 (from the Gangwan Square station to Zhongshan Square station), they managed to analyze groups of data collected under different shield driving conditions using the partial least squares regression method. Based on a trial section of the Beijing Underground Diameter Line shield tunnel, Wang et al. (2013) systematically analyzed the causes and mechanism of the abrasion of a large-diameter shield construction tool for a water-rich sand and gravel stratum using in situ monitoring data. The adaptability of different types of shield cutters under this formation condition is also investigated. Yuan et al. (2009) carried out an in-site test on the wear condition of tools during a construction process, using sand and gravel stratum shield tunneling. Afterward, the authors worked out the wear modulus of various cutting tools during the construction of an earth pressure balance shield in a sandy gravel stratum. Finally, by using statistical regression analysis, the relationship between tool wear and cutting trace length and the longest distance of shield tunneling without changing its blades in the sand and gravel stratum is determined.

Although some of the aforementioned studies investigated on the tool wear problem, in variable compound strata, this issue is more difficult to handle. Therefore, based on the shield tunnels built in the Line 1 phase 1 project of the NanNing Rail Transit, a section from the MinZu University Station to the QingChuan Station, this study analyzed the configuration and wear condition of an EPB shield machine in a variable compound stratum. In this paper, the tool wear data collected from in-site have been studied, the tool wear condition is discussed, and then suggestions are provided for construction under similar conditions. This paper could serve as a reference for EPB shield construction in similar layers in NanNing.

2 TOOL CONFIGURATION

The Line 1 phase 1 project of the NanNing Rail Transit, a section from the MinZu University Station to QingChuan Station, involved layers that mainly consist of silty clay, silt, fine sand, medium sand, gravel sand, gravel, pebbles, mudstone and silty mudstone, together with compound strata confronted with the same tunneling section. The shield cutter-head style used in this project is web-plate cutter head, with a head aperture ratio of 35% and a cutter diameter of 6280 mm. The configuration of the cutter-head is shown in Table 1.

3 TOOL WEAR PROBLEM ANALYSIS

3.1 *Disk cutter wear count*

As can be seen in Table 1, large numbers of disk cutters are equipped in the cutter-head to break up the rocks and stones, in the case of the stratum of this section being solid in its strength and hardness. Systematical analysis is put on the disk cutter wear problem. The configuration of disk cutters in this project is spiral, the details of which can be seen in Figure 1. The amount of tool wear from the MinZu–QingChuan section to tool changing in 3# connecting passages has been counted. Taking No. 13 and 36 blades as examples, the result is shown in Table 2 and the photos are shown in Figure 2.

Table 1. Cutter head configuration.

Section	Disk cutter	Scraper tool	Tool changing instruction
Before tool changing	2 double-blade disk cutters (18 inches) 2 scraper tools (18 inches) 30 single-blade disk cutters (18 inches) 2 single-blade disk cutters (17 inches)	six-hole scraper tool (38) edge-preserving knife (4) edge knife (8)	Tool changing in 3# connecting passage
After tool changing	2 double-blade disk cutters (18 inches) 2 scraper tools (18 inches) 30 single-blade disk cutters (18 inches) 2 single-blade disk cutters (17 inches)	six-hole scraper tool (38) edge-preserving knife (4) edge knife (8)	

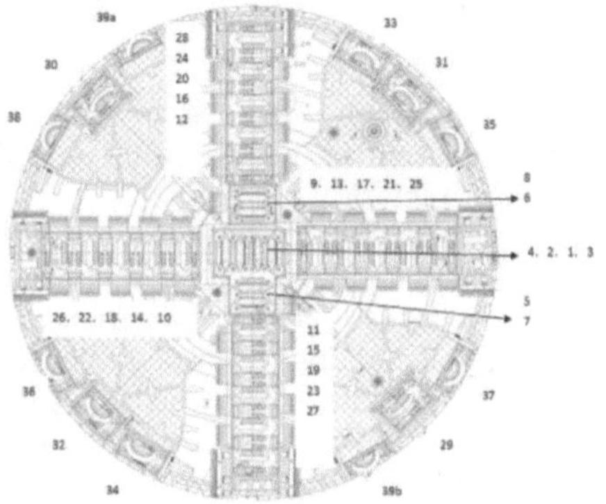

Figure 1. Number for the disk cutters.

Table 2. Tool wear account.

Tool number	Right tunnel wear condition	Left tunnel wear condition
13	The maximum eccentric wear is 19 mm. Another one is 16 mm and elsewhere it is 8.5 mm.	There are three partial wears; the biggest one is 45 mm and the other two are 35 mm and 40 mm. The rest are 3 mm. The blade width ranged from 30–50 mm.
36	9 mm in average	10 mm in average, blade sharpened to 10 mm

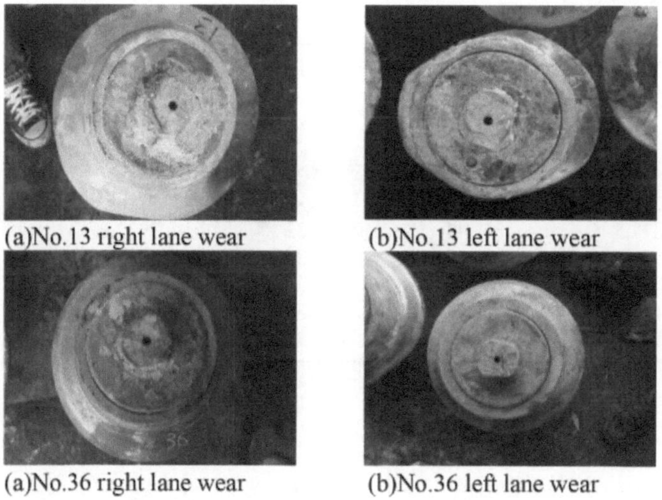

Figure 2. Tool wear in disk cutters.

By investigating the wear condition of 33 disk cutters in the left lane and 14 disk cutters in the right lane, the wear shield disk cutters appear to be uniform wear and cutter ring eccentric wear. In detail, for the 33 disk cutters counted in the left lane, 12 of them have shown different degrees of eccentric wear, with the rest showing uniform wear. For the 14 disk cutters counted in the right lane, 3 of the front disk cutters have shown different levels of eccentric wear, with the rest showing uniform wear; for those which have uniform wear, 2 of them are normal failure (a normal failure of a disk cutter refers to the uniform wear that the edge width of a cutter ring (20 mm usually) exceeds the allowed amount).

3.2 *Disk cutter wear analysis*

3.2.1 *Disk cutter normal wear analysis*

The configuration of the disk cutter in this project is spiral; the wear condition may vary under different conditions when the uniform wear appears in different distances from the shield center. The result is shown in Table 3.

From Table 3, it can be seen that as the distance of the tool go further from the center of its cutter-head, the uniform wear becomes larger at first, but minimizes later. The uniform wear of disk cutters from 10#-30# increases as it goes further from the center of a cutter-head. The main reason explaining this problem is that the distance between the tool and the center of the cutter-head increases by 100 mm increments; as the distance increases, the rotation becomes quicker, the stress alternating time increases together with the effective working area and the tool wear gets faster. For disk cutter 31#-39#, the uniform wear minimizes as the distance between the tool and the center of the cutter-head increases, because as the growing speed decreases, the effective working area decreases but the stress alternating time is on the rise, thereby the tool wear becomes slower.

3.2.2 *Disk cutter eccentric analysis*

The load-bearing condition of the disk cutter while boring can be seen in Figure 3. While rotational torque T_{rot} is provided by excavation face, resistance torque T_{res} consists of scrape resistance torque in muck soil, resistance torque in the cutter box and starting torque from the disk cutter (30–50 N·m). When T_{res} is bigger than T_{rot}, the disk cutter fails to rotate, but

Table 3. The normal wear condition of disk cutters in different places.

Number	Distance from the center of cutterhead/mm	Uniform wear/mm	Number	Distance from the center of cutterhead/mm	Uniform wear/mm
10	900	5	14	1300	5
17	1600	2	18	1700	3–4
20	1900	2.8	21	2000	3
22	2100	3–5	25	2400	5.5
26	2500	4–5	28	2700	9
29	2761.27	10	30	2820.51	11–12
31	2877.5	11	32	2927.67	10
33	2974.53	11	34	3017.48	9
35	3053.18	8	36	3084.67	9–10
37	3111.63	7–8	38	3130.41	7–10
39a	3140	4–6.5	39b	3140	4–6

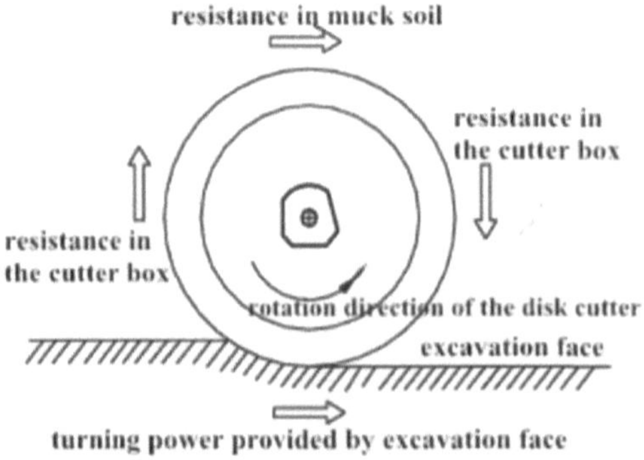

Figure 3. Force diagram of disk cutter while boring.

it slides. Serious eccentric wear occurs due to long-time friction on one side of the disk cutter. The causes for eccentric wear are listed here: 1) excavation face is so loose that rotation torque fails to move the disk cutter; 2) blocking of waste in the cutter box increases the resistance torque during the rolling process.

3.3 *Construction advice*

From the analysis on tool wear it can be seen that the disk cutter mainly consists of eccentric wear and uniform wear. In order to minimize the disk cutter eccentric wear, and thus make sure that all front disk cutters are basically the same regarding under the wear condition, the following advice is given in terms of shield construction in similar compound strata in NanNing:

1. For future projects, amplifying the rotation torque and minimizing the resistance torque are suggested to cut down the eccentric wear for the cutter ring on the disk cutters: 1) use double-blade disk cutters; 2) improve the mobility of muck; 3) improve the wear resistance configuration according to its layout; 4) use a wear resistance material surfacing on the knife drum of the disk cutters; 5) use a wear-resistance leading blade and sharp scraper tools.
2. Increase the use of disk cutters as their distances from the center of cutter-head grow up to ensure that the wear conditions of all front disk cutters are basically the same. Reducing the height of the general layout or raise the wear resistance of tools may also help.

4 CONCLUSION

Taking the Line 1 phase 1 project of the NanNing Rail Transit, a section from the MinZu University Station to the QingChuan Station, as an example, this study analyzes the tool configuration and wear conditions of an EPB shield machine in a variable compound stratum. After discussion and in-site investigation, it can be found that the current problems of tool wear in disk cutters is in the eccentric wear and uniform wear of the cutter ring. Based on the analysis, the following conclusions can be drawn: measures such as enlarging the rotation torque, minimizing the resistance torque and improving the configuration of the layout of disk cutters can be taken to minimize the eccentric wear to reach the ideal wear condition in future projects alike in NanNing.

REFERENCES

Pei, Ruiying. 2009. Research on the tool wear condition and layout of shield machine. *Tianjin University*.

Wang, Zhenfei & Zhang, Chengping & Zhang, Dingli & Sun, Yi. 2013. Wearing and adaptability of cutting tools for large diameter shield in water-rich sandy gravel stratum. *Journal of Beijing Jiaotong University* 37(3): 62–67.

Wu, Jun & Yuan, Dajun. 2015. Analysis and prediction on composite shield cutters wear in extremely hard rock in Dalian metro. *China Civil Engineering Journal* 48(S1): 250–255.

Yuan, Dajun & Hu, Xianpeng & Sun, Li. 2009. Tests and analysis upon shield tool wear in sandy cobble stratum. *Urban Mass Transit* 05: 48–51.

Zhang, Mingfu. 2007. Research on dynamic tool wear of boring in EPB shield. *Beijing Jiao Tong University*.

Applied Mechanics and Civil Engineering VI – Xie (Ed.)

Anchor cable encircled modes of unbonded annular anchors for pre-stressed tunnel lining

Jin Pi
School of Transportation Science and Engineering, BeiHang University, Beijing, China

Ruilang Cao, Yufei Zhao & Zhixin Jia
China State Key Laboratory of Simulation and Regulation of Water Cycle in River Basin, China
Institute of Water Resources and Hydropower Research, Beijing, China

ABSTRACT: Currently, single circle with single layer and double circle with double layer are the main types of anchor cable encircled modes of pre-stressed unbonded annular anchors in tunnel lining engineering. The new encircled mode, double circle with single layer, is proposed as the design technique for the Songhua river water supply tunnel in China. The equivalent load method and the solid modeling method are combined in Flac^{3d} 3D modeling. A numerical simulation of the pre-stressed effects of the single circle with single layer and double circle with single layer modes on the tension and operation of the tunnel lining was analyzed. The results indicated that the lining had a better stress distribution and pre-stressing effects under the double circle with single layer mode.

1 INTRODUCTION

At present, annular anchor pre-stressed tunnel lining is the main design method of poor-quality surrounding rocks, thin covering and high internal water pressure tunnel. Based on the contact of tendons and concrete, the annular anchor lining can be classified into bonded and unbonded (Kang, 2005). In bonded pre-stressed concrete lining, metal conduits are installed according to the desired shape in the template before pouring the concrete, and then the pre-stressing strand bundles that penetrate the pipes are tensioned and anchored until the concrete reaches the target strength. Finally, the high pressure slurry is injected into the hollow pipe grout. In unbonded pre-stressed concrete lining, the external pre-stressing strand bundle is wrapped in a PE plastic sleeve, internal voids are filled with grease, and the pre-stressed and non-pre-stressed reinforcements are laid in the same template before pouring the concrete. Two ends of the cable are locked and tensioned in the same anchorage until the concrete reaches the target strength. A closed ring stress system is formed after backfilled block-out. Compared with other pre-stressed lining structures, the advantages of unbonded pre-stressed concrete lining can be summarized as follows: more uniform distribution of stress within the reinforced concrete of lining; reduced loss of pre-stress; reduced steel demand; less area of adverse pre-stress conditions within the lining; convenient construction, etc. (Sun, 2014 and Li, 2012). The unbonded circular anchored tendon pre-stressed tunnel lining technology is applied in hydraulic pressure tunnels such as Crimsel Tailrace Tunnel in Swiss and Presenzano Tunnel in Italy (Trucco, 1978 and VSL, 1990), which use the Single Circle with Single layer (SCS) encircled mode. This technology is also applied in Xiao Langdi Desilting Tunnel, Da Huofang Water Tunnel and Xi Longchi Storage Plant in China (Lin, 1999, Jin, 2006 and Jin, 2005), which use the Double Circle with Double layer (DCD) encircled mode.

The encircled mode of unbonded pre-stressed anchor cables can affect the lining structure construction, block-out size and anchor choice, which are very important factors for structural design and construction. The encircled modes of the projects developed at home and abroad are SCS and DCD. The SCS mode is used to arrange the anchor cables outside the lining, but the anchor cables inside the lining are not arranged, and the anchor cables cross the anchor slot after around a week. The advantage of the SCS mode is that a sufficient anchor tensile force is used and lining pre-stressing is larger and more uniform. It can also help reduce the lining thickness and cost. The disadvantage of the SCS mode is that the concrete pouring between anchors is difficult, and hence the distance of the anchor and block-out width should be increased, which may cause the weaker parts of the region near the block-out to increase. The DCD mode is used to arrange the anchor cables both outside and inside the lining with two layers, and the two layers are tied to the inside of the non-pre-stressed reinforcement in parallel with a 10 cm spacing, so it can reduce the amount of banding erect ribs. The advantage of the DCD mode is that it is easy to tension with smaller deflection angle and the size of block-out is smaller than that of the SCS mode. The disadvantage of the DCD mode is that the pre-stressing is small due to the positioning of the anchor, which is too close to the inside lining. Also, the DCD mode has a large lining thickness that results in more concrete quantities. Therefore, the two types of anchor encircled modes have an optimized space.

The Songhua river water supply tunnel in the Jilin Province of China has an inner diameter of 6.8 m and an operation pressure of 0.55 MPa, for which pre-stressed annular anchor design and construction techniques are used. Due to low water pressure, high anchor tension is not required, and hence the encircled mode of single-layer anchors may be more appropriate. In this study, the advantages of both SCS and DCD modes were combined to form a new type of encircle mode, namely the Double Circle with Single layer (DCS), which was used as the design technique for the Songhua river water supply tunnel. The equivalent load method and solid modeling method were combined in Flac^{3d} 3D modeling. A numerical simulation of the pre-stressed effects of the SCS and DCS modes on the tension and operation of the tunnel lining was analyzed. The results indicated that the DCS lining had a better stress distribution than the SCS lining.

2 NUMERICAL SIMULATION METHODS AND PROGRAMS

2.1 *Numerical simulation method*

The equivalent load method (Kang, 2006), temperature stress method (Xiang, 2014), the initial strain method (Guo, 2015) and solid modeling method (Qu, 2008) are the common methods to simulate the stress state of a closed curve pre-stressing strand. In circular anchor tension simulation, the equivalent load method can only calculate the value of a constant initial pre-stress, but not simulate the variation of pre-stressing causes by the lining stress. The temperature stress method and initial strain method also cannot simulate the un-bound strand because the un-bound strand does not meet the deformation coordination. The solid modeling method can establish a solid model or a structural unit model of pre-stressing tendons. Although its mechanism of action is consistent with the unbonded system, it can accurately simulate single circle encircled pre-stressed tendons, but the presence of cross-cutting issues of pre-stressed tendons is encircled by a double circle method. The pre-stressed tendons cannot be tensioned in the numerical simulation; therefore, the above methods all have defects in simulation of closed curve anchors. In addition, in order to achieve the tension of pre-stressed anchors, a lining structure is required to set a fixed block-out for anchorage before the concrete pouring. Block-out micro-expansion concrete backfill can generate expansion stress to make up the lack of pre-stressed parts, thus the stress should not be overlooked in the numerical calculation. Because of the pre-stressing, internal water pressure, external water pressure, weight and stress of rock pressure together, the touching relationship of the lining and surrounding rock is varied during the construction, operation and maintenance period. For example, in the maintenance period, pre-stressed lining appears retracted with the dissipation of internal water pressure, and as the surrounding rock is in plastic state residual deformation, micro-gaps will appear between the lining and the surrounding rock

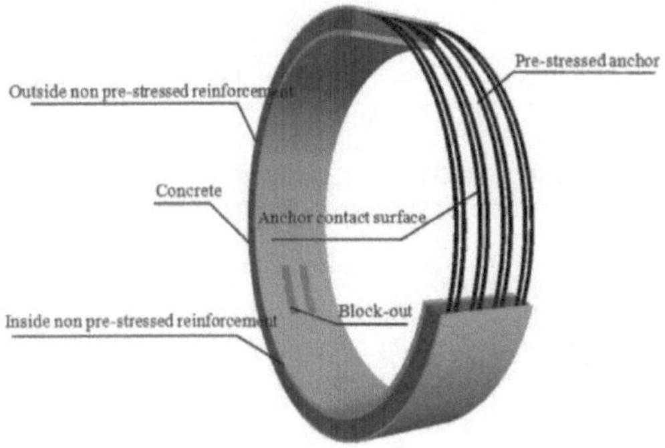

Figure 1. Numerical model diagram.

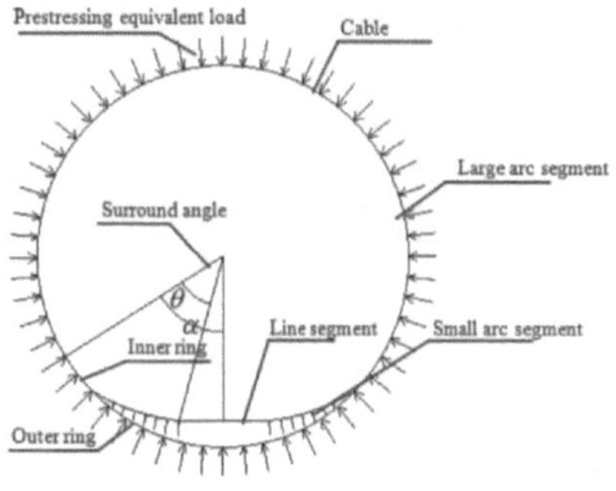

Figure 2. Equivalent load and solid model simulation.

with unloading, which are locally in the disengaged state. Therefore, numerical modeling is required to establish a reasonable relationship for the surrounding rock binding effect of lining, not only to ensure that when the surface of the lining and the surrounding rock at the junction of the maximum tensile stress is larger than the normal bond strength, both automatic disengaging no longer transfers contact stress, but also to ensure that the surrounding rock and lining share the internal water pressure again after re-fitting.

The unbonded pre-stressed concrete lining of the numerical simulation is shown in Figure 1. It is combined with the equivalent load and solid modeling method that simulate the transfer process of the unbonded pre-stressed steel strand. According to the unbonded pre-stress steel strand stress characteristics in concrete lining, the stress of the steel strand is divided into the same pre-stress component and the variation of the stress component. In the numerical simulation process, first of all, the interface property is established in the anchor and concrete delivery interface to simulate the friction and slip relationship of steel wire and casing pipe. And then, as shown in Figure 2, the equivalent method is used to simulate constant pre-stress after the steel strand is anchored, and the solid model of passive force is used to simulate varied non-pre-stressing of the steel strand. Finally, stress and strain states at different stages of the steel strand are calculated by force super position.

2.2 *Numerical scheme and parameter selection*

1. Numerical scheme

The encircled modes of anchor cables are SCS of eight steel strands and DCS of four steel strands. The calculation scheme is shown in Table 1. The internal lining and the annular anchor model diagram is shown in Figure 3.

2. Calculation parameters

The inner diameter of the lining is 6.8 m, its wall thickness is 0.50 m, operating water pressure is 0.55 M Pa and dome above the minimum overburden thickness is 18 m. The lining is poured with C40 concrete, a single beam of unbonded pre-stressed anchor is composed by a 7 × Φ5 mm high-strength low-relaxation strand, the tensile strength standard value is 1860 MPa, the strand friction coefficient is 0.032, and the swing factor is 0.0007. A slot with a size of 1.20 m (length) × 0.20 m (width) × 0.20 m (depth) is created in the lower anchorage lining 45° position, and the adjacent row crossing angle is 90°. In the operation period numerical simulation, taking into account the complexity of the model, the number of units is high, and if the external water load applies as the penetration, the convergence rate of the model will be significantly reduced. Also, the external water pressure is small, so the external water pressure is applied by equivalent surface stress. The rock is class IV weak weathered rock of tuff, and its physical and mechanical parameter values are shown in Table 2.

Table 1. Calculation model size.

Scheme	Block-out position	Spacing	Lining thickness	The number of anchor
SCS	The side 45°	45 cm	50 cm	8
DCS	The side 45°	45 cm	50 cm	4

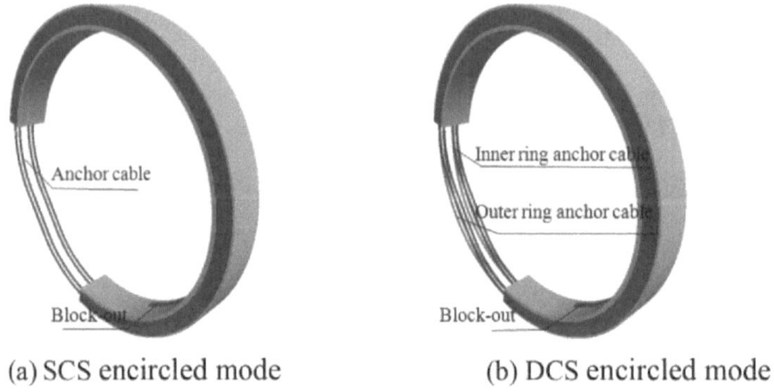

Figure 3. Numerical model.

Table 2. Rock mechanics parameters.

Lithology	Rock classification	Density kg/m³	Deformation modulus GPa	Poisson's ratio	Cohesion MPa	Friction angle°
Weak weathered tuff	IV	2500	2.1	0.34	0.3	29.7

3. Monitoring scheme

The inner and outer surfaces of the lining and block-out area stress state are focused for monitoring in numerical simulation. The inner and outer surfaces annularly arrange uniformly 16 monitoring points; the anchor slot site monitoring line layout is shown in Figure 4.

3 NUMERICAL RESULTS AND ANALYSIS

3.1 *Pre-stressed effect analysis of post-tensioned lining*

The minimum principal stress (Smin) distribution of lining concrete under two encircled modes after anchor tension is shown in Figures 5 and 6. It can be found from Figures 5 and 6 that under the two schemes, the overall effects of pre-stressed lining are all relatively uniform and there is stress concentration only near the block-out. The pre-stressing value increases from outside to inside along the direction of lining thickness. The stress laws of both encircled modes are similar. After the anchor is tensioned, the maximum pre-stress of the inner lining in SCS is −6.5 MPa while that in DCS is −5.9 MPa, and they are all located in the lining surrounding angle close to 90° and 270°. The minimum compressive stress of the inside lining in SCS is −4.1 MPa while that in DCS is −3.8 MPa, and they are all located at the top of the lining. The maximum pre-stress of the outer lining in SCS is −4.1 MPa while that in DCS is −4.6 MPa, and they are all located in the lining surrounding angle close to 135° and 225°. The minimum compressive stress of the outer lining is −3.0 MPa in SCS while that in DCS is −3.8 MPa, and they are all located in the surrounding angle close to 90° and 270°. In the inner lining, the pre-stressing of SCS is larger compared to DCS, the maximum difference being 9.23% and the value is 0.6 MPa. In outside lining, the pre-stress of DCS is larger

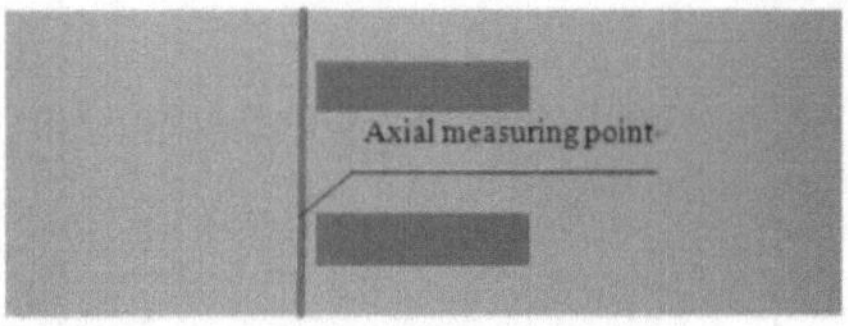

(a) Axial measuring point arrangement

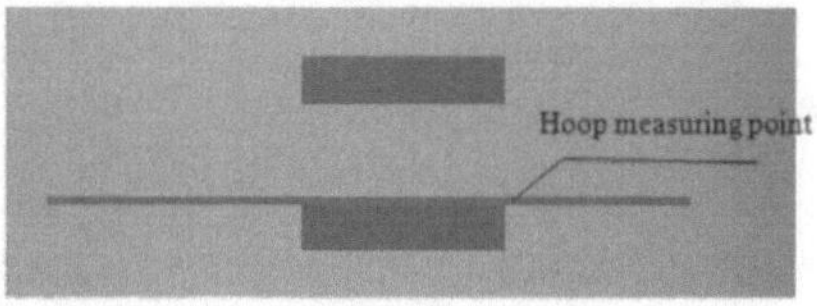

(b) Hoop measuring point arrangement

Figure 4. Concrete monitoring points arrangement near the block-out.

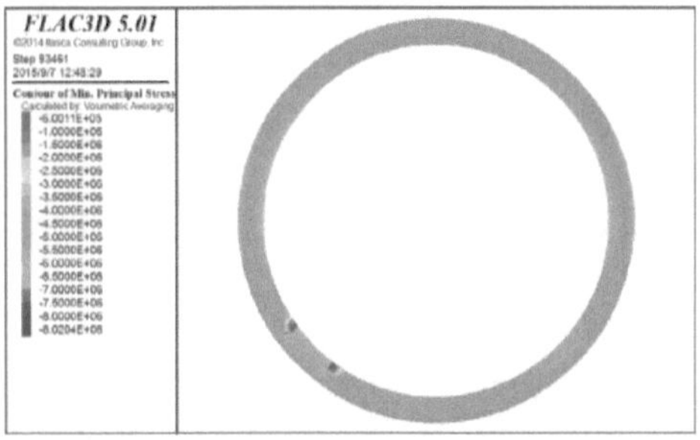

Figure 5. Concrete lining cloud Smin of SCS (Pa).

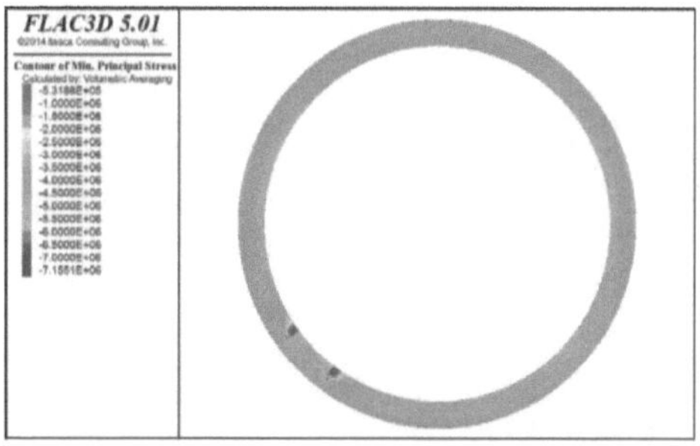

Figure 6. Concrete lining cloud Smin of DCS (Pa).

compared to SCS; the maximum difference being 10.8% and the value is 0.5 MPa. From the pre-stressed effects of post-tensioned lining, there is no great difference was observed between the two encircled modes.

Analysis results show that the lining concrete near the block-out exhibits a complex stress state, and becomes the most weak area of the lining due to the stress localization around the block-out, resulting in the maximum tensile stress being close to the hoop-free face of the block-out, and the maximum compressive stress is located between the adjacent block-outs. More reinforcement in the vicinity of the block-out should be strengthened. The stress diagram near the block-out of SCS and DCS is shown in Figures 7 to 10. After the anchor is tensioned, the maximum pre-stress near the block-out in DCS is −7.16 MPa while that in SCS is −8.08 MPa; the maximum tensile stress near the block-out in DCS is 0.5 MPa while that in SCS is 0.7 MPa. The DCS had a better stress state than SCS.

3.2 *Pre-stressed effect analysis of operation*

The minimum principal stress distribution of lining concrete under two encircled modes in operation at 55 m water pressure is shown in Figures 11 and 12. The minimum principal stress and maximum principal stress (Smax) of feature points along surrounding angle variety in lining are shown in Figures 13 and 14.

The axis at the crown is set to 0°, clockwise argument and the anchor slot center position angle is 225°. It can be seen from Figures 11 and 12 that both encircled modes of pre-stressed anchor lining are always compressed in operation. The maximum and minimum stress location of the inner and outer lining surface is consistent with the tension period. It can be seen from Figure 13 that the pre-stress is more uniform in DSC. In DSC, the outside pre-stress of lining is larger than in SCS. As can be seen from Figure 14, under the two encircled modes in operation, the inner lining has no tensile stress, the outer lining is tensioned slightly, the maximum tensile stress is only about 0.2 MPa, and the lining is in safe condition.

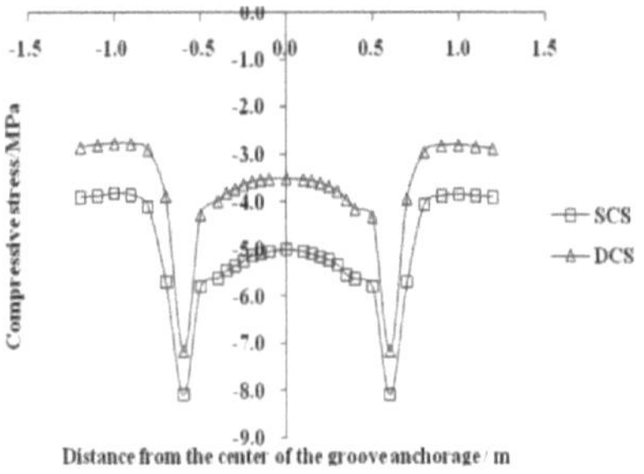

Figure 7. The hoop σ_1 graph near the block-out.

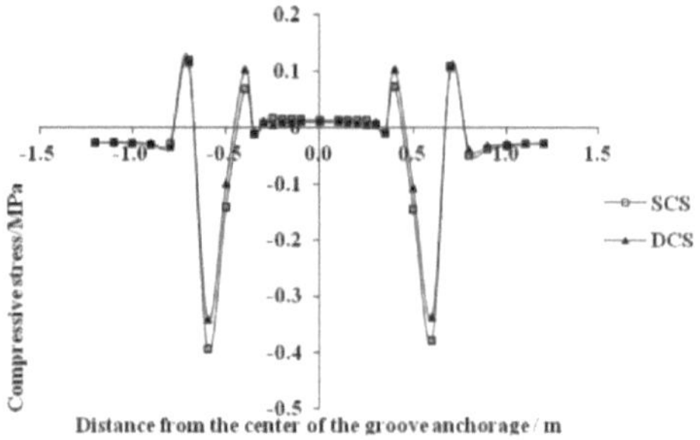

Figure 8. The hoop σ_3 graph near the block-out.

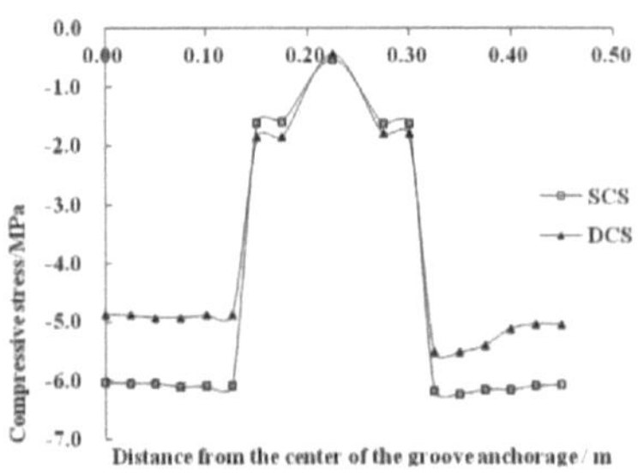

Figure 9. The axial σ_1 graph near the block-out.

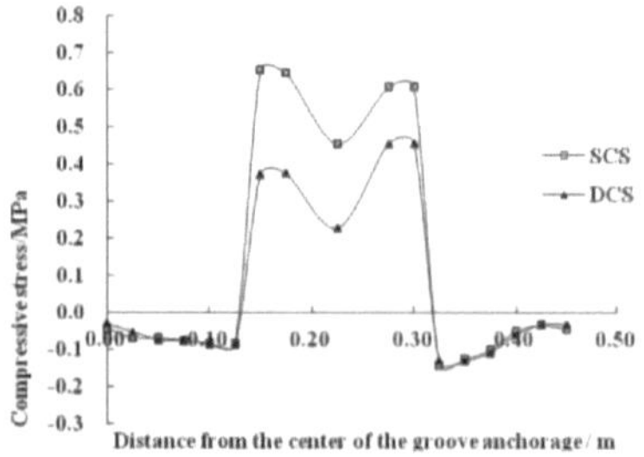

Figure 10. The axial σ_3 graph near the block-out.

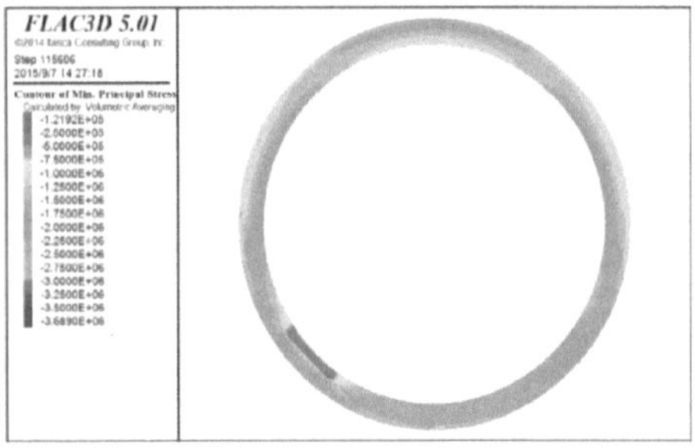

Figure 11. Concrete lining cloud Smin of SCS (Pa).

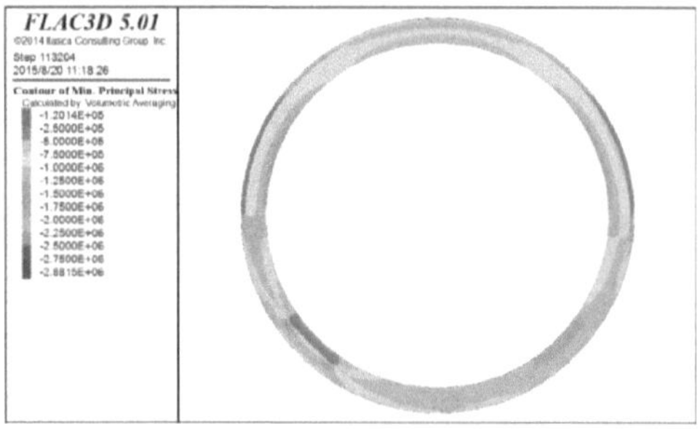

Figure 12. Concrete lining cloud Smin of DCS (Pa).

3.3 *Pre-stressed effect analysis of exceed pressure*

The two encircled modes of pre-stressed lining under water pressure at 55 m, 65 m, 75 m and 100 m were calculated, and the results are summarized in Table 3.

From Table 3, at a water pressure of 55 m and 65 m, the lining is in a compressed state. The lining concrete tensile stress area extends outward from the anchor slot position, gradually expanding the scope. The anchor slot position had a greater tensile stress with 75 m water pressure. And the lining with the SCS mode has a trend destruction. The lining tensile stress in DCS is from 0.4 to 2.0 MPa at a water pressure of 100 m, and in SCS it is from 0.75 to 2.25 MPa. In SCS, the anchorage groove portion tensile stress is weak, and pre-stressed lining has been destroyed. The rear of anchorage groove concrete is the weak pre-stressed area of the lining in SCS; this area is difficult to form a uniform pre-stress distribution after the block-out backfill. So, in the case of overload, this part of the concrete stress state is poor; therefore, SCS pre-stressed lining is more susceptible to damage under high pressure.

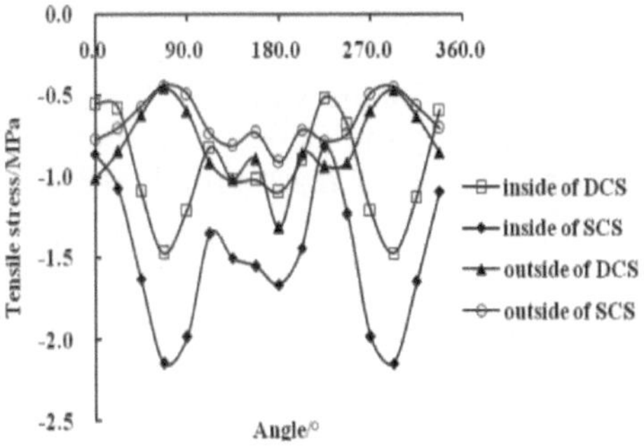

Figure 13. Smax imagery of concrete lining.

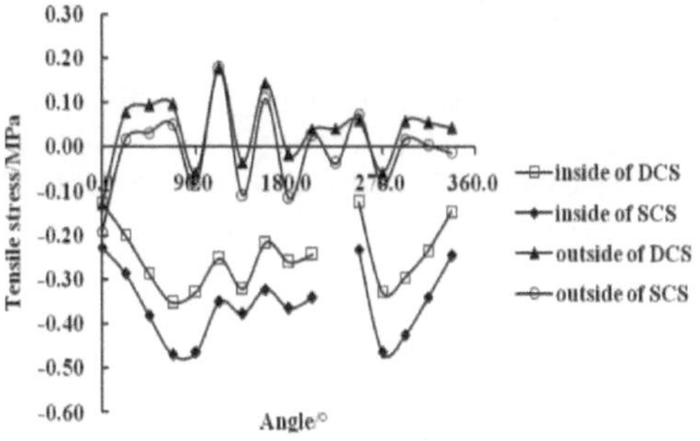

Figure 14. Smin imagery of concrete lining.

Table 3. The calculation results of different anchor encircled modes.

Calculation condition	Distribution location	Calculation results	DCS/MPa	SCS/MPa
55 m	Lining	Compressive stress	−0.75~−2.0	−0.75~−2.1
	Near the block-out	Tensile stress	small	small
65 m	Lining	Compressive stress	−0.5~−0.9	−0.6~−1.0
	Near the block-out	Tensile stress	small	small
75 m	Lining	Compressive stress	−0.4~−0.8	−0.5~−0.8
	Near the block-out	Tensile stress	0~1.25	0~1.50
100 m	Lining	Compressive stress	small	small
	Near the block-out	Tensile stress	0.4~2.0	0.75~2.25

4 CONCLUSIONS

The new encircled DCS mode was proposed as the design technique for the Songhua river water supply tunnel. The equivalent load method and the solid modeling method were combined in $Flac^{3d}$ 3D modeling. The numerical simulation of the pre-stressed effects of the SCS and DCS modes on the tension and operation of the tunnel lining was analyzed. From the above analysis, the following conclusions were drawn:

1. After the anchor cable is tensioned, the pre-stressing of SCS is larger in the inner lining compared to the DCS; the maximum difference is 9.23% and the value is 0.6 MPa. In the outer lining, the pre-stressing of DCS is larger compared to SCS; the maximum difference is 10.8% and the value is 0.5 MPa. The pre-stressed effect of DCS is more uniform than that of SCS, but from the pre-stressed effect of post-tensioned lining, there is no great difference between the two encircled modes.
2. The lining tensile stress area is mainly distributed in the hoop-free face of the block-out after the anchor cable is tensioned. The tensile stress area and the tensile stress value of SCS are larger than those of DCS. Also, a partial stress concentration appeared. DCS had a better stress distribution of lining than SCS.
3. Lining concrete compressive stress decreased and tensile stress increased with water pressure applied during the operation. The main discontinuity advantageous location is near the block-out, and the stress of the lining structure under the two schemes meets the requirements. With the increase in overload water pressure, the lining concrete tensile stress area extends from the block-out portion outwardly, with the scope gradually expanding. The rear of anchorage groove concrete is the weak pre-stressed area of the lining in SCS. In this area, it is difficult to form a uniform pre-stress distribution after the block-out was backfilled. So, in the case of overload, this part of the concrete stress state is poor, and therefore SCS pre-stressed lining is more susceptible to damage under high pressure.

ACKNOWLEDGMENTS

This study was supported by the China Institute of Water Resources and Hydropower Research Project (GE0145B452016), the National Key Research and Development Project (2016YFC0401804) and the Natural Science Foundation of China Project (Youth Science Fund Project, 51409283).

REFERENCES

Guo Qiaohang, Zheng Huang, Li Wei etc. Finite element simulations on mechanical self-assembly of helical shapes driven by initial strains [J]. Chinese journal of applied mechanics, 2015, 32(2): 192–197 (in Chinese).

Jin Zhaohui. Design and research on un-boned pre-stressed concrete of pressure tunnels [D]. Nanjing: HoHai university, 2006 (in Chinese).

Jin Qiulian. Study on the numerical simulation analysis of pre-stressed concrete tunnel liner with 3D FEM [D]. Tianjin: Tianjinuniversity, 2005 (in Chinese).

Kang Jingfu, Hu Yuming. Techniques and performance of post-prestressed runnel liner [J]. ASCE Practice Periodical on Structural Design and Construction, 2005, 10(2): 102–108 (in Chinese).

Kang Jingfu, Liang Yuehua, Zhang Qincheng. 3-D finite element analysis of post-pre-stressed concrete lining [J]. Journal of Tianjin University, 2006, 39(8): 968–972 (in Chinese).

Li Xiaoke. Investigation on the deformation feature and crack restricting design of circular lining of hydraulic pressure tunnel [D]. Dalian University of Technology, 2012 (in Chinese).

Lin Xiushan, Shen Fengsheng. Technology and practice study on pre-stressed tunnel Lining with unbonded annular anchors of Xiao langdi project [M]. Zhenzhou: The yellow river water conservancy press, 1999 (in Chinese).

Qu Weilian, Lu Lijun, Li Ming. Solid modeling method for structure with 3-D straight through crack [J]. Journal of Wuhan University of Technology, 2008, 30(1): 87–90 (in Chinese).

Sun Zhenchuan, Zhong Shengxing. Key points in implementation of bonded circumferential pre-stressing of Yellow River crossing tunnel on south-to-north water diversion project [J]. Tunnel Construction, 2014, 34(1): 73–77 (in Chinese).

Trucco G., O. Zeltner. Grimsel-Oberrar pumped storage system [J]. Water Power & Dam Construction, 1978 (2).

VSL Job Report. Offtake tunnels for the Presenzano hydroelectric facility [R]. L' Industria Italiana del Cemento, 1990(4).

Xiang Yan. Influence of temperature stress on internal force and deformation of retaining structures for deep excavations [J]. Chinese Journal of Geotechnical Engineering, 2014, 36(S2): 64–69 (in Chinese).

Applied Mechanics and Civil Engineering VI – Xie (Ed.)
© 2017 Taylor & Francis Group, London, ISBN 978-1-138-62631-7

Simulation of oil transport after oil spill in the Bohai Sea

L. Lei
China Waterborne Transport Research Institute, Beijing, China

W. He
China Transport Telecommunications and Information Center, Beijing, China

ABSTRACT: In this study, a hybrid model was developed by coupling a MIKE 21 and OILMAP model to simulate oil transport after oil spill. The coupled model was applied in the Bohai Sea. This model was able to reproduce hydrodynamics processes. Moreover, the oil transport trace modeled by the hybrid model was also compared with the observed data. It was found that the developed model could predict oil transport.

1 INTRODUCTION

With rapid economic development, the demand for petroleum is rising. It will increase maritime traffic flow and lead to the release of oil during ship accidents. To better prepare for the emergency response and mitigation of such spills, there is a need to predict the fate and transport of oil (Afenyo et al. 2016). OILMAP, developed by Applied Science Associates (ASA), is one of the useful models to predict the trajectory of oil. This model coupled several main processes, such as spreading, evaporation, emulsification, dispersion, advection, photo-oxidation, dissolution, encapsulation and sedimentation of oil (Applied Science Associates Inc. 2004). It is able to predict the movement of oil on a water surface and its distribution in the environment (evaporated, in the water column, on the shoreline). For these calculations, the model relies on environmental data such as wind and currents. HYDROMAP (also developed by ASA) is used to calculate the current field for OILMAP. However, HYDROMAP operates with a rectangular regular mesh. A regular grid has a disadvantage, i.e. it does not conform to nearly any desired shoreline (Applied Science Associates Inc. 2006). Thus, HYDROMAP cannot model the flow field accurately at the boundaries. The inaccuracy of the flow field will increase the deviation between modeled and measured oil trace. Keeping this in mind, MIKE 21 (DHI 2004), which can resolve flow field by using an unstructured body-fitted grid, is applied to couple with OILMAP to better forecast the trajectory of oil spill.

2 MATERIALS AND METHODS

2.1 *Study area*

Bohai Sea is located at the North-Eastern region of China (Figure 1). It is a shallow, semi-enclosed marginal sea (surrounded by mainland and Liaodong peninsula) of the West Pacific Ocean (Ding et al. 2016). The area of the Bohai Sea is about 77,000 km^2 (East-West width is about 300 km and North-South length is about 550 km). Bohai Sea, hosting over 60 seaports (including Tianjin, Dalian, Yingkou, and Qinhuangdao seaports), is one of the world's most prosperous shipping networks with dozens of ship routes (Guo et al. 2016). All ship routes and seaports are potential origin of oil slicks.

Ocean University of China (OUC) has conducted an oil spill emergency exercise in the Bohai Sea. Engine oil (500 liters; density 0.85 kg/L) was leaked at 122°14′48″E, 38°00′35″N,

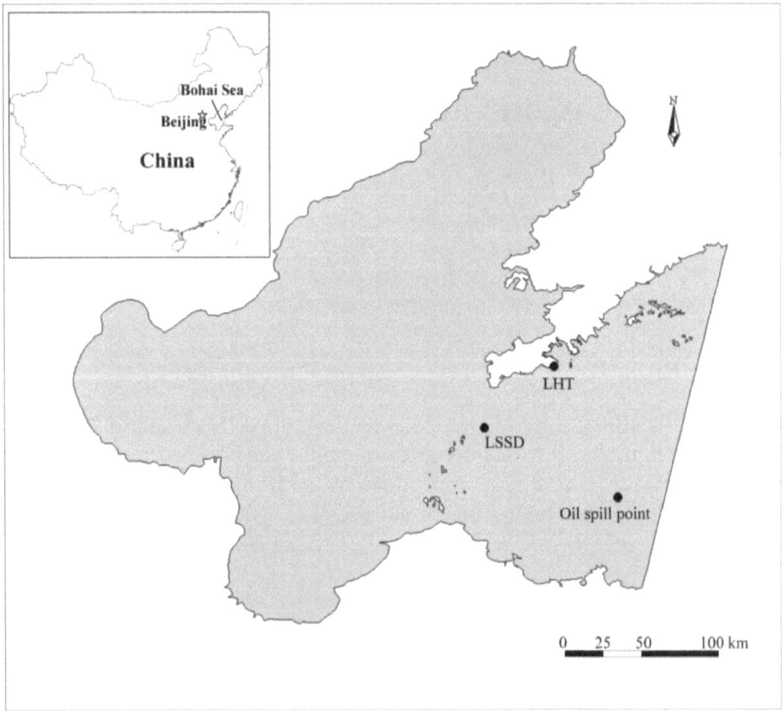

Figure 1. General localization of the Bohai Sea and the sampling stations.

on May 17, 2001. A tracer was released and tracked by a ship after oil spill. During the emergency exercise, the tracer position was recorded for every hour. More information about the exercise can be found in Maritime Safety Administration of the People's Republic of China (2004).

2.2 *Model development*

2.2.1 *MIKE 21*

The MIKE 21 modeling system is based on a flexible mesh approach and it has been developed for applications within oceanographic, coastal and estuarine environments. Hydrodynamics models, containing momentum and continuity equations, are constructed based on the numerical solution of two-dimensional incompressible Reynolds averaged Navier–Stokes equations invoking the assumptions of Boussinesq and of hydrostatic pressure. More information about governing equations can be found in DHI (2004).

2.2.2 *OILMAP*

The OILMAP system, containing oil trajectory and the fate model, is intended for rapid, first-order estimates of spill movement. It predicts the surface trajectory of spilled oil while maintaining a mass balance of the fate of oil. The user may specify either instantaneous or continuous release spills. The spill is initially represented by a number of spillets, each of which represents an equal fraction of the total oil spilled. Spillets are advected in response to wind and current force with random turbulent dispersion superimposed. These forces combine to affect the trajectory and final destination of the oil. At the same time, the oil evaporates, spreads, is entrained into the water column, emulsifies and can become stranded on the shore in coastal regions. Together, these processes affect the physico-chemical state of the oil and its distribution in the environment, and they are termed the "weathering" and fates of the oil. More information about the OILMAP system can be found in Applied Science Associates Inc. (2004).

However, hydrodynamic models are not developed in OILMAP. Currents, which transport oil, may be calculated by other hydrodynamic models. MIKE 21 is applied to simulate flow field in this study. Then, the modeled flow field is used as the input into the OILMAP system for oil transport simulation.

3 RESULTS AND DISCUSSION

3.1 *Hydrodynamics*

The flow field in the Bohai Sea is simulated by MIKE 21. Figure 2 shows the comparison of simulated and observed tidal elevation at station LHT. Differences between model results and field data are less than 0.8 m. The computed tidal elevations are satisfactory. Figure 3 presents the comparison of calculated and measured tidal velocity at station LSSD. Generally, the model results agree well with the observed tidal velocity.

3.2 *Transport of oil*

The calculated flow field is stored as a current NetCDF file, and then inputted into OILMAP. Figure 4 shows the comparison of the modeled and recorded trace after oil spill. The average Lagrangian separation distance between the simulated and observed position was 3.1 km. To evaluate the performance of the oil spill model, the skill score *ss* is used (for more details, see Liu and Weisberg (2011)):

$$ss = \begin{cases} 1 - \dfrac{s}{n} & (s \le n) \\ 0 & (s > n) \end{cases} \tag{1}$$

$$s = \sum_{i=1}^{N} d_i \Big/ \sum_{i=1}^{N} l_{oi} \tag{2}$$

where s is the non-dimensional index; n is the tolerance threshold; d_i is the separation distance between the modeled and observed endpoints of the Lagrangian trajectories at time step i;

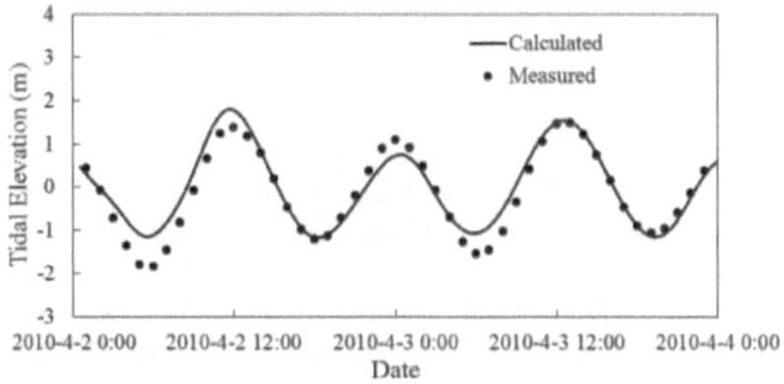

Figure 2. Comparison of the calculated and measured tidal elevations at station LHT.

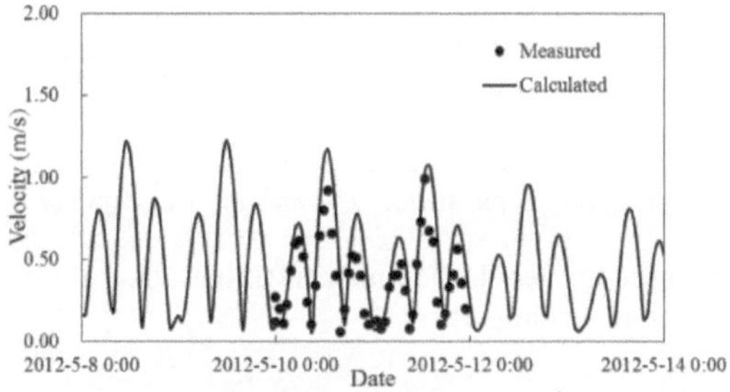

Figure 3. Comparison of the calculated and measured velocity at station LSSD.

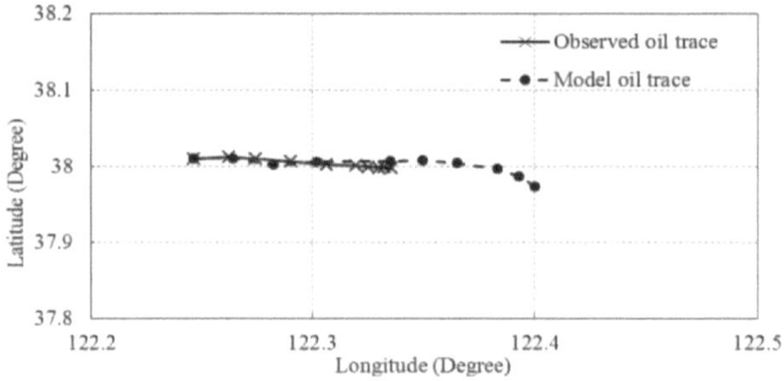

Figure 4. Comparison of the predicted and observed oil trace.

and l_{oi} is the length of the observed trajectory. The tolerance threshold n is chosen as 1 and the skill score ss is equal to 0.48. Generally, the coupled model reproduces main features of oil transport in the Bohai Sea.

4 CONCLUSION

A hybrid model, which coupled the MIKE 21 model and OILMAP model, was developed to simulate oil transport in this study. The coupled model was applied in the Bohai Sea. Rising and ebbing tide processes were successfully reproduced. Then, the calibrated flow field was inputted in the OILMAP system. Compared with the oil spill emergency exercise, the predicted oil trace agreed well with the observed data. It was found that the main features of oil transport after oil spill could be modeled by the proposed model.

ACKNOWLEDGMENTS

The authors acknowledge the financial support from the Science Foundation for Young Scientists of China Waterborne Transport Research Institute (Grant No. 2016 QN-HB04105).

REFERENCES

Afenyo, M., Veitch, B., Khan, F. 2016. A state-of-the-art review of fate and transport of oil spills in open and ice-covered water. *Ocean Engineering* 119, 233–248.

Applied Science Associates Inc. 2004. *Technical manual OILMAP for windows.* Narragansett, Rhode Island.

Applied Science Associates Inc. 2006. *HYDROMAP user manual.* Narragansett, Rhode Island.

DHI. 2004. *MIKE 21 HD, flow model, hydrodynamic module, user guide.* Denmark: Danish Hydraulic Institute (DHI), Water & Environment

Ding, Y., Cao, C.H., Huang, J., Yan, S., Liu, G.Y., Wu, L.J., Wan, Z.W. 2016. Origins and features of oil slicks in the Bohai Sea detected from satellite SAR images. *Marine Pollution Bulletin* 106 (1–2), 149–154.

Guo, W.J., Wu, G.X., Liang, B.C., Xu, T.J., Chen, X.B., Yang, Z.W., Xie, M.X., Jiang, M.R. 2016. The influence of surface wave on water exchange in the Bohai Sea. *Continental Shelf Research* 118, 128–142.

Liu, Y.G., Weisberg, R.H. 2011. Evaluation of trajectory modeling in different dynamic regions using normalized cumulative Lagrangian separation. *Journal of Geophysical Research* 116 116(C9), 1527–1540.

Maritime Safety Administration of the People's Republic of China. 2004. *Tutorial for oil spill emergency response.* Beijing: China Communications Press.

Applied Mechanics and Civil Engineering VI – Xie (Ed.)
 ISBN 978-1-138-62631-7

The construction technology of Earth Pressure Balance shield in sand stratum section

Yitao Li
Key Laboratory of Road and Traffic Engineering of Ministry of Education, Tongji University, Shanghai, China

Shuguang Qiao
Nanning Rail Transit Co. Ltd., Nanning, China

Zhaohui Zhang & Langming Yong
China Railway Construction Bridge Engineering Bureau Group Co. Ltd., Tianjin, China

ABSTRACT: Based on the shield tunneling project of the first phase project of Nanning rail transit line 1 in the section from Xixiangtang passenger depot station to University for Nationalities station, this paper discusses the main construction and technology measure for Earth Pressure Balance (EPB) shield tunneling in sand stratum section, including the repair and maintenance of shield machine before it enters the sand stratum, the boring posture and parameter control and other technical measures. By summarizing the construction experience in this section, this paper can provide reference for problems that can be encountered in sand stratum section EPB shield tunneling, such as the difficulty in establishing a balance model and the surface settlement exceeding standard, meanwhile it has guiding significance for EPB shield construction in similar stratums in Nanning district.

1 INTRODUCTION

With the development of our urban construction, shield tunneling technique is widely used in tunnel engineering in energy, traffic, water conservancy and other fields. The earth pressure balance shield is especially well adopted in subway constructions in many cities for its characteristics like strong stratum adaptability, fast driving speed, and little impact on the environment. But for earth pressure balance shield construction in sand stratum section, many problems are yet to solve. For example, muck conditioning can be difficult, balance model can be difficult to establish and surface settlement can exceed the standard easily. Therefore, construction can be challenging and risky.

Many domestic scholars and engineers have also discussed the problem of earth pressure balance shield construction in sand stratum section. Chen et al. (2010) studied the case of shield construction in Guangzhou–Foshan segment Chongleigang station–Qiandeng Lake station–Financial Hi-Tech Zone station section of pearl river delta intercity rapid transit, the technical measures of earth pressure balance shield construction in this region's watery sand stratum were analyzed. Based on TA15 section construction in Nanjing Metro, Jiang et al. (2006) analyzed the EPB excavating in watery sand stratum from aspects of shield design, soil stability in each construction phase and passing through shallow foundation buildings, discussed the crucial and important points in this method. Zeng et al. (2006) took the shield construction in Zhujiang Xincheng station to Kecun station section of Guangzhou metro line 3 as an example, discussed the construction technology of earth pressure balance shield going through sand stratum section. Wang (2012) aimed at the difficulties in EPB construction in watery sand stratum, based on practical engineering, systematically summarized and studied the crucial technology of EPB construction in watery sand stratum, from aspects of the adaptability of EPB shield, settlement control method and muck conditioning measure.

It can be seen that the problems of EPB shield construction in sand stratum section in different region has been thoroughly discussed by scholars and engineers. But sand stratums in different cities have different characteristics. As an important city in southwest China, Nanning has started its rail transit construction relatively late, and the experience in EPB shield construction in sand stratum section is not sufficient. Based on the shield tunneling project of the first phase project of Nanning rail transit line 1 in the section from Xixiangtang passenger depot station to University for Nationalities station, this paper discusses the main construction and technology measure for Earth Pressure Balance (EPB) shield tunneling in sand stratum section, and provides reference for EPB shield construction in similar stratums in Nanning district.

2 PROJECT BACKGROUND

The overall length of the section from Xixiangtang passenger depot station to University for Nationalities station of the first phase project of Nanjing rail transit line 1 is 1.1 km, with a connected aisle and a wastewater pump house. The section tunnel starts at University for Nationalities station, left line enters the sand stratum at the point which is 145 m from the starting point, and the height of the tunnel top is increasing gradually, right line enters the sand stratum at the point which is 105 m from the starting point. The tunnel depth in the sand stratum section is about 10–15 m, the largest depth of sand stratum intrusion is 8 m. The stratigraphic section of the sand stratum section is shown in Figure 1.

The structures in the shield tunnel range in this section are mainly houses located at both sides of the road and 10 kV, 110 kV pylons. Though these structures are in the theoretical influence range of the tunnel, they are all at a great distance from it. There is basically no structure within 20 m distance of each side of the road.

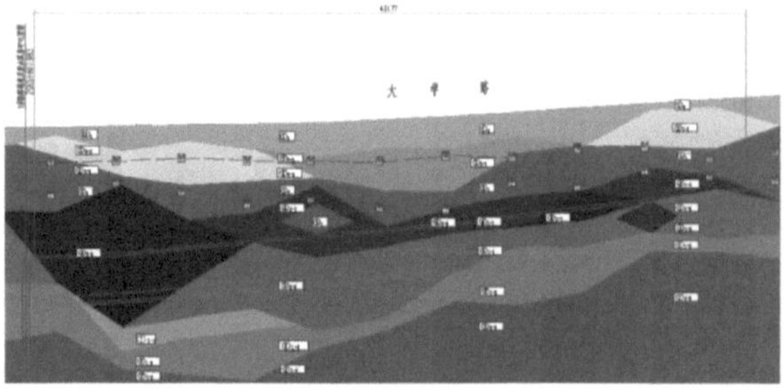

(a) left line

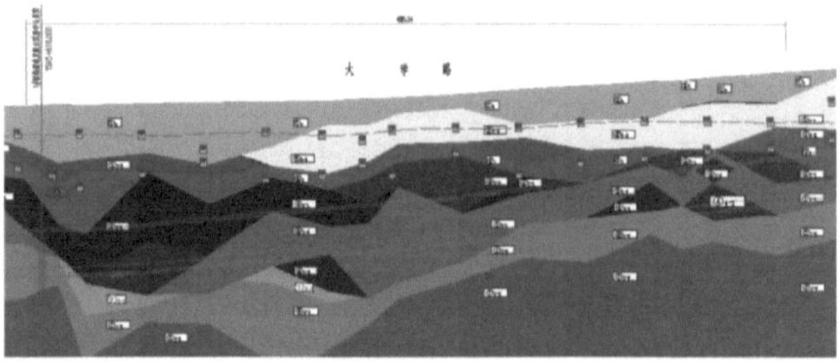

(b) right line

Figure 1. Stratigraphic section.

3 MAIN CONSTRUCTION TECHNOLOGY AND MEASUR

Keeping the shield machine boring continuously and quickly is an effective way to pass the sand stratum section using EPB shield method. To achieve this, a serious of technical measures must be adopted before and during construction.

3.1 *The maintenance and repair of shield machine before entering the sand stratum*

3.1.1 *The maintenance of tail sealing*

The sand stratum section is rich in groundwater, to insure the effectiveness of the tail sealing, high quality shield tail sealant should be used while passing the sand stratum section, the quantity of sealant usage should be increased and the sealant should be able to be infused incessantly during the process of tunneling. When the time length of parking is too long, in order to avoid the sealing element to be consolidated by the infused mortar, the tunnel boring machine should be moved several centimeters forward, and this interspace should be filled up with tail sealant. Meanwhile the posture of the shield machine should be adjusted to avoid the shield tail to make an oversized turn.

3.1.2 *The maintenance and repair of other parts*

Before the shield machine arrives at the sand stratum, the hydraulic system, the electrical system, the water circulation system, the lubrication system and the grouting system should be overhauled. The key points are the maintenance and repair of the air compressor, the screw conveyor gate, the greasing system, the segment erector and the foam system, and the state monitoring of the hydraulic system.

3.2 *The boring posture and parameter control of the shield machine when it tunnels through the sand stratum*

In this section, when the shield machine is tunneling though the sand stratum, two circumstances are most common. The shield tunnel section could be entirely below the sand stratum, or partly in the sand stratum. Either way, the boring posture and parameter should be controlled.

3.2.1 *Strict control upon the boring posture of the shield machine*

During the tunneling process, the axis of the shield machine can't be parallel to the Designed Tunnel Axis (DTA) all the time. The shield machine's posture would shift while going forward. When the shield machine deviates, it should be adjusted gradually, sudden correction is strictly forbidden. In practice, the deviation can be adjusted according to the difference between the value displayed on the VMT screen, which is obtained by automatic measurement system, and designed tunnel axis.

3.2.2 *Boring parameter in the sand stratum*

In Nanning district, when adopting EPB shield to tunnel through the sand stratum, the boring parameters mainly include the followings: the soil chamber top pressure is 1.3~1.4 bar, the cutter head rotation speed is 1~1.2 rpm, the boring speed is controlled at 40~50 mm/min, the cutter head torque is 3000~5000 kN · m, the thrust is 8000~13000 kN, the foam infused into the front of the cutter head is 20~24 m^3/ring, the bentonite infused is 3~6 m^3/ring, the volume of the residue discharge should not be bigger than 58 m^3, the weight of it should be 88~89t. In construction practice, the boring parameters above should be adjusted by technological explanation according to the actual geological conditions and measurement and monitoring results.

3.3 *Synchronized grouting and secondary grouting management*

For synchronized grouting, because the permeability of sand is relatively good, the actual grouting volume should be larger than the calculated amount to ensure the grouting quality.

If necessary, the mortar mixing ratio should be adjusted by increasing the portion of cement. By this measure, the initial setting time of the mortar would be shorter, the consolidation of the soil around the lining segment would be accelerated, therefore the overrun surface settlement can be avoided.

The principle of grouting must be followed, which is "simultaneous grouting, quick setting, information feedback and adequate supplement". Concretely speaking, the grouting material should have great groutability and durability, high calculus strength, and no pollution for groundwater and surrounding environment. It should be a relatively inexpensive kind of cement mortar. The initial setting time of it should be 4 to 8 hours. The slurry should be dynamically managed and constantly adjusted based on information during the tunneling process. The actual grouting pressure should be slightly bigger than the theoretical value. According to the geological status and the thickness of earth covering, the grouting pressure should be controlled at 0.2 to 0.4 MPa. To make up for the ground loss, the synchronized grouting volume behind every lining ring segment should be no smaller than 6.0 m^3, at the same time the grouting volume should be constantly adjusted and dynamically managed according to the monitoring condition of the surface hogging and sinking.

For secondary grouting, while the shield machine is tunneling, binary slurry should be backfilled into the back of every lining segment through the lifting hole for the 4 to 5 rings of segments that left the shield. On-site secondary grouting is controlled by grouting pressure, the grouting should be stopped once the pressure is over the controlled value. The grouting value should be controlled at 4–5 bar, at most 5 bar, to avoid excessive pressure around the lining segment damaging the segment. Secondary grouting must insure that the gaps behind the lining segments are filled up densely so that the surface settlement would not be excessive.

3.4 *Residue discharge and groundwater control measure*

When tunneling through sand stratum section, the shield machine's residue discharge per ring should be strictly controlled, the discharge amount should be overseen and recorded by special-assigned person to avoid over excavation causing damage to nearby buildings. At the same time, the stratum over the tunnel in this section is rich in groundwater, and losing a large amount of water in the stratum can cause settlement, which can lead to building inclination. Therefore, to avoid water losing in the boring construction, measures as followings should be adopted: (1) When the boring for one ring is finished and the machine is stopped waiting for segment assembling, the soil chamber should be filled with residue earth to prevent the groundwater from the soil chamber; (2) Secondary grout should be backfilled into every lining segment through the lifting hole for the 4 to 5 rings of segments that left the shield. Therefore, the route for water to gather to the soil chamber would be sealed.

3.5 *Other technical measures*

1. Using bentonite to condition muck, the volume per ring is 3–6 m^3. The waterproof effect would be stronger and the surface settlement could be controlled. Meanwhile, the usage amount of foaming agent should be enlarged.
2. If allowed, the boring speed should be as fast as possible. Therefore, the time length of the disturbance in the stratum caused by cutter head rotation can be shortened, and the liquefaction of the sand above can be avoided.
3. Exit passageway should be reserved and kept unblocked. In case of an emergency, people should be evacuated immediately.
4. Surface settlement monitoring should be treated seriously. The boring parameter should be adjusted in time and the surface settlement should be controlled to insure the safety of the pipelines and buildings nearby.

4 CONCLUSION

Based on the shield tunneling project of the first phase project of Nanning rail transit line 1 in the section from Xixiangtang passenger depot station to University for Nationalities station, this paper discusses the main technology measures for Earth Pressure Balance (EPB) shield tunneling in sand stratum section. These measures include the repair and maintenance of shield machine before it enters the sand stratum, the boring posture and parameter control, synchronized grouting and secondary grouting management, residue discharge and groundwater control and others. By summarizing the construction experience in this section, this paper provides reference and guidance for follow-up EPB shield construction in sand stratums in Nanning district.

REFERENCES

Chen, Hongjie & Wei, Xing. 2010. The Construction Technology of Earth Pressure Balance Shield in Watery Sand Stratum. *Urban Mass Transit*, 02: 50–52.

Jiang, Zhaosheng, Guo, Guangcai & Huang, Weiran. 2006. The Technical Measures of Earth Pressure Balance Shield in Watery Sand Stratum. *Urban Mass Transit*, 07: 58–60.

Wang, Huaizhi. 2012. Study on Construction Technology of Earth Pressure Balance Shield in Watery Sand Stratum. *South China University of Technology*.

Zeng, Huapo, Yu, Weiya & Wu, Laiming. 2006. A Discussion of Construction Technology of Earth Pressure Balance Shield in Sand Stratum. *Pearl River*, 04: 22–24.

Applied Mechanics and Civil Engineering VI – Xie (Ed.)
© 2017 Taylor & Francis Group, London, ISBN 978-1-138-62631-7

Effect of chemical–mechanical activation on the cementitious activity of iron ore tailings

Chun-ai Piao & Dongmin Wang
China University of Mining and Technology, Beijing, China

Li-ran Zhang
Department of Civil Engineering, Tsinghua University, Beijing, China

ABSTRACT: In this study, an orthogonal experiment was carried out to explore the effect of grinding time, the type and dosage of activators on the cementitious property of iron ore tailings. The following were found to be the optimal scheme: the grinding time to be 2.0 h, $CaSO_4 \cdot 2H_2O$ as an activator and its dosage as 0.5%. Subsequently, the results indicated that the size, which is less than 5.13 μm, reached the highest value when the $CaSO_4 \cdot 2H_2O$ dosage was 0.5%; meanwhile, SEM observation also confirmed the finding. The flexural strength ratio and activity index of 3*d*, 7*d* and 28*d* reached the maximum values correspondingly. IR and hydration heat measurement indicated that compared with the mechanical activation, the chemical–mechanical activation can significantly increase the content of disordered substances. While the hydration of the aluminum phase was accelerated, it suggested that the cementitious activity of iron ore tailings was higher during the chemical–mechanical activation than during the mechanical activation.

1 INTRODUCTION

With the rapid development of the steel industry, the scale of iron ore mining and iron tailings is increasing. Mass accumulation of the tailing discharge leads not only to occupation of the land but also to threats to the environment of the local mining area. Therefore, it is highly necessary to carryout comprehensive utilization of iron tailings (Li, 2012; Huang, 2013; Dong, 2013).

The main components of iron tailings are calcium, silicon, aluminum and iron oxides. The composition is similar to the chemical composition of the cementitious material, but it is an inert component. If the corresponding modification is not carried out, then it does not have the potential of volcanic ash activity. Iron tailings as cementitious materials in concrete mainly depend on whether it has a volcanic ash activity and the size of its activity (Yang, 1998). Activation methods for improving the activity of volcanic ash can be activated by mechanical activation (Wu, 2001), chemical activation (Yang, 2006a; Yang, 2006b), and thermal activation (Feng, 2007). Mechanical energy can induce to produce a chemical reaction which can hardly be induced by heat energy. It is not required to be calcined at high temperatures. Also, the process is simple, and thus it saves energy and resources (Yang, 2000a; Yang, 2000b). Therefore, mechanical and chemical activation methods are effective activation measures, in particular the combination of the two methods. The combination of mechanical and chemical activation provides the maximum activity of the powder released (Gao, 2002), and hence it is necessary to develop a combined process.

According to the principle of mechanical and chemical activation, the combination of mechanical and chemical activation methods (referred to as chemical–mechanical activation) makes the activation of iron tailings, so as to improve the reaction activity. The variation of particle size distribution and the activity index of activated iron tailings powder was studied. The chemical–mechanical activation mechanism was analyzed by infrared spectroscopy and heat release experiments. In order to achieve bulk harmless utilization of iron ore tailings, the modified iron tailings as a concrete mineral admixture will prove to be a more environmentally friendly and efficient new way for rational utilization of iron ore tailings.

2 EXPERIMENTAL

2.1 *Materials*

Iron tailings used in the experiment are from Qian'an, Hebei. The main chemical and mineral compositions of iron tailings are listed in Tables 1 and 2. The screening results of iron ore tailings are shown in Table 3. Through the screening results, the fineness modulus of iron ore tailings is 1.4, which belongs to the superfine sand. We used P · I42.5 reference cement, which is in accordance with the relevant national standards (GB 2008-8076). Its chemical and mineral compositions are shown in Table 4. The sand used in the experiment is standard sand. In the experiment, activators include Na_2SiO_3, NaOH, Na_2SO_4 and $CaSO_4 \cdot 2H_2O$, whose purity is greater than or equal to 96%.

Table 1. Chemical composition of iron ore tailings (*wt.*%).

SiO_2	CaO	MgO	Fe_2O_3	Al_2O_3	SO_3	Other
64.71	3.69	6.89	12.17	7.59	0.19	4.76

Table 2. Mineral composition of iron ore tailing (*wt.*%).

Types and contents of minerals/%						Total amount of clay minerals/%
Quartz	Plagioclase	Hornblende	Potash feldspar	Calcite	Dolomite	
53.8	27.3	7.0	5.9	1.9	1.6	2.5

Table 3. Screening results of iron ore tailings.

Sieve analysis/cumulative sieve residue%							Fineness modulus	Clay content%	Clod content%
4.75 mm	2.36 mm	1.18 mm	0.5 mm	0.3 mm	0.15 mm	0.075 mm			
0	3	8	20	40	66	100	1.4	0.3	0.1

Table 4. Composition of Portland cement (*wt.*%).

Chemical composition								
SO_3	SiO_2	Fe_2O_3	Al_2O_3	CaO	MgO	Na_2O_{eq}	f-CaO	Cl^-
2.42	21.82	3.45	4.04	61.70	2.59	0.60	0.83	0.007
Mineral composition								
C_3S	C_2S	C_4AF	C_3A					
56.01	27.15	8.42	7.06					

2.2 *Methods*

2.2.1 *Grinding of iron ore tailing*

The cement test mill of 5 kg in the laboratory is commonly used, whose model is SYMØ500 × 500 (Wuxi Xiyi Construction Machine Factory). At 105°C, the iron tailings were made to dry at a constant temperature and chemical–mechanical grinding was carried out. Then different chemical activators (Na_2SiO_3, NaOH, Na_2SO_4, $CaSO_4 \cdot 2H_2O$) were added at a certain proportion (the dosage is 0.5%, 1.0%, 1.5% and 2.0%, respectively) to the test mill. Finally, by mixing and grinding for 1.0 h, 1.5 h, 2.0 h and 2.5 h, we prepared iron tailing powder with different fineness.

2.2.2 *Testing and characterization*

1. Test of particle size distribution of iron tailings

The particle size distribution was detected using an LS-C (IIA) laser particle size analyzer. Then the particle size distribution of iron tailings was measured.

2. Activity evaluation test of iron tailings

The cement with 30% mass fraction was replaced with iron tailing powder obtained from chemical–mechanical activation. The compressive strength and flexural strength of the rubber sand specimens were measured at 3d, 7d and 28d. The level of activity was evaluated using GB/T2847-2005 "Pozzolanic" materials for cement production.

3. The analysis of the iron tailing powder microstructure

The microstructure of the samples was analyzed using a Hitachi S-3400 scanning electron microscope. The sample preparation method is as follows: after grinding, the samples are dried in a vacuum dryer. After vacuum gold plating, we observed the microstructure of the specimen and took photos using a scanning electron microscope. The sample is pasted on the copper sample holder by conductive adhesive.

4. Infrared spectrum analysis of iron tailing powder

The change of the Si-O bond with different fineness of iron tailing powder was determined using a spectrophotometer by FT-IR. The infrared spectrometer instrument model used in this test is NICOLET iS10, whose resolution is 4 cm^{-1}. With KBr grinding, the sample is pressed.

5. Hydration heat test

The hydration heat of the chemically and mechanically activated iron tailings by adding 10%, 30% and 50% of cement was determined using a microcalorimeter (Thermometrics TAM air). The ratio of water to cement is 0.4 in this experiment. The test temperature is 25°C.

3 RESULTS AND DISCUSSION

3.1 *Orthogonal test*

Through an orthogonal experiment, the effect of milling time, the type of activator and its content on the properties of iron tailings were studied. The design of the orthogonal experiment is shown in Table 5. Taking 3, 7 and 28d compressive strength as the evaluation index, the optimum scheme of grinding and the verification test data of each age are obtained as shown in Table 6.

The analysis of Table 6 data shows that for the optimal group of 3d, its 7d and 28d compressive strength values are 27.8 MPa and 34.7 MPa, which are the lowest of the three groups.

Table 5. $L16(4)^3$ orthogonal design.

	Factor		
Level	Grinding time/h	Activator dosage/%	Activator type
1	1.0	0.5	Na_2SiO_3
2	1.5	1.0	NaOH
3	2.0	1.5	Na_2SO_4
4	2.5	2.0	$CaSO_4 \cdot 2H_2O$

Table 6. Comparison of optimal method in each age test.

	Grinding time t/h	Activator dosage d/%	Activator type	3d/MPa	7d/MPa	28d/MPa
3d	2.0	0.5	Na_2SO_4	21.6	27.8	34.7
7d	1.5	0.5	Na_2SiO_3	19.6	30.6	36.5
28d	2.0	0.5	$CaSO_4 \cdot 2H_2O$	19.7	28.2	39.1

Table 7. Effect of $CaSO_4 \cdot 2H_2O$ dosage on the differential distribution of iron ore tailings at the grinding time of 2h.

Dosage/ %	Differential distribution/μm						
	≤1.00	1.00~3.00	3.00~5.00	5.00~10.00	10.00~16.00	16.00~24.00	>24.00
0	1.91	44.08	31.71	20.82	1.44	0.03	0.00
0.5	3.31	56.17	21.77	17.67	1.08	0.00	0.00
1.0	2.13	44.08	28.71	22.89	2.19	0.00	0.00
1.5	0.87	40.43	29.66	19.89	7.20	1.95	0.00
2.0	1.33	38.20	25.20	20.39	6.65	7.90	0.33

For the optimal group of 7d, its 3d compressive strength is 19.6 MPa. The value is the lowest in the three groups. Through a comprehensive analysis of the optimal scheme, the late strength optimum group and the optimal group of 28d for the final optimal solution are determined: t = 2.0 h, d = 0.5% and the activator is $CaSO_4 \cdot 2H_2O$.

3.2 *Effect of $CaSO_4 \cdot 2H_2O$ content on the powder particle distribution of iron tailings*

Under the action of chemical–mechanical force, the particle size will be reduced gradually, and the distribution of the particles will also change. When the mechanical grinding time is 2 h, the effect of the content of $CaSO_4 \cdot 2H_2O$ on powder particle distribution of iron tailings is shown in Table 7. It can be seen from the table that with the increase of the content of $CaSO_4 \cdot 2H_2O$ under mechanical action, the particles of size 1–3 μm gradually increased. When the content is 0.5%, the particle content of size less than 3 μm is more than 59%. Iron tailings of 16–24 μm gradually decreased and then increased gradually. It shows that the increase of content makes particle agglomeration.

When the content of $CaSO_4 \cdot 2H_2O$ is 0.5%, the influence of mechanical grinding time on the particle size distribution of iron tailings is shown in Table 8. As can be seen from the table, with the increase of grinding time, the particle content of 1–3 μm first gradually increased and then decreased. When the grinding time is 2 h, the particle content of less than 3 μm is more than 59%. Iron tailings of size 16–24 μm gradually decreased and then increased gradually. It shows that the increase of content give rise to particle weak agglomeration.

The effect of the $CaSO_4 \cdot 2H_2O$ content by chemical–mechanical grinding on particle accumulation distribution is shown in Figure 1. With the increase of dosage, the accumulation content of iron tailing grains of less than 5.13 μm is 77.70%, 81.25%, 74.92%, 70.96% and 64.93%, respectively. When the dosage is 0.5%, the grain accumulation of less than 5.13 μm is the highest.

3.3 *The morphology analysis of iron tailing powder by chemical–mechanical activation*

The morphology of iron tailing powder by chemical–mechanical activation is shown in Fig. 2. From Fig. 2(a), it can be seen that the iron tailing particles are irregular. Particles of size 150–300 μm are mainly found. By using mechanical force to grind for 2 h, the particles are refined, which is shown in Fig. 2(b).

The morphology of particles mixed with 0.5% $CaSO_4 \cdot 2H_2O$ after grinding for 2 h is shown in Fig. 2(c). Compared with the mechanical grinding, the particles are refined again. The conclusion is consistent with the conclusion in Section 2.2.

Table 8. Effect of grinding time on the differential distribution of iron ore tailings at the $CaSO_4 \cdot 2H_2O$ dosage of 0.5%.

Grinding time/h	Differential distribution/μm						
	≤1.00	1.00~3.00	3.00~5.00	5.00~10.00	10.00~16.00	16.00~24.00	>24.00
1.0	1.29	36.79	26.98	20.62	7.02	6.86	0.44
1.5	1.67	41.03	28.97	19.75	5.19	3.26	0.13
2.0	3.31	56.17	21.77	17.67	1.08	0.00	0.00
2.5	1.87	42.07	30.57	20.93	3.51	1.05	0.00

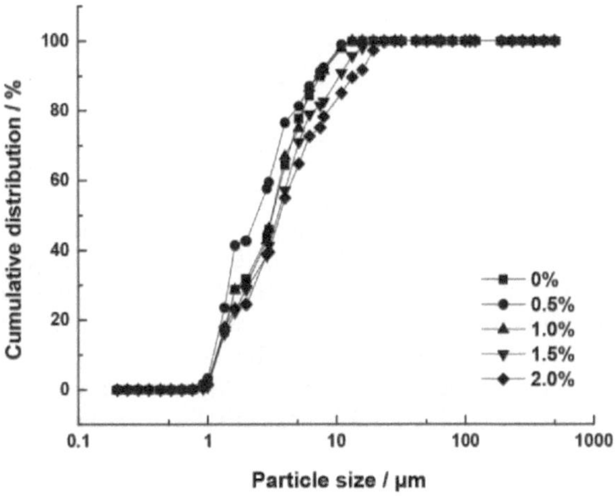

Figure 1. Effect of $CaSO_4 \cdot 2H_2O$ dosage on the cumulative distribution of iron ore tailing powder.

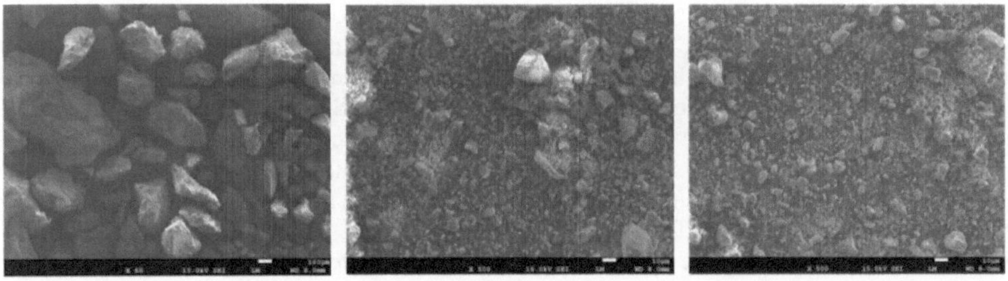

(a) (b) (c)

Figure 2. Microstructure of iron ore tailings at different grinding times: (a) raw ore; (b) mechanical activation 2 h; (c) chemical–mechanical activation 2 h.

3.4 *Effect of $CaSO_4 \cdot 2H_2O$ content on the activity of iron tailing powder*

Different dosages of $CaSO_4 \cdot 2H_2O$ and iron tailing powder obtained by chemical–mechanical grinding for 2 h are blended with reference cement in the proportion of 30%. The law of the flexural strength ratio (the flexural strength of the composite system/the flexural strength of reference cement) and mortar activity index (the compressive strength of the composite system/the compressive strength of reference cement) in 3d, 7d and 28d is shown in Fig. 3.

Fig. 3(a) shows that when the $CaSO_4 \cdot 2H_2O$ content increases, the flexural strength of the composite cementitious system first increases and then decreases. It can be seen in Fig. 3(b) that the activity index has the same law. When the dosage of CaSO4 · $2H_2O$ is 0.5%, the flexural strength ratio and activity index of 3d, 7d and 28d reach the highest value.

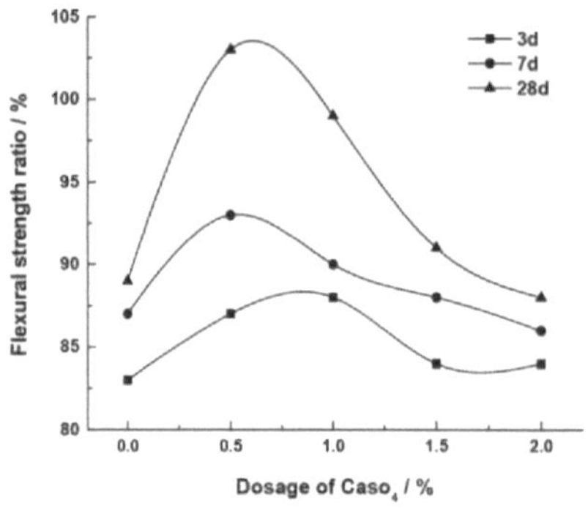

(a)

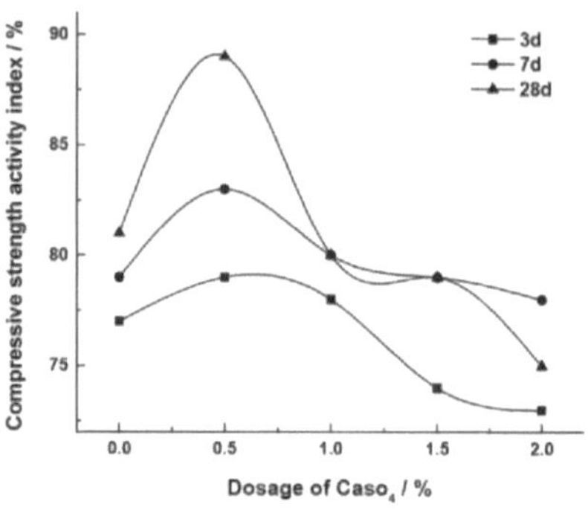

(b)

Figure 3. Effect of $CaSO_4 \cdot 2H_2O$ dosage on (a) the flexural strength ratio and (b) the activity index of iron ore tailing powder.

After chemical–mechanical activation, the grain refinement degree of iron tailings blended with 0.5% $CaSO_4 \cdot 2H_2O$ is very high. New fault ion polarization distortion and rearrangement are performed to make the surface lattice distortion. The number of fracture bonds on its surface increases. The mortar strength increases with the reaction degree of $Ca(OH)_2$, so the flexural strength ratio and the activity index are relatively high.

According to GB/T2847-2005 "Mixed materials for the use of volcanic ash in cement", the individual results of the activity index are determined. With the iron tailings at the content of 30% by chemical–mechanical grinding for 2 h, the activity index was more than 65%. It can be judged as an active mixed material.

3.5 *IR analysis of chemical–mechanical activation on iron tailing powder*

Through the infrared spectral changes of iron tailings by different grinding methods, the change of the internal structure of powder is studied. As shown in Fig. 4, the absorption peaks of 1101 cm^{-1}, 1040 cm^{-1}, 995 cm^{-1}, 965 cm^{-1}, 882 cm^{-1} and 527 cm^{-1} (Mako, 2001) are caused by the change of Si-O and Si-O-Si bonds. The 1101 cm^{-1} and 1040 cm^{-1} absorption peaks are caused by the asymmetric stretching vibration of Si-O. After mechanical grinding for 2 h, these two absorption peaks gradually degenerate and weaken. But these two absorption peaks completely disappear after chemical–mechanical grinding for 2 h. This phenomenon shows that the continuous grinding caused the destruction of the integrity and symmetry of mineral crystals. The Si-O bond is broken, which makes the surface of powder particles charged and forms the terminal Si-O bond effect (Hao, 2001). 995 cm^{-1} and 965 cm^{-1} correspond to symmetric stretching vibration of the Si-O bond. The two peaks by mechanical grinding for 2 h nearly disappear, and the peaks by chemical–mechanical grinding for 2 h completely disappear, degenerate and widen. This phenomenon is caused when the effect of the terminal Si-O bond makes the new fracture surface of powder charged. And then, it makes adsorption polymerization of cross-section and forms a new O-Si-O or Si-O-Si bridge. At the same time, compared with the mechanical activation, chemical–mechanical activation makes the disordered material to increase. The vibration can be improved and the activity is improved.

3.6 *Effect of chemical–mechanical activated iron tailing powder on cement hydration law*

The two kinds of chemical–mechanical and mechanical activated (2 h) iron tailings as raw materials are used. The effect of different grinding activation and iron tailing powder dosages on cement hydration law is studied, the results of which are shown in Fig. 5. It can be seen from Fig. 5(a) that with the increase of two kinds of iron tailing powder content, the maximum heat release peak gradually decreases.

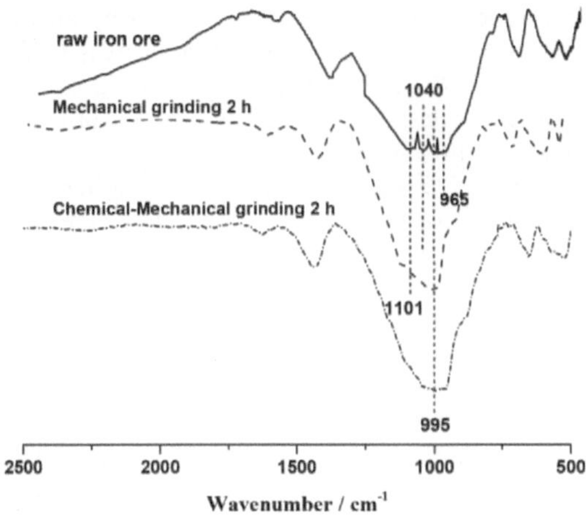

Figure 4. Effect of chemical–mechanical activation on the absorption of infrared spectra of iron ore tailing powder.

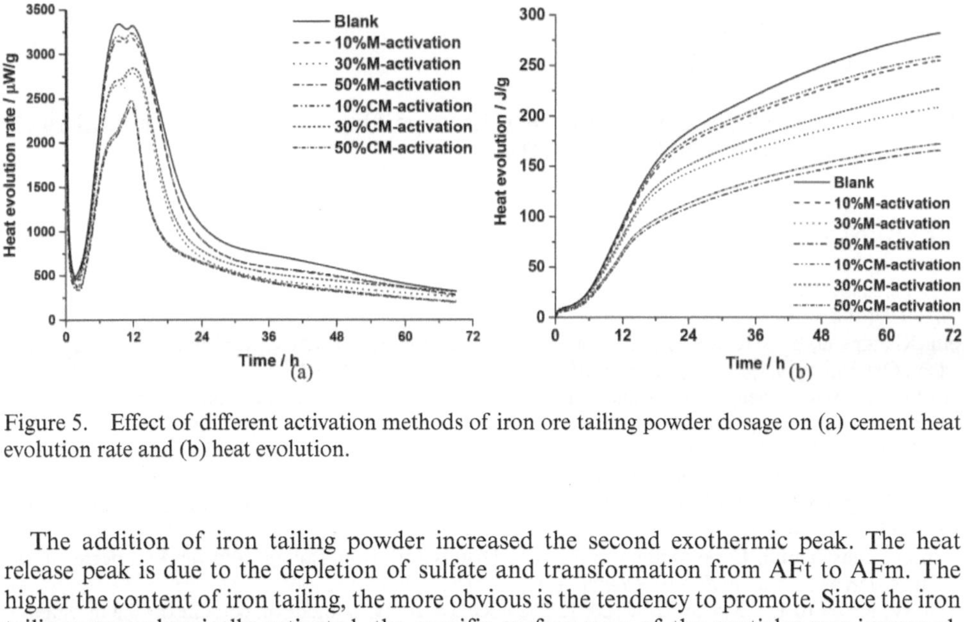

Figure 5. Effect of different activation methods of iron ore tailing powder dosage on (a) cement heat evolution rate and (b) heat evolution.

The addition of iron tailing powder increased the second exothermic peak. The heat release peak is due to the depletion of sulfate and transformation from AFt to AFm. The higher the content of iron tailing, the more obvious is the tendency to promote. Since the iron tailings were chemically activated, the specific surface area of the particles was increased. The increase of the number of surface breaking bonds promotes the second hydration of the aluminum phase. It can be observed that the iron tailing powder decreased the hydration heat release, which is shown in Fig. 5(b). Compared with mechanical activation, the maximum heat release peak and accumulated heat release of iron tailings by chemical–mechanical activation are higher. This shows that the hydration activity of chemical–mechanical activated iron tailing is higher than that of mechanical activation.

4 CONCLUSIONS

This study takes iron tailings of Hebei Qian'an as the main raw material. The chemical–mechanical activation of iron tailings is studied. The method can improve the reaction activity of iron tailings.

1. The optimal solution of chemical–mechanical activation by an orthogonal experiment was obtained: t = 2.0 h, d = 0.5% and the activation agent is $CaSO_4 \cdot 2H_2O$. When the content of $CaSO_4 \cdot 2H_2O$ is 0.5%, the particle accumulation of less than 5.13 μm is the highest. The particle content of less than 2 μm was greater than 40%. The SEM observation also showed the same result.
2. When the dosage of $CaSO_4 \cdot 2H_2O$ is 0.5%, the flexural strength ratio and the activity index of 3d, 7d and 28d are highest. For this content, iron tailing is maximally activated, which belongs to the active admixture. Through IR analysis, it can be found that the chemical–mechanical activation makes the disordered materials to increase compared with the mechanical activation. The vibration can be improved and then the activity is improved.
3. The iron tailing powder decreased the heat release rate of cement hydration and cumulative heat release, and the hydration of the aluminum phase was promoted. The maximum heat peak and cumulative heat release of iron tailing by chemical–mechanical activation are higher than those by mechanical activation. This shows that the hydration activity of chemical–mechanical iron tailing is higher than that of mechanical activation.
4. The appropriate chemical–mechanical activation can better stimulate the potential activity of iron tailings. It can increase the content of cement concrete. In order to reduce the amount of cement at the same time, a new type of mineral admixture was actively studied and developed. It is not only to reduce the amount of cement, but also to actively study and develop a new type of mineral admixture.

ACKNOWLEDGMENTS

The authors gratefully acknowledge the financial support from the National Natural Science Foundation of China (Grant No. 51572293).

REFERENCES

Dong Fengzhi. 2013. Study on comprehensive utilization of iron tailings. *Mining Research and Development* 33(01): 59–61.

Feng Xiangpeng, Sun Henghu, Zhang Na, et al. 2007. Study of Activity Optimization Mechanism of Iron Ore Tailing. *Express Information of Mining Industry*. (06): 21–24.

Gao Shujun, Wu Qisheng, Zhang Shaoming. 2002. Study on mechano-chemical activity of the slag treated by high-energy ball milling. *Journal Nanjing University of Technology* 24(06): 61–65.

Hao Baohong. 2001. IR analysis of the chemical bond changes in quartz powder during superfine milling. *Mining and Metallurgical Engineering* 21(04): 65–66.

Huang Yonggang. 2013. Utilization status and outlook of China's iron ore tailings. *Resoures and Industries* 15(03): 40–44.

Li Jinjin, Ding Yimin, Lu Wenxiong. 2012. The Disposal of Iron Tailings and its Advancement of Utilization as Resource. *Fly Ash Comprehensive Utilization* (06): 52–56.

Mako E., R.L. Frost, J. Kristof, et al. 2001. The effect of quartz content on content on the mechano-chemical activation of kaolinile. *Journal of colloid and interface science* 244, 359–364.

Wu Qisheng, Zhang Shaoming, Zhou Yongming. 2001. The Research and Development of Mechanochemistry in Inorganic Materials Preparation. *Materials Science and Engineering* 19(01): 137–142.

Yang Nanru. 1998. Making full use of resources and development of new cementitious materials. *Journal of Building Materials* 1(01): 19–25.

Yang Nanru. 2000a. Processes and effects of mechanochemistry (I)—Chemical effects of mechanochemistry. *Journal of Building Materials* 3(1): 19–26.

Yang Nanru. 2000b. Processes and effects of mechanochemistry (II)—Chemical effects of mechanochemistry. *Journal of Building Materials* 3(2): 93–97.

Yang Nanru, Zeng Yanwei. 2006. Research topics of chemical activating cementitious materials (Part one)—The necessity and feasibility of the research and development of chemical activating cementitious material (I). *Development Guide of Building Materials* (01): 44–47.

Yang Nanru, ZengYanwei. 2006. Research topics of chemical activating cementitious materials (Part two)—The necessity and feasibility of the research and development of chemical activating cementitious material (II). *Development Guide of Building Materials* (02): 42–46.

Applied Mechanics and Civil Engineering VI – Xie (Ed.)
 ISBN 978-1-138-62631-7

The design value of vertical earth pressure on hydraulic tunnels

Z.Y. Zhou & G. Wang
Shanghai Municipal Engineering Design and Research Institute (Group) Co. Ltd., Shanghai, China

ABSTRACT: The calculation model of vertical earth pressure on a tunnel has been argued for a long time, and it becomes more complex when the tunnel is filled with water because of the inner water pressure. This paper discusses several vertical earth pressures on tunnels and pipes, and analyzes the application conditions, respectively. It also provides the principle of assigning earth pressure values in the hydraulic tunnels' design. However, there are still many influencing factors of vertical earth pressure. The design of hydraulic tunnels should be both safe and economic, which still needs further analysis and research.

1 INTRODUCTION

1.1 *The background of hydraulic tunnels*

Most hydraulic tunnels are important municipal projects, which include urban water supply and drainage pipes, deep water storage and draining tunnels, etc. In recent years, the up-ground infrastructure has become increasingly crowded since the city scale keeps expanding, and the land paving has modified the original ground feature, making water draining enormously difficult in the downtown. The super large draining tunnel is an important solution for urban waterlogging. At the same time, long-distance water supply pipes with super large diameter have also been adopted extensively. There are already dozens of large hydraulic tunnels in China, such as the Qingcaosha water supply tunnel, Bailonggang waste water drainage tunnel, North water transfer tunnel, etc. Hydraulic tunnel construction can use either the shield method or the pipe-jacking method, and it is necessary to seriously consider the load model, seam tightness and long-term service ability.

1.2 *Calculation theories of vertical earth pressure*

Trenchless construction is usually used in super large hydraulic tunnels, where the vertical earth pressure is different from that of buried pipes. When the surrounding earth of a trenchless tunnel is unperturbed, and the tunnel is deeply set without swamp soil, silt or embankment, there will be a soil arching effect above the tunnel.

1. Terzaghi ideal earth pressure theory

Loose earth pressure is normally calculated by the Terzaghi formula. Since the above earth will slip down during the excavation, there will be an active slipping surface beside and above the tunnel, which extends to the ground, and then the Terzaghi formula is derived by balancing all the forces on the earth.

As shown in Figure 1, take an earth element (width $2B_1$, thickness dz) at the depth of z, and an equation will be derived according to the force balance.

$$2B_1\gamma dz = 2B_1(\sigma_y + d\sigma_y) - 2B_1\sigma_y + 2\tau dz \tag{1}$$

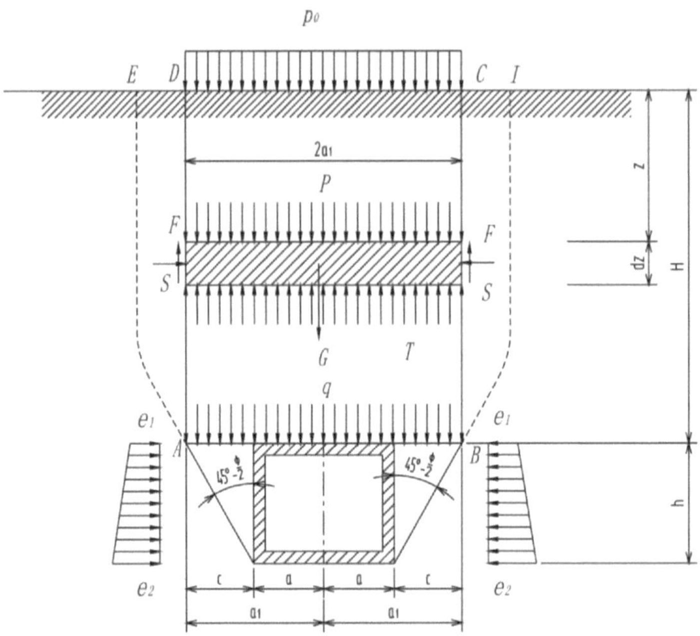

Figure 1. Terzaghi earth pressure calculation model.

B_1 is half of the influence width of tunnel excavation, which is calculated as given below:

$$B_1 = R\cot\left(\frac{45° + \varphi/2}{2}\right) \tag{2}$$

can be derived by Coulomb earth failure criteria:

$$\tau = c + \sigma_x \tan\varphi = c + K_0 \sigma_y \tan\varphi \tag{3}$$

K_0 is an experienced parameter, which is normally set as 1.
Take 3 into 1:

$$\frac{d\sigma_y}{dz} = \gamma - \frac{c}{B_1} - K_0 \sigma_y \frac{\tan\varphi}{B_1} \tag{4}$$

Considering the boundary condition of $\sigma_y = p_0$ at $z = 0$:

$$\sigma_y = \frac{B_1\left(\gamma - \frac{c}{B_1}\right)}{K_0 \tan\varphi}\left(1 - e^{-K_0 \tan\varphi \cdot \frac{z}{B_1}}\right) + p_0 e^{-K_0 \tan\varphi \cdot \frac{z}{B_1}} \tag{5}$$

As a result, the above earth pressure σ_v can be deduced:

$$\sigma_v = \frac{B_1\left(\gamma - \frac{c}{B_1}\right)}{K_0 \tan\varphi}\left(1 - e^{-K_0 \tan\varphi \cdot \frac{H}{B_1}}\right) + p_0 e^{-K_0 \tan\varphi \cdot \frac{H}{B_1}} \tag{6}$$

2. M.M. Promojiyfakonov theory

Consider the earth with a height of h_1 below the earth arching as the vertical earth pressure, as shown in Figure 2.

h_1 can be calculated as given below:

$$h_1 = \frac{a_1}{f_{KP}} \tag{7}$$

$$a_1 = \frac{D}{2}\left[1 + \tan\left(45° - \frac{\varphi}{2}\right)\right] \tag{8}$$

$$G = \gamma_1 h_1 D_0 \tag{9}$$

where

h_1—arching height (m);
f_{KP}—$\tan\varphi$;
D—tunnel height, which is the outer diameter for circle pipes.

When using the M.M. Promojiyfakonov theory, there should be enough depth, and a nature balancing earth arching should be formed after excavation. Generally, it can be used under two conditions: $f_{KP} \geq 0.6$ (which means $\varphi \geq 30°$) and the above earth height $H \geq 2h_1$.

3. The structural design code for pipelines of water supply and waste water engineering (Chinese code: GB50332-2002)

The vertical earth pressure on the pipe constructed by the jacking method can be calculated as given below:

$$F_{sv,k} = C_j \gamma_s B_t D_1 \tag{10}$$

$$B_t = D_1\left[1 + \tan\left(45° - \frac{\varphi}{2}\right)\right] \tag{11}$$

$$C_j = \frac{1 - \exp\left(-2K_a \mu \cdot \frac{H_s}{B_t}\right)}{2K_a \mu} \tag{12}$$

where

C_j—earth pressure coefficient of non-slotting construction;
B_t—influence width of the above earth pressure at the top of the pipes;

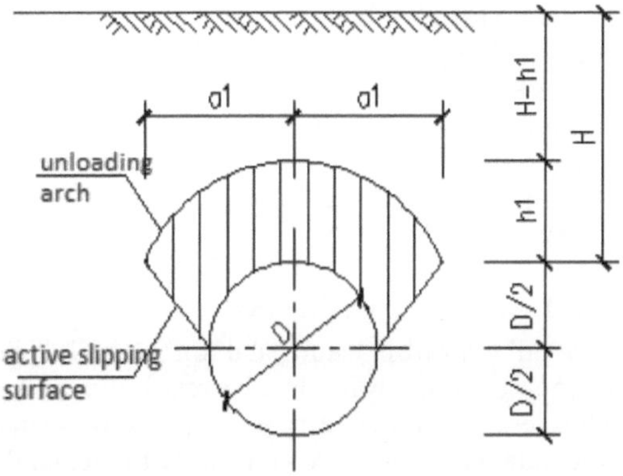

Figure 2. M.M. Promojiyfakonov theory.

γ_s—earth gravity;

$K_a\mu$—the product of the active earth pressure coefficient and inner friction coefficient of the above original earth, which can be set as 0.19 for general earth;

H_s—depth from ground to the top of the pipe;

φ—inner friction angle of the earth beside the pipe, which can be set as 30°, if there is no experiment conclusions.

4. Shanghai Road Tunnel Design Code (DG/TJ08-2033-2008, J11197-2008)

There are two rules in the Shanghai Road Tunnel Design Code: 1) the vertical load should be designed as all the above earth pressure in shallow tunnels; 2) when the height of the above earth exceeds 2 times the diameter, the vertical load can use the unloading arch theory or all earth pressure according to specific construction conditions.

2 THE DESIGN VALUE OF VERTICAL EARTH PRESSURE UNDER DIFFERENT CONSTRUCTION CONDITIONS

There have been many research studies and experiments on the design value of water and earth pressure. However, arguments always exist on the design value of the earth pressure on the tunnel. Ohta et al. analyzed the earth and underground water pressure distribution in 14 shield tunnel projects and concluded that the vertical earth pressure nearly equals the weight of the above soil column in soft clay, while in hard clay, the actual vertical earth pressure is less than the soil column weight, and in sand, it is much less than the soil column weight because of the soil arching effect [1]. Mashimo et al. analyzed the surrounding earth pressure distribution of shield tunnels by site monitoring, which showed that the actual earth pressure is more close to the Terzaghi earth pressure [2]. In recent years, many research and monitoring results showed that the actual earth pressure is very different from the theoretical value [3][4][5]. Li compared the earth pressure calculation results and the monitoring data, and derived that, in large and deep shield tunnels, the vertical earth pressure is between loose earth pressure and soil column weight, but more close to the loose earth pressure [7].

3 THE CONTROL LOAD COMBINATION IN HYDRAULIC TUNNELS

The load on a hydraulic tunnel is shown in Figure 3, where

H_0—underground water depth (m);

H_1—above net water head (m);

H—above earth thickness (m);

q_1—above vertical earth pressure (kPa);

q_2—bottom subgrade reaction (kPa);

q_G—arching earth pressure (kPa);

q_w—static water pressure (kPa);

e_1—top lateral earth pressure (kPa);

e_2—bottom lateral earth pressure (kPa);

PP—lateral triangular earth pressure (kPa);

R_H—calculated radius (m);

g—lining gravity (kPa);

α—angle of the calculated surface and vertical axis.

The hydraulic tunnels are different from traditional tunnels mainly because of inner water pressure. In normal conditions, the inner water head is relatively small in urban water supply and sewage tunnels, which is less than outer earth water pressure; therefore, the worst case is still the combination of the outer earth pressure and empty tunnel, and the calculation mode is the so-called "Load-Structure Method". While the tunnel is shallow, and the inner water pressure is more than the outer water pressure, the control case may be the combination of

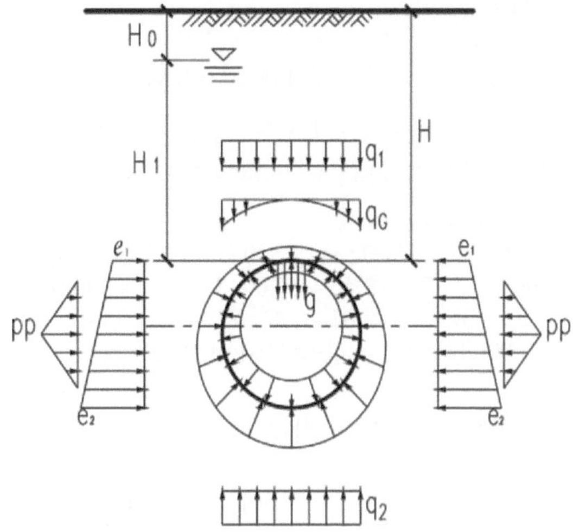

Figure 3. Calculation diagram of hydraulic tunnels.

Table 1. The design value of earth pressure parameters.

Earth classification		Diluvial clay	Diluvial sand	Alluvial clay
Nature earth gravity γ		18	19	17
Effective earth gravity γ'		8	9	7
External loads under maximum earth pressure	Calculation model for soil and water	Divided	Jointed	Jointed
	Maximum vertical earth pressure	All of the earth	$1.5D_0$	All of the earth
	Lateral earth pressure coefficient λ	0.6	0.4	0.75
External loads under minimum earth pressure	Calculation model for soil and water	Divided	Jointed	Jointed
	Minimum vertical earth pressure	$0.175D_0$	$0.175D_0$	
	Lateral earth pressure coefficient λ	0.6	0.4	0.5
Coefficient of subgrade reaction k(MN/m^3)		10	50	0
The above earth		Diluvial clay	Diluvial clay/Sand	Alluvial clay
Crossed earth layer		Diluvial clay	Hard diluvial clay/sand	Alluvial clay

minimum outer earth pressure and maximum inner water pressure. In this case, the normal force of the tunnel will be much less, and even there could be tension is some structure area. As a result, the earth pressure calculation can use the "Stratum-Structure Method", and consider the adverse effect of the arching effect.

The earth pressure is complexly influenced by earth properties and depth for shield tunnels, especially for the tunnels with inner water pressure. For now, there is no systematical minimum earth pressure calculation principle in China. In Japanese shield tunnel design manual with inner water pressure, when the smaller outer earth pressure is adverse to the structure, the earth of 0.175D above the tunnel is used as the vertical pressure, so as to mainly consider the influence of maximum inner water pressure.

The Japanese shield tunnel design manual with inner water pressure provides the calculation model and value determination basis for earth pressure, as shown in Table 1.

4 CALCULATION CASE

In this paper, a large water supply tunnel in Shanghai is taken as an example.

1. Construction condition 1: no inner water pressure

The above earth height H is 12.335 m and the lateral earth pressure coefficient, λ, is 0.700;

- The vertical earth pressure on the tunnel:
  $Parc = 0.2146 \times Rc \times \gamma i = 0.2146 \times 2.975 \times 18.500 = 17.717$ (kN/m);
  $P1 = (H \times \gamma i + P0 + Parc) = (12.33 \times 18.50 + 30.0 + 17.717) = 390.013$ (kN/m)
  $Pg = \pi \times \gamma c \times T = \pi \times 25.00 \times 0.45 = 53.014$ (kN/m)
  $Pw = -\pi \times Rc \times \gamma w/2 = 0.000$ (kN/m) (soil-water-jointed)
- The counter vertical force:
  $P2 = P1 + Pg + Pw = 390.013 + 53.014 + 0.000 = 443.027$ (kN/m)
- Lateral earth pressure:
  $Q1 = \lambda \times (P1\text{-}Parc) = 0.70 \times (390.0128 - 17.7166) = 260.607$ (kN/m)
  $Q2 = Q1 + 2 \times Rc \times \gamma i \times \lambda = 260.6074 + 2 \times 2.975 \times 18.50 \times 0.700 = 376.186$ (kN/m)

  The results are shown in Figure 4.

2. Construction condition 2: inner water pressure and outer load

When the tunnel is in normal use, the inner water pressure should be calculated. The calculation process is shown below.

The central inner water pressure Hw = 7 + 10 = 17(m), and the calculation results under single inner water pressure is shown in Figure 5.

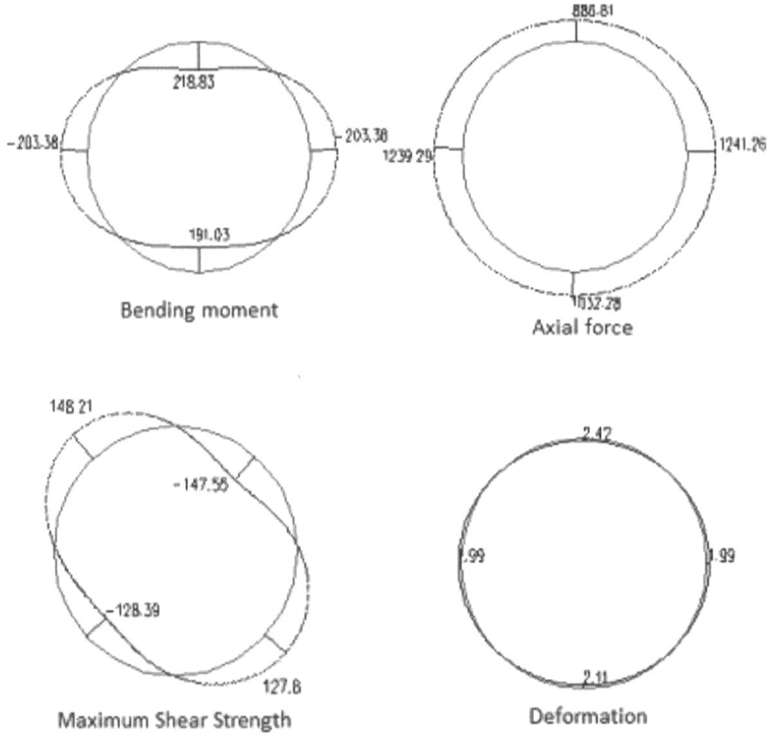

Figure 4. Calculation results under no inner water pressure.

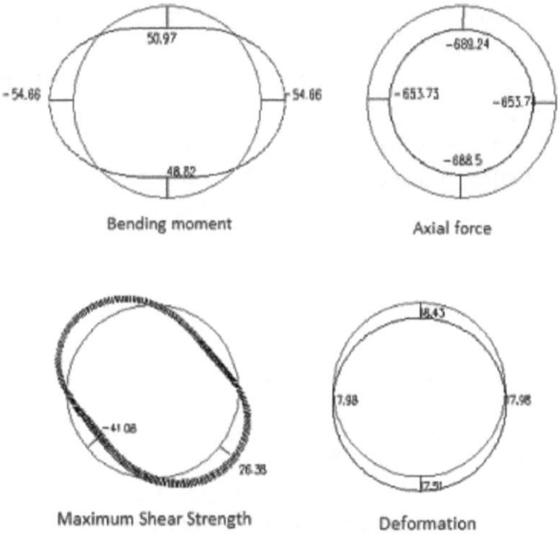

Figure 5. Calculation results under single inner water pressure.

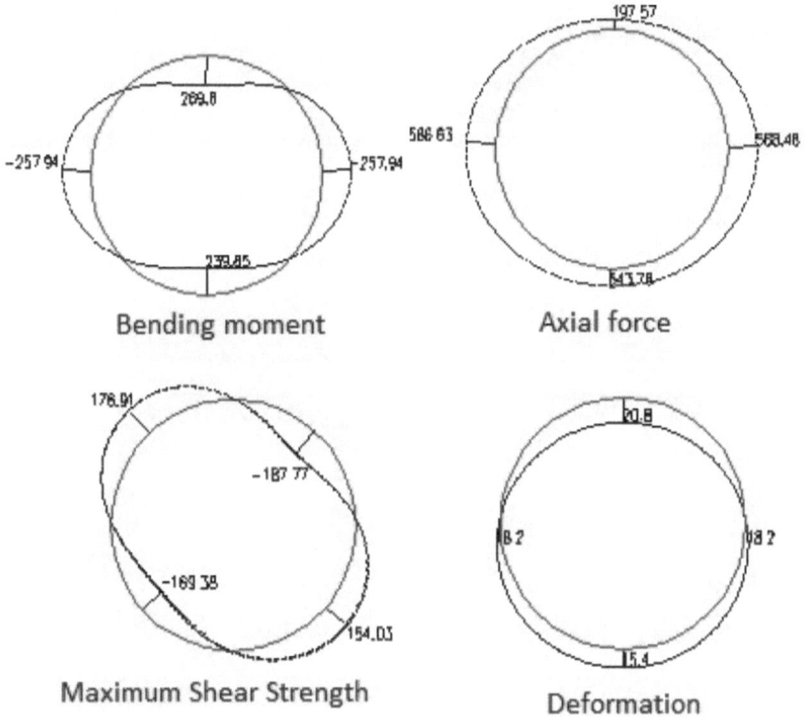

Figure 6. Calculation results under both inner water pressure and outer force.

The results of the inner water pressure combined with the calculation results in construction condition 1 are shown in Figure 6.

5 CONCLUSIONS

In the tunnel design and construction, a specific calculation model has already been proposed for the pipe-jacking method. However, the above vertical earth pressure on larger and deeper tunnels is usually calculated using the soil column theory. In fact, whether in clay or sand, there are still many problems when using loose earth pressure theories or any theory considering the soil arching effect, such as the arching effect weakening by seismic earth liquidation or long-term service.

Nevertheless, as the inner water pressure can play an important role in the hydraulic tunnel load combination, it is not safe to use only the soil column theory for calculation. Therefore, the earth arching effect should be considered when calculating the vertical earth pressure.

REFERENCES

[1] Japanese shied tunnel design manual with inner water pressure [M], Japan.

[2] Koizumi A. Segment Design of Shield Tunnel Design [M], Japan.

[3] Li X., Tian Z., Yang Z., et al. Observed ground pressures and internal forces of the Shanghai Yangtze River tunnel[C]/Tunneling and Underground Construction. [S. l.]: [s. n.], 2014: 984–992.

[4] Li X., Zhou S., Gong Q., Chen C., Evaluation of earth pressure around a deeply buried metro shield tunnel with a large cross-section under high water pressure conditions, Rock and Soil Mechanics, Vol. 36, No. 5, 2015.

[5] Mashimo H., Ishimura T. Evaluation of the load on shield tunnel lining in gravel [J]. Tunnelling and Underground Space Technology, 2003, 18(2): 233–241.

[6] Ohta H., Takeuchi T., Nishiwaki, Yohta H., et al. Performance of linings for shield driven tunnels—A survey on Japanese shield tunneling [C]/International Symposium on Underground Construction in Soft Ground. New Delhi: A. A. Balkema. 1995: 367–374.

[7] Technical specification for pipe jacking of water supply and sewerage engineering [M], China.

[8] Wu S., Wang Z., Wang L. Monitoring and analysis of the force and deformation of large section crossing-river tunnel during operation period [J]. Journal of Zhejiang University (Engineering Science), 2013, 47(4): 595–608.

[9] Zhang H., Zhang L., Ma G., et al. Experimental study on in-situ monitoring of ground pressures acting on shield-driven tunnels [J]. Tunnel Construction, 2006, 26(Supp. 2): 8–11.

Applied Mechanics and Civil Engineering VI – Xie (Ed.)
 ISBN 978-1-138-62631-7

Numerical simulation of sediment erosion and transport in the diversion tunnels of a hydropower station

Xu Wang, Ping Yu & Jing Liang
College of Water Conservancy Engineering, Tianjin Agricultural University, Tianjin, China

ABSTRACT: To reveal sediment erosion and transport processes in the diversion tunnels of a large hydropower station in Southwestern China, a three-dimensional numerical model is established to simulate sediment movement. The calculation results indicate that the current will cause sediment erosion near the tunnel entrance, which leads to sediment transport to the downstream in the water diversion process. The speed of sediment erosion is faster at the beginning stage. When the residual sediment is less, the sediment erosion speed is obviously slower, and it takes more time to wash away the residual sediment. The calculation results of the sediment transport rate and the average sediment concentration indicate that the sediment transport rate increases with the increase in the flow rate. Although the sediment concentration change trend probably increases with the flow rate, an abrupt concentration decrease is found at an intermediate flow rate. This study can provide the scientific basis for the design of hydropower stations and the treatment of sediments.

1 INTRODUCTION

Sediment deposition is always an important problem in the operation management of hydropower stations. Reservoir sediments will not only reduce the effective capacity, but also produce adverse effects on water diversion and power generation in hydropower stations, such as unit wear, cooling water blocking and sluice gate clogging. Therefore, a reasonable solution to the problem of sediments is an important guarantee for the safety in flood season and the safe power generation of hydropower stations.

The sediment deposition in the water diversion process is an important problem in water conservancy and hydropower engineering. There are abundant research results in this area. A lot of theoretical innovation and successful experience in engineering have come into being, which has important guiding significance for the water and sediment movement near the water intake. Many scholars have made a lot of physical model research on sediment deposition near the water intake of hydropower stations. Li Yinan (2002) studied the scouring and deposition of water intake of the Jining Shandong canal power plant under different flow conditions by using a physical model. Zhang Xujin (2013) studied water intake of the Huaneng Jiangjin power plant by using a physical model experiment. An analysis of sediment deposition and sediment effects on water intake was performed to compare water intake with respect to the built conditions of Jiangjin station. However, the physical models are sometimes difficult to be widely carried out due to the limitation of the experimental area and the high cost of the experimental study. As a convenient research method, numerical simulation has been widely favored by a majority of the scholars. Zhou Fayi (1997) established a three-dimensional numerical simulation method of suspended sediments near a water intake area by using a large eddy turbulence model. Li Mengguo (2006) studied the annual sediment deposition of the Huangdao power plant in different periods by using irregular triangular meshes based on the wave and the crushing effect. Hu Xiaozhang (2009) establishes a two-dimensional sediment mathematical model to research on the sediment thickness of water intake of the Hainan Dongfang Power Plant. However, the mathematical model cannot completely replace the physical model; therefore, it is necessary to use a physical model to accomplish its complement and validation. Lee (2000, 2002) developed a series of studies on the movement of sediment particles based on the sediment movement mechanism.

In this study, a large range of sediment deposition will appear in the diversion tunnel inlet, which can bring the sediment into the water turbine unit during the operation of the diversion and power generation system, leading to the risk of improper operation of the turbine. Therefore, it is necessary to make an in-depth study on sediment distribution and movement in hydropower station water inlet and water diversion tunnels, as well as the influence on the hydropower station. The results of the analysis and conclusions will provide the scientific basis for the safe and efficient operation and management of hydropower stations.

2 MODEL SETUP

The paper mainly studies the diversion process of sediment deposition in the entrance of the diversion tunnel and the calculation of sediment erosion and transport. From the water inlet of the tunnel to the generator, water will flow through about 16 km distance. A three-dimensional model is established to simulate the process. The model is shown in Figures 1 and 2. The figures show only about 300 m range of the whole model, because it is too long; there are nearly 16 km downstream of the tunnel not shown up.

In the model, the grid scale in vertical and horizontal spacing is 1 m, the longitudinal grid scale is 2 m, and the longitudinal length of the model is 16 km. The upper and lower boundary conditions are set to the flow rate boundary. According to the experimental results,

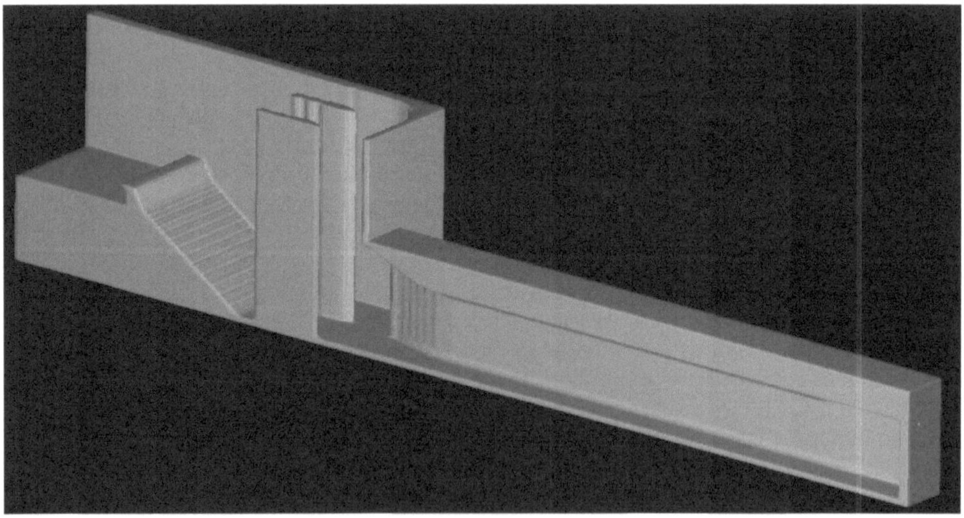

Figure 1. Three-dimensional model of symmetric processing.

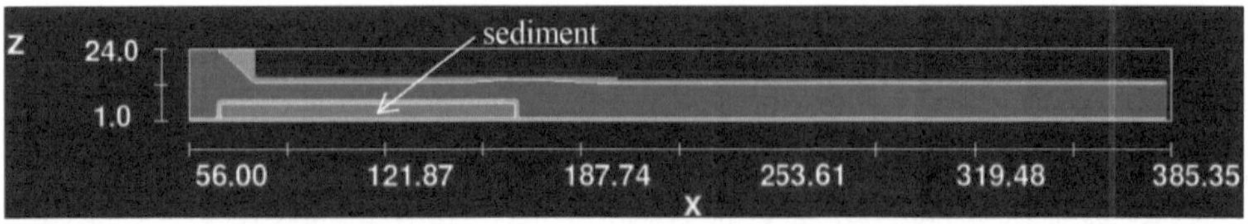

Figure 2. Sediment distribution in the front section of the water diversion tunnel in condition 1 (0 s time).

Table 1. Sediment gradation.

Particle size composition								
Particle size (mm)	1~0.5	0.5~0.25	0.25~0.1	0.1~0.05	0.05~0.025	0.025~0.01	0.01~0.005	<0.005
Proportion	7%	16%	52.5%	18%	2%	2%	2%	0.5%

Table 2. Calculation condition.

Condition number	Flow rate (m^3/s)	Sediment laying height (m)	Condition number	Flow rate (m^3/s)	Sediment laying height (m)
1	228.6	5	11	117	3
2	228.6	4	12	117	2
3	228.6	3	13	66	5
4	228.6	2	14	66	4
5	170	5	15	66	3
6	170	4	16	66	2
7	170	3	17	14.7	5
8	170	2	18	14.7	4
9	117	5	19	14.7	3
10	117	4	20	14.7	2

the sediment packing density is 1512.2 kg/m^3 and the critical volume fraction of sediment deposits is 0.57. The sediment gradation in the calculation is shown in Table 1. The calculation conditions are shown in Table 2.

This study adopts the k-ε turbulence model to solve the flow field and uses the sediment empirical model to calculate sediment erosion and transport. The sediment model includes two basic models: the flow model and the sediment incipient motion model. When the sediment is suspended, the model of the flow model is started, and the sediment will deposit under the action of gravity and other forces.

The sediment incipient motion model is triggered at the contact surface between the bed and the water. When the shear stress of the flow is greater than the critical shear stress, the bed sediment moves. This sediment model has been successfully applied in many practical projects. Compared with the experimental results, the calculation result accuracy is high, and the error can be controlled within 10%.

3 CALCULATION RESULTS

3.1 *Variation of sediment scour in the entrance of the diversion tunnel*

Due to space constraints, only the typical working condition calculation results and figures are listed. Figures 1–7 shows sediment distribution in the typical times of working condition 1. In the figures, only the front section of the tunnel is expressed; the remaining part of the sediment distribution characteristics will be given in the form of a concentration curve and time variation curve. It can be clearly seen from the figures that with the passage of time, the sediment erosion gradually produces at the entrance of the tunnel. The overall scouring process takes 6400 s, but the sediment is not linearly decreased during the scouring process. As shown in the figures, before 3000 s, the sediment has been mostly washed away, and the rest of the sediment is washed in the time range of 3000 s~6400 s. Similarly, under other conditions, there are similar processes. However, in the case of different flow rates, the erosion time is different.

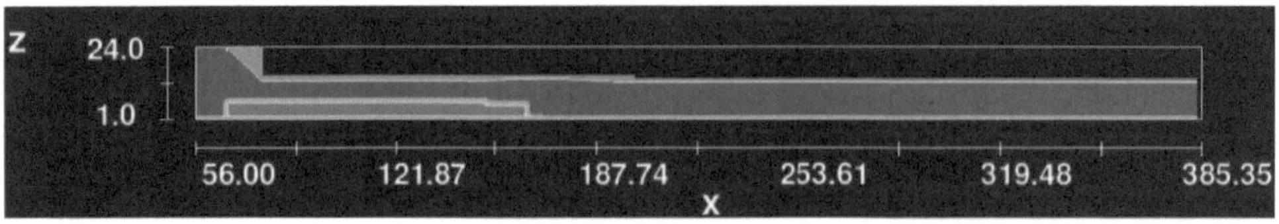

Figure 3. Sediment distribution in the front section of the water diversion tunnel in condition 1 (300 s time).

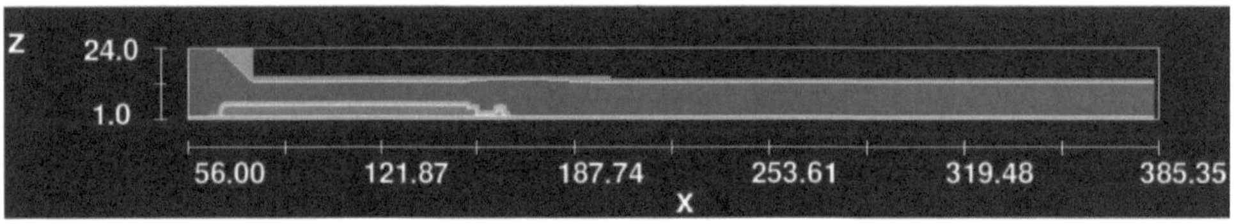

Figure 4. Sediment distribution in the front of the water diversion tunnel in condition 1 (1000 s time).

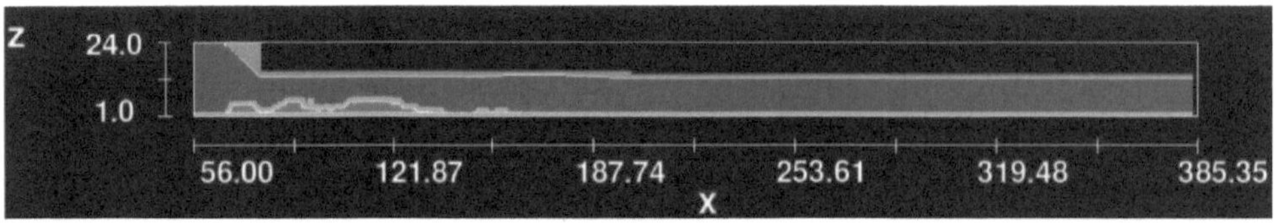

Figure 5. Sediment distribution in the front of the water diversion tunnel in condition 1 (1700 s time).

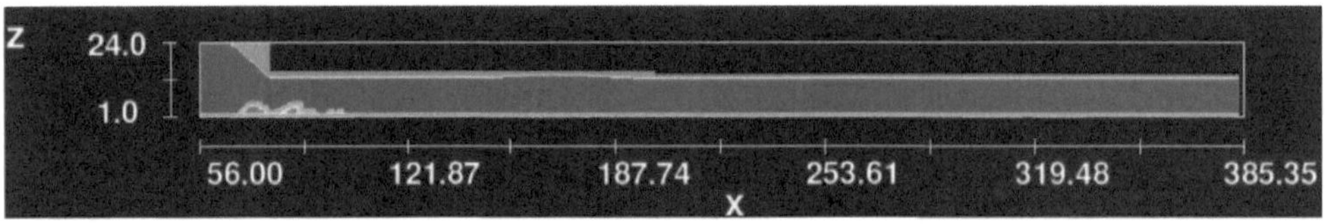

Figure 6. Sediment distribution in the front of the water diversion tunnel in condition 1 (2500 s time).

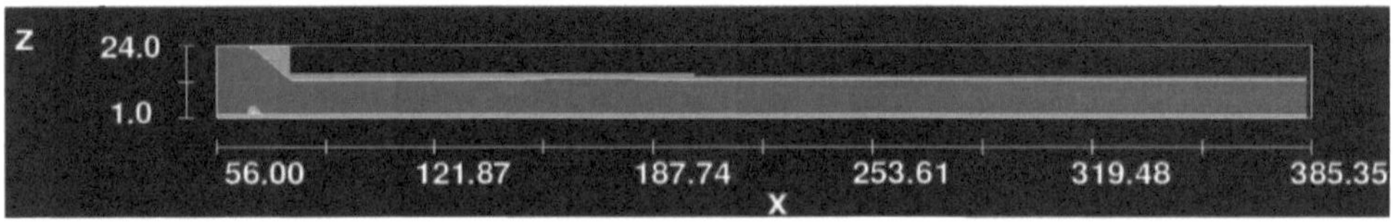

Figure 7. Sediment distribution in the front of the water diversion tunnel in condition 1 (6000 s time).

3.2 *Sediment transport rate and sediment concentration on typical cross sections*

In the water diversion process, sediment transport is the focus of this study, especially investigating whether the whole mass of sediment transport exists. Movement of large group sediments may lead to unit damage, resulting in huge economic losses. Therefore, analysis of transient characteristics of sediment transport is very important. In this paper, the characteristics of the time course of sediment movement are analyzed from the point of the instantaneous sediment transport rate of the typical cross- sections. In the study area, 4 typical cross-sections are selected, which are 1 km, 6 km, 11 km and 16 km.

Taking the working condition 1 as an example, it can be seen from Figure 8 that the sediment transport rate increases in the 1 km cross-section at 400 s time, which indicates

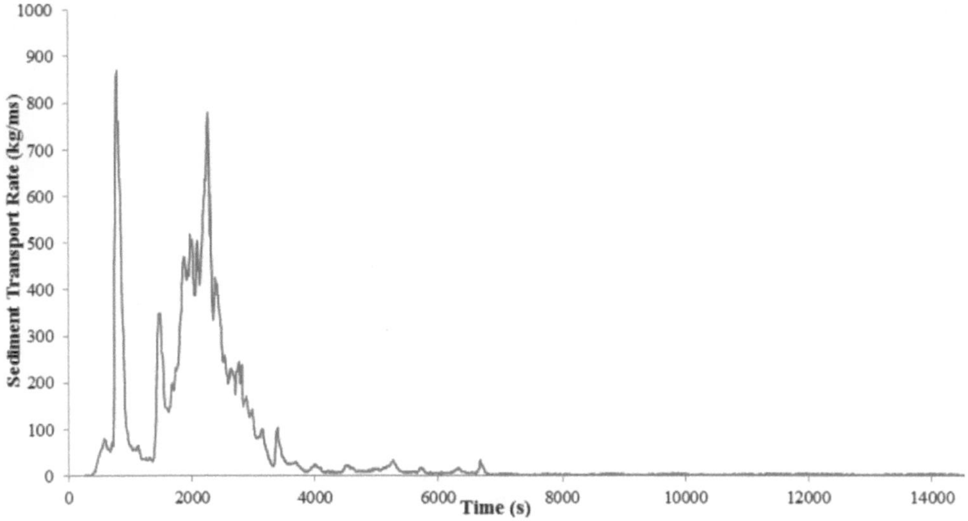

Figure 8. Sediment transport rate at 1 km section in condition 1.

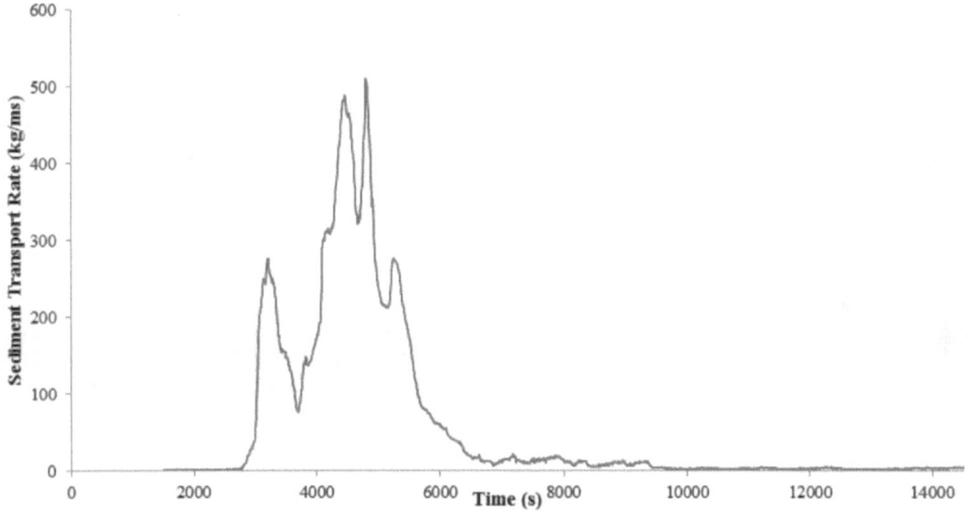

Figure 9. Sediment transport rate at 6 km section in condition 1.

that the sediment has reached the section. The sediment transport rate curve shows that the process of sediment erosion is not linear with time, but the characteristic of fluctuation. It can be clearly seen from the figure that there are two peak values in the vicinity of 800 s and 2000 s. The appearance of the peak indicates that a large amount of sediment begin to pass through the section. A higher sediment transport rate has been maintained to about 3500 s, until the 7000 s sediment rate changes as the minimum value, indicating that the upstream scour sediment has been basically all through the 1 km section. Figures 9–11 shows that other sections of the sediment transport process is similar with the 1 km cross-section. The main difference is that the peak of the sediment transport rate appears at different times.

In other conditions, the variation trend of the sediment transport rate is similar to condition 1. In the condition of a small flow rate, the time required for the sediment to pass through the tunnel should be increased. In each condition, the peak value of the sediment transport rate appears in the 1 km section. When the sediment transports to the end of the

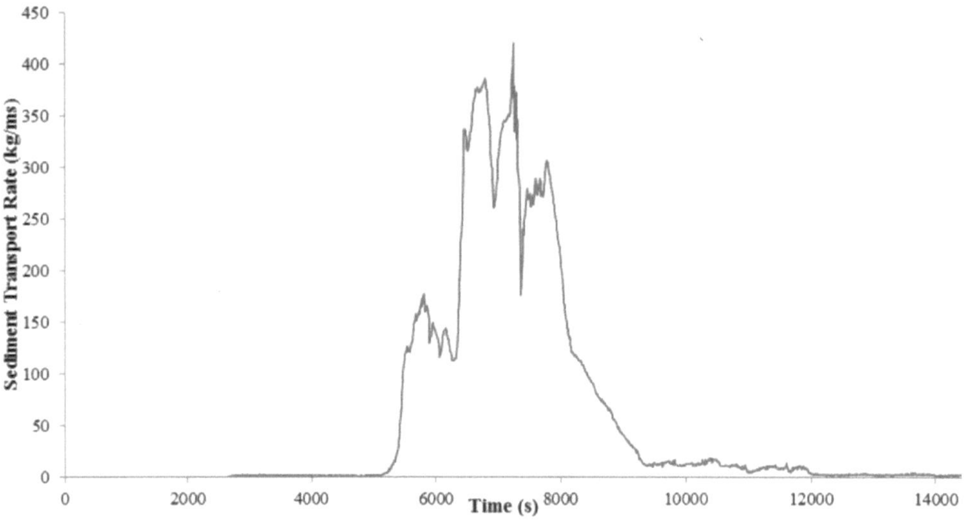

Figure 10. Sediment transport rate at 11 km section in condition 1.

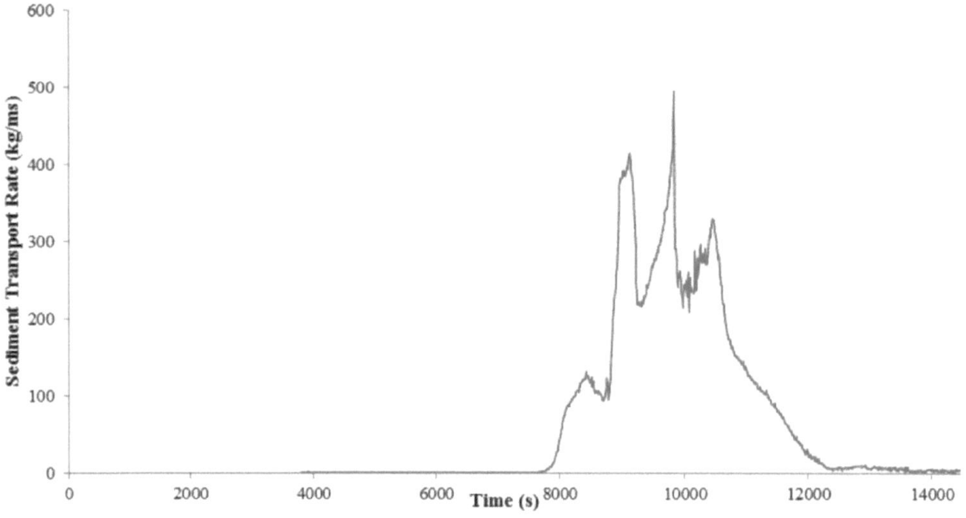

Figure 11. Sediment transport rate at 16 km section in condition 1.

tunnel, the peak value is significantly reduced. This shows that the homogenization of the sediment transport rate appears in the sediment movement process.

The calculation results of the single wide sediment transport rate and sediment concentration of each condition are shown in Table 3. The average sediment concentration is calculated from the single wide sediment transport rate divided by the vertical height and the average flow velocity. The physical meaning of it can be considered sediment content in unit volume of water when the sediment body passes through some cross-section.

The sediment transport rate increases along with the flow rate on the condition of 5 m sediment laying height, as shown in Figure 12. However, the average sediment concentration does not show the same trend with the transport rate. From the data in Table 3 and Figure 13, it can be found that with the increase in the flow rate, the sediment concentration begins to increase, but at the flow 117 m^3/s, the sediment concentration decreases abruptly, and then continues to increase. The reason for this situation is the effect of the flow velocity. The average sediment concentration is calculated by dividing the flow velocity. When the flow velocity value is less than 1, the result is enlarged.

Table 3. Sediment transport rate and average sediment concentration.

Condition number	Flow rate (m^3/s)	Average velocity (m/s)	Sediment laying height (m)	Sediment transport rate (kg/s · m)	Average sediment concentration (kg/m^3)
1	228.6	2.09	5	179	8.56
2	228.6	2.09	4	88	4.21
3	228.6	2.09	3	50	2.39
4	228.6	2.09	2	43	2.06
5	170	1.55	5	55	3.54
6	170	1.55	4	32	2.06
7	170	1.55	3	27	1.74
8	170	1.55	2	19	1.22
9	117	1.07	5	15	1.40
10	117	1.07	4	14	1.31
11	117	1.07	3	9	0.84
12	117	1.07	2	4	0.37
13	66	0.60	5	11	1.82
14	66	0.60	4	10	1.66
15	66	0.60	3	9	1.49
16	66	0.60	2	7	1.16
17	14.7	0.13	5	2	1.49
18	14.7	0.13	4	1.85	1.38
19	14.7	0.13	3	0.83	0.62
20	14.7	0.13	2	0.65	0.48

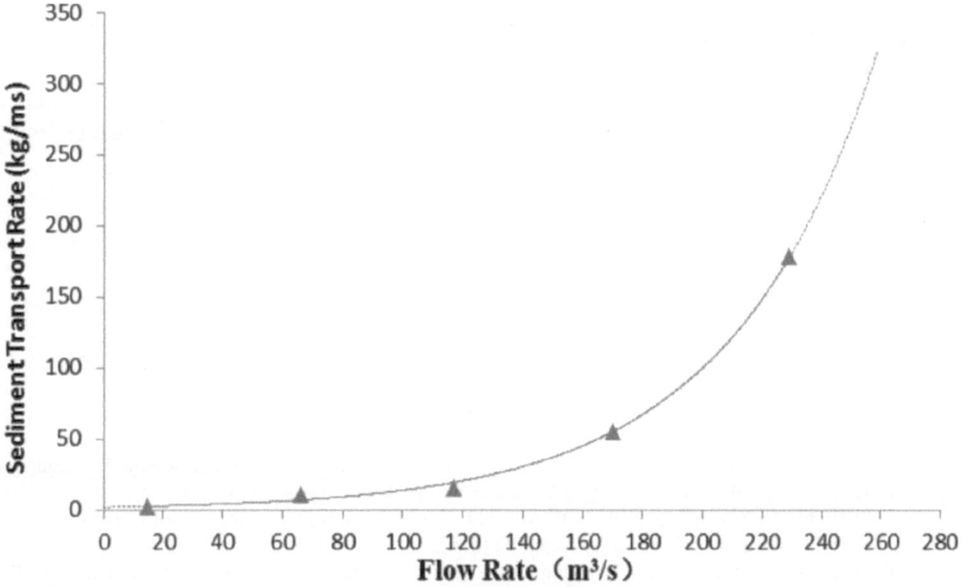

Figure 12. Relationship of the flow rate and sediment transport rate on 5 m sediment laying height.

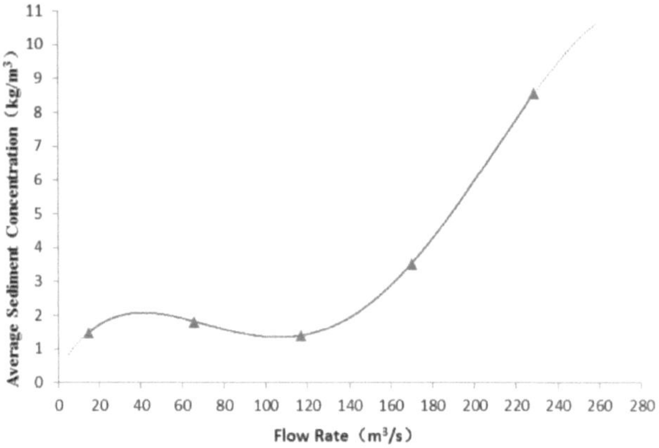

Figure 13. Relationship of the flow rate and average sediment concentration on 5 m sediment laying height.

4 CONCLUSION

In this paper, a numerical calculation method was used to simulate the sediment movement in the diversion tunnel of a hydropower station. The calculation results were analyzed, and the calculation method was considered to be in line with the actual situation.

In condition 1, the sediment erosion time and transport time were shortest, and sediment deposition in the entrance of the tunnel was washed away within two hours of time. Also, about 3.5 hours later, the sediment transported out of the tunnel. In the minimum flow rate conditions, the sediment erosion time was nearly 148 hours and the sediment discharge time was about 190 hours.

At a flow rate of 117 m^3/s, the maximum average sediment concentration in the water tunnel was 1.40 kg/m^3. When the flow rate was greater than this value, the average sediment concentration increased significantly, and such high transport strength may cause an adverse impact on station units.

ACKNOWLEDGMENTS

This study was supported by the Science and Technology of Tianjin Agricultural University (2013N13) and Open Cooperation Funds of State Key Laboratory of Hydraulics and Mountain River Engineering Sichuan University (2016gzzx1-1).

REFERENCES

Hu Xiaozhang. 2009. Mathematical simulation study of sediment intake and outfall of Dongfang power plant in Hainan at north bay. Pearl River 18(6): 18–23.

Lee Hongyuan, Chen Yenhsu, You Jiingyun, et al. 2000. Investigations of continuous bed load saltating process. Journal of Hydraulic; Engineering, ASCE 126(9): 691–700.

Lee Hongyuan, You Jiingyun, Lin Yingtien. 2002. Continuous saltating process of multiple sediment particles. Journal of Hydraulic; Engineering, ASCE 128(4): 443–450.

Li Mengguo, Shi Zhong. 2006. Numerical simulations of tidal current and sediment for the Huangdao power plant water intake project. Coastal Engineering 25(2): 13–20.

Li Yinan, Li Xianli, Yu Chunhua, et al. 2002. Model test of surface water intake of Jining canal power plant. Journal of Shandong Institute of Architecture and Engineering, 17(1): 67–71.

Zhang Xujin, Zhang Jinlin et al. 2013. Influence of reservoir sediment deposition on water-intake of power plant. Journal of Chongqing Jiaotong University. Natural Science 32(5): 1035–1039.

Zhou Fayi, Chen Bihong. 1997. 3-D simulation of the flow and sediment transport near a water intake. Journal of Hydraulic Engineering 28(12): 30–37.

Applied Mechanics and Civil Engineering VI – Xie (Ed.)
 ISBN 978-1-138-62631-7

Lattice Boltzmann modeling of two-phase flows in complex porous media considering surface hydrophobicity

Hao Chen
Applied Mechanics and Structure Safety Key Laboratory of Sichuan Province, School of Mechanics and Engineering, Southwest Jiaotong University, Chengdu, China
Department of Civil and Mechanical Engineering, University of Missouri-Kansas City, Kansas City, MO, USA

ZhiQiang Chen
Department of Civil and Mechanical Engineering, University of Missouri-Kansas City, Kansas City, MO, USA

ABSTRACT: Two-phase flows in porous media with surface hydrophobicity are an important hydro-geomechanical phenomenon in many practical applications, such as irrigation in farming land where soil hydrophobicity impacts efficiency of water use. In this study, a Lattice Boltzmann Method (LBM) is attempted to simulate 2D pore-scale two-phase flows in porous media. Assuming the availability of 2D profiling of the porous media structures (e.g. X-ray scanning), 2D digital images coupled with the bounce-back scheme are used as the input to deal with complex geometric boundaries. Based on this method, the detailed information of flow in pore spaces is obtained, and droplet or bubbles spreading and the influences of different wetting properties can be simulated successfully. The numerical method established in this work provides a promising approach to pore-scale flow simulation in porous media with complex geometrical boundaries, which could be applied to predict flow behavior in hydrophobic porous media. (e.g. soils).

1 INTRODUCTION

Global weather changes have cast significant uncertainties upon the proportion of water supplied to agriculture by precipitation (Reilly et al., 2003, Turral et al., 2011). Therefore, one of the grand challenges faced by global population is securing water for food (USAID). Soil hydrophobicity in general has adverse impacts on water use in agriculture. Particularly, soil hydrophobicity changes the flow patterns of the subsurface water by creating preferential flows (Ritsema and Dekker, 1994). In some cases, more than 80% of the percolating water is transported through preferential flows bypassing the soil matrix that may further enhance leaching or groundwater contamination. In the research community, experimental methods are usually applied for studying water percolation and preferential flows (Allaire et al., 2009). For example, Wessolek et al. experimentally showed that soil hydrophobicity could diminish flow region sand result in more significant preferential flows (Wessolek et al., 2008).

On the other hand, modeling preferential flow in general soil media is a recognized challenging problem as it involves multi-phase flows (water and air), complex geometric boundaries, possible unsaturated soil-matrix deformation, and finally soil hydrophobicity (Meakin and Tartakovsky, 2009). Therefore, few computational mechanics methods are reported for modeling preferential flows in soil (Gomez et al., 2013, Vogel et al., 2010); moreover, soil hydrophobicity is rarely found in these existing studies. From a computational mechanics view, how to simulate the fluid behavior in porous media considering complex geometric structures at the pore-scale remains a great challenge. To begin with, macroscopic approaches are often adopted to describe the flow phenomena in porous media by averaging pore-scale processes over the representative volume, such as the well-known macroscopic description of the flow based on Darcy's equation. But the evolution of the permeability during the process is required in addition to Darcy's law, and the key coefficient is commonly determined from laboratory, which makes its application much limited in a computational framework. In reality the pore-scale characteristics (geometric boundaries and surface hydrophobicity or hydrophilic properties) determine the macro-scale transport process. Therefore the understanding of flow at the pore-scale is of significant importance and is critical in the analysis of the macro-scale behavior (Chen et al., 2012, Blunt et al., 2013). In short, the pore-scale study provides a bridge between the pore-scale and macro-scale description, and it has attracted much attention in recent years.

Numerical investigation of pore-scale porous media flows requires a method that can easily handle flow within complex geometries and multi-phase fluid dynamics (such as bubble or droplet coalescence and breakup). The traditional computational fluid dynamics methods (e.g. finite volume or element based methods) hardly meet these requirements. A new and promising method to characterize multiphase flow in porous media is the Lattice Boltzmann Method (LBM). In recent years, this method has reached a high degree of maturity for solving both complex geometries and multiphase flow problems thanks to advances in modern computers. One of the attractive features of the LBM is its ability to reproduce the dynamics of an immiscible multiphase mixture naturally without making use of any interface tracking algorithms. Due to the LBM's discrete kinetic equations defined in both spatial and velocity spaces for the particle distribution functions, this intrinsic kinetic nature makes the LBM particularly suitable for multiphase flows (Lee and Lin, 2005, Yuan and Schaefer, 2006). Three popular multiphase LBM models are usually used. The first type is the color-gradient-based LBM (Gunstensen et al., 1991), which is based on the Rothman–Kellerlattice gas model (Rothman and Keller, 1988). The second type is the original Shan-Chen model (Shan and Chen, 1993). The third type is the free-energy-based LBM (Swift et al., 1995). The last one is proposed in (He et al., 1999), which uses the idea of a level set.

Among these multiphase flow LBM models, the Shan-Chen model is widely used due to its simplicity. It can easily handle fluid phases with different densities, viscosities, and more importantly, surface hydrophobicity (or wettability). The major drawbacks of the original Shan-Chen model are that it is not well-established thermodynamically and the maximum gas-liquid ratio is limited to 58 (Pan et al., 2004). The disadvantage can be overcome by incorporating different kinds of equations of state into the original Shan-Chen model (Yuan and Schaefer, 2006). Furthermore, the wettability with various contact angles of the liquid/gas to the solid surface can be conveniently implemented by changing a parameter named the virtual density of wall. Therefore, the Shan-Chen model was chosen to simulate the multiphase flow in porous media in this study. In this paper, we would focus on numerical study of two-phase flows in complex porous media as a preliminary endeavor. Particularly, emphasis is on introducing the effect of surface hydrophobicity or its opposite, surface hydrophilicity, in a LBM framework. Realistic simulation of preferential flow (e.g. finger flow) will be explored in our future efforts.

2 SHAN-CHEN LATTICE BOLTZMANN MODELING

The incompressible Shan-Chen type multiphase LBM model (Shan and Chen, 1993, Shan and Chen, 1994) is adopted to predict the two-phase flows in this work. This model involves a single relaxation time in the Bhatnagar-Gross-Krook (BGK) collision operator (Qian et al., 1992). It allows calculating the temporal and spatial evolution of the density distribution functions f_i^σ for σ number of components. Each distribution function satisfies the Lattice Boltzmann equation.

$$f_i^\sigma(x + c_i\Delta t, t + \Delta t) - f_i^\sigma(x, t) = -\Delta t / \tau^\sigma [f_i^\sigma(x, t) - f_i^{\sigma,eq}(x, t)] \tag{1}$$

where Δt is the time step, $f_i^\sigma(x,t)$ is the density distribution function of the σth component in direction c_j at $x \cdot f_i^{\sigma,eq}(x,t)$ is the equilibrium function of the σth component. τ^σ is the relaxation time of the σth component characterizing the collision processes, by which the distribution functions relax towards their equilibrium distributions.

The equilibrium distribution function $f_i^{\sigma,eq}(x,t)$ depends only on local density and velocity and can be chosen as having the following form,

$$f_i^{\sigma,eq} = \omega_i \rho^\sigma \left[1 + (c_i \cdot u^{\sigma,eq}) / c_s^2 + (c_i \cdot u^{\sigma,eq})^2 / 2c_s^4 - (u^{\sigma,eq} \cdot u^{\sigma,eq}) / 2c_s^2\right] \quad (2)$$

In the two-dimensional nine-velocity model the discrete velocities are,

$$[c_0, c_1, c_2, c_3, c_4, c_5, c_6, c_7, c_8] = c \begin{bmatrix} 0 & 1 & 0 & -1 & 0 & 1 & -1 & -1 & 1 \\ 0 & 0 & 1 & 0 & -1 & 1 & 1 & -1 & -1 \end{bmatrix} \quad (3)$$

And the weight factors $\omega_i = 4/9 (i = 0)$, $\omega_i = 1/9 (i = 1, 2, 3, 4)$, $\omega_i = 1/36 (i = 5, 6, 7, 8)$, and $c_s = c/\sqrt{3}$, where $c = \Delta x / \Delta t$ is the ratio of lattice spacing Δx and time step Δt.

The mass density ρ^σ and momentum density $\rho^\sigma u^\sigma$ of each component fluid are calculated for each node via the following equations.

$$\rho^\sigma = \sum_i f_i^\sigma \quad (4)$$

$$\rho^\sigma u^\sigma = \sum_i c_i f_i^\sigma \quad (5)$$

The relaxation time tunes the kinematic viscosity of the σth component by

$$\nu^\sigma = (\tau^\sigma - 1/2) c_s^2 \Delta t \quad (6)$$

The interactions between gas, liquid and skeleton control the flow and distribution of fluids in porous media (Xu et al., 2005, Saraji et al., 2010, Genty and Pot, 2013). In the Shan-Chen multiphase model, the fluid-fluid interaction, which is proportional to the product of a function of the particle number densities, can be written as,

$$F_l^\sigma(x) = -G_c \rho^\sigma(x,t) \sum_i \omega_i \rho^{\bar{\sigma}}(x + c_i \Delta t, t) c_i \quad (7)$$

where $F_l^\sigma(x)$ is the fluid-fluid interaction force, and the strength of two-phase interaction is denoted by G_c which is negative and represents the attraction between the two phases. σ and $\bar{\sigma}$ denote two different fluid components, respectively. It is simple to describe the interaction between a fluid and a wall by introducing an extra interaction force, and this method is first used (Martys and Chen, 1996). The idea is to create an analogue to the particle-particle interaction force used to induce phase separation, and it is described as,

$$F_{ads}^\sigma(x) = -G_{ads}^\sigma \rho^\sigma(x,t) \sum_i \omega_i s(x + c_i \Delta t, t) c_i \quad (8)$$

where s = 0 or 1 for the fluid and the solid phases, respectively. The interaction strength between each fluid and solid surface was adjusted by the parameter G_{ads}^σ. In the simulation, the hydrophilic and hydrophobic regions of the solid particles have different adhesion parameter.

The macroscopic velocity $u^{\sigma,eq}$ of the σth component in the multiphase model is defined by

$$u^{\sigma,eq} = u' + \frac{\tau^\sigma (F_l^\sigma + F_{ads}^\sigma + G)}{\rho^\sigma} \quad (9)$$

where G a steady body force and u' is the lattice velocity, defined as

$$u' = \frac{\sum_\sigma \frac{1}{\tau^\sigma} \sum_i c_i f_i^\sigma}{\sum_\sigma \frac{1}{\tau^\sigma} f_i^\sigma} \quad (10)$$

The averaged fluid momentum before and after collision is

$$\rho u = \rho u^{\sigma} + \frac{1}{2}\sum_{\sigma} F^{\sigma} \tag{11}$$

where u is the overall fluid velocity, and $\rho = \sum_{\sigma} \rho^{\sigma}$ is the total mass density of the whole fluid. The pressure of the whole fluid is given by

$$P(x) = c_s^2 \sum_{\sigma} \rho^{\sigma}(x) + \frac{G_c}{3} \rho^{\sigma}(x) \rho^{\bar{\sigma}}(x) \tag{12}$$

The Shan-Chen model modifies the equilibrium velocity in the collision operator to include an interactive force, and thus it can realize phase separation and surface tension effects. The wettability of the solid surface can be obtained by adjusting the liquid-solid interaction. Different contact angles are obtained through adjusting G_{ads}^{σ}, which introduces the hydrophilic surface condition and the hydrophobic surface condition.

3 NUMERICAL RESULTS AND DISCUSSION

Numerical simulation for two-phase flows driven by the gravity in porous media is carried out to test the validity of the LBM model. Porous structure generation is not a topic of our concern herein. The porous medium in our simulation is the same as that in the literature of (Arab et al., 2009) as shown in Fig. 1, wherein the white zones are flow domains and the black are the solids.

Furthermore, binary processing of the image is adopted to obtain geometric data of the simulation. We marked the solid site of value 1 and the flow domain as 0 in the binary image. Based on the data, the bounce-back scheme is easily accepted to deal with the complex geometry boundary of porous media, and the Shan-Chen type LBM model is implemented using the open source solver Palabos (FlowKit Ltd., 2016). In the numerical example, the full domain includes 1042 × 512 pixels, which are used to form the 1042 × 512 lattice units. For the LBM part, the relaxation basic time unit is set to be 1 ms. The upper boundary is applied with the velocity boundary condition, the lower boundary is applied with non-slip boundary condition, and the left and right boundaries are applied with periodic boundary condition.

Initially, the liquid phase is set to lie in upper half of the domain and the gas phase lies in lower half. The body force is gravity only, which only acts on the liquid phase. Fig. 2 presents contours of density field of liquid phase at three different time, which show that the liquid movement in the porous media with surface hydrophilic (Fig. 2a, b, c) and hydrophobic properties (Fig. 2d, e, f), respectively. It is found that small droplet and bubbles can be generated in both cases, which is a common phenomenon in gas-liquid two-phase flows. For the hydrophilic surface condition, the gas-liquid interface moves slowly. Due to the adhesion force caused by the hydrophilic porous structure, the gas-liquid interface just reaches the bottom

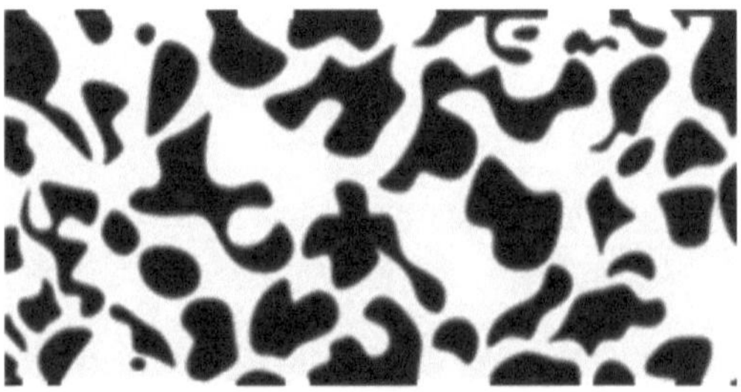

Figure 1. Image of 2D porous media [Courtesy of (Arab et al., 2009)].

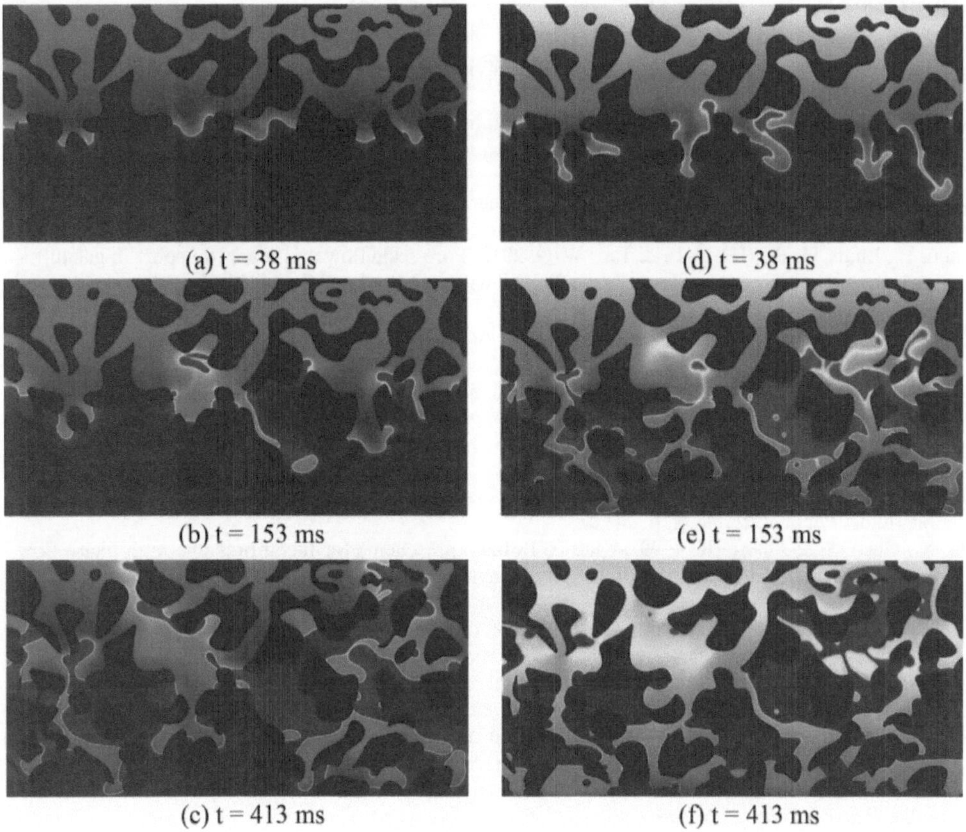

Figure 2. Evolution of density field of liquid phase: (a), (b) and (c) in hydrophilic condition; (d), (e) and (f) in the hydrophobic condition.

after 413 ms, as shown in Fig. 2(c). For the hydrophobic surface condition, the gas-liquid interface moves fast. After 153 ms the gas-liquid interface reaches the bottom, as shown in Fig. 2(e). And after 413 ms the lower half domain is mostly occupied by the liquid phase, as shown in Fig. 2(f). The gas-liquid interface moves fast because the adhesion force is not dominant under the hydrophobic surface condition and the liquid is not attached to the porous structure any more.

4 CONCLUSIONS

In this paper Shan-Chen type LBM is presented in view of application for the two-phase flows in complex porous media on the pore scale. The method is coupled with 2D digital image to deal with the complex porous structure. In numerical example the LBM model is proved to have a good advantage in simulating two-phase flows in very complex porous media due to its particulate nature and simplicity. The effects of hydrophobicity and hydrophilicity are tested in the numerical experiment, which shows obvious influences on the two-phase flows in complex pore space.

ACKNOWLEDGEMENTS

Funding from a University of Missouri System's Intercampus Interdisciplinary Grant and a USDA-NIFA Grant (Award No. 2015-68007-23214) for both authors is greatly appreciated.

REFERENCES

Allaire, S.E., Roulier, S. & Cessna, A.J. 2009. Quantifying preferential flow in soils: A review of different techniques. *Journal of Hydrology*, 378, 179–204.

Arab, M.R., Semma, E., Pateyron, B. & El ganaoui, M. 2009. Determination of physical properties of porous materials by a lattice Boltzmann approach. *FDMP: Fluid Dynamics & Materials Processing*, 5, 161–176.

Blunt, M.J., Bijeljic, B., Dong, H., Gharbi, O., Iglauer, S., Mostaghimi, P., Paluszny, A. & Pentland, C. 2013. Pore-scale imaging and modelling. *Advances in Water Resources*, 51, 197–216.

Chen, L., Luan, H.-B., HE, Y.-L. & Tao, W.-Q. 2012. Pore-scale flow and mass transport in gas diffusion layer of proton exchange membrane fuel cell with interdigitated flow fields. *International Journal of Thermal Sciences*, 51, 132–144.

Flowkit Ltd. 2016. *Palabos, CFD, Complex Physics* [Online]. Available: http://www.palabos.org/ [Accessed].

Genty, A. & Pot, V. 2013. Numerical simulation of 3D liquid–gas distribution in porous media by a two-phase TRT lattice Boltzmann method. *Transport in porous media*, 96, 271–294.

Gomez, H., Cueto-Felgueroso, L. & Juanes, R. 2013. Three-dimensional simulation of unstable gravity-driven infiltration of water into a porous medium. *Journal of Computational Physics*, 238, 217–239.

Gunstensen, A.K., Rothman, D.H., Zaleski, S. & Zanetti, G. 1991. Lattice Boltzmann model of immiscible fluids. *Physical Review A*, 43, 4320.

He, X., Chen, S. & Zhang, R. 1999. A lattice Boltzmann scheme for incompressible multiphase flow and its application in simulation of Rayleigh–Taylor instability. *Journal of Computational Physics*, 152, 642–663.

Lee, T. & Lin, C.L. 2005. A stable discretization of the lattice Boltzmann equation for simulation of incompressible two-phase flows at high density ratio. *Journal of Computational Physics*, 206, 16–47.

Martys, N.S. & Chen, H. 1996. Simulation of multicomponent fluids in complex three-dimensional geometries by the lattice Boltzmann method. *Physical review E*, 53, 743.

Meakin, P. & Tartakovsky, A.M. 2009. Modeling and simulation of pore-scale multiphase fluid flow and reactive transport in fractured and porous media. *Reviews of Geophysics*, 47.

Pan, C., Hilpert, M. & Miller, C.T. 2004. Lattice-Boltzmann simulation of two-phase flow in porous media. *Water Resources Research*, 40.

Qian, Y., D'Humières, D. & Lallemand, P. 1992. Lattice BGK models for Navier-Stokes equation. *EPL (Europhysics Letters)*, 17, 479.

Reilly, J., Tubiello, F., Mccarl, B., Abler, D., Darwin, R., Fuglie, K., Hollinger, S., Izaurralde, C., Jagtap, S. & Jones, J. 2003. US agriculture and climate change: new results. *Climatic Change*, 57, 43–67.

Ritsema, C.J. & Dekker, L.W. 1994. How water moves in a water repellent sandy soil: 2. Dynamics of fingered flow. *Water Resour. Res.*, 30, 2519–2531.

Rothman, D.H. & Keller, J.M. 1988. Immiscible cellular-automaton fluids. *Journal of Statistical Physics*, 52, 1119–1127.

Saraji, S., Goual, L. & Piri, M. 2010. Adsorption of asphaltenes in porous media under flow conditions. *Energy & Fuels*, 24, 6009–6017.

Shan, X. & Chen, H. 1993. Lattice Boltzmann model for simulating flows with multiple phases and components. *Physical Review E*, 47, 1815.

Shan, X. & Chen, H. 1994. Simulation of nonideal gases and liquid-gas phase transitions by the lattice Boltzmann equation. *Physical Review E*, 49, 2941.

Swift, M.R., Osborn, W. & Yeomans, J. 1995. Lattice Boltzmann simulation of nonideal fluids. *Physical Review Letters*, 75, 830.

Turral, H., Burke, J.J. & Faurès, J.M. 2011. *Climate change, water and food security*, Food and Agriculture Organization of the United Nations Rome.

Vogel, T., Brezina, J., Dohnal, M. & Dusek, J. 2010. Physical and numerical coupling in dual-continuum modeling of preferential flow. *Vadose Zone Journal*, 9, 260–267.

Wessolek, G., Schwärzel, K., Greiffenhagen, A. & Stoffregen, H. 2008. Percolation characteristics of a water-repellent sandy forest soil. *European Journal of Soil Science*, 59, 14–23.

Xu, Y., Liu, Y. & Huang, G. 2005. Lattice Boltzmann simulation of momentum and energy transfer in a porous medium. *Modern Physics Letters B*, 19, 1531–1534.

Yuan, P. & Schaefer, L. 2006. Equations of state in a lattice Boltzmann model. *Physics of Fluids (1994-present)*, 18, 042101.

Applied Mechanics and Civil Engineering VI – Xie (Ed.)
© 2017 Taylor & Francis Group, London, ISBN 978-1-138-62631-7

Optimization of primary support parameters for the main structure of the tunnel of Beijing

Ruilang Cao & Lifeng Zheng
China State Key Laboratory of Simulation and Regulation of Water Cycle in River Basin, China Institute of Water Resources and Hydropower Research, Beijing, China

Xingzu Xue
JiLin Province Water Resource and Hydropower Consultative Company of P.R. China, Changchun, China

Yufei Zhao
China State Key Laboratory of Simulation and Regulation of Water Cycle in River Basin, China Institute of Water Resources and Hydropower Research, Beijing, China

ABSTRACT: On the basis of the main structure of the tunnel of Beijing metro line 14, this study focused on the optimization of tunnel support parameters under complicated geological conditions. This study first evaluates the engineering geological conditions related to construction risk. Then, it applies finite difference software FLAC3D in the simulation of tunnel excavation. In the process of Cross Diaphragm (CRD) excavation, the characteristics of the surrounding rocks with and without supporting structures were analyzed. The length and spacing of anchors, row spacing, and the thickness of spray layer were set as variables in this study. The optimum scheme was obtained by the orthogonal tests. The results not only provide technical guidance for the construction of Beijing metro line 14, but will also provide valuable references for similar future projects.

1 INTRODUCTION

With the development of the society, the exploitation and utilization of the underground space are increasing, which simultaneously bring some new geotechnical problems. Large-scale land subsidence induced by the construction of urban metro tunneling systems can lead to the damage of road surface, buried pipelines, and buildings in the area (Richard, 2006). It may also bring great loss of the life and property of people, especially in densely populated areas (Shin, 2006). Therefore, it is a critical problem that should be solved.

The following are some problems and risks that may occur in tunnel projects under different construction conditions:

1. Complicated geological conditions with weak water strata;
2. Large cross-sectional excavation;
3. Intensive buildings and pipelines nearby.

However, previous studies have shown that research on the tunnel projects under these construction conditions is relatively rare (Alessio, 2007). The tunnel of Beijing metro line 14 is a typical engineering case in which studies on the construction methods and main design parameters are necessary (Attewell, 1986). Based on the theoretical studies and numerical simulation, this study discussed the optimization of the tunnel support parameters. The results will be useful for reducing the project cost.

2 PROJECT BACKGROUND

2.1 *Project profile*

Beijing subway line 14 was an "L"-shaped backbone line of Beijing subway network, throughout the south and east of Beijing. It was one of the important transport hubs connecting the two directions of the city. This study analyzed the 11th interval of the project (Figure 1). The main construction method applied in the project interval was mine construction method, and the conditions of site environment were as follows:

1. The surface buildings along the subway line were generally one- or two-storey houses and warehouses. There were high-rise and multi-storey residential buildings about 18–20 m away from the tunnel, such as six 18-storey residential buildings and two 6-storey residential buildings.
2. This interval mainly crossed under one road, which was about 45 m wide and with high traffic flow. Along the line, there was a bridge measuring 70.0 m × 50.0 m across the river. The existing road and bridge were the influencing factors considered in the designing and construction process.
3. There was a river about 50 m wide and 6.30 m deep with water depth of 0.60–1.00 m in the south of the site. The river edge was 10–12 m away from the interval. The tunnel which closed to the river bank and the bottom was reinforced by cement solid lining and ecological lining, respectively. The water exchange between the river water and the groundwater had an important impact on the project.

Thus, it was extremely necessary to control ground surface settlement in the site during the construction to reduce its influence on the surrounding buildings, the road, and the bridge.

2.2 *Project geological condition*

In the urban areas of Beijing, the Quaternary river alluvium was thick in the west while thin in the east, and the landscape was transformed from mountain to plain in a step-by-step manner. The west urban area of Beijing was just on the top and the middle-upper part of the alluvial-proluvial fan, which was formed from the Quaternary ancient river alluvium.

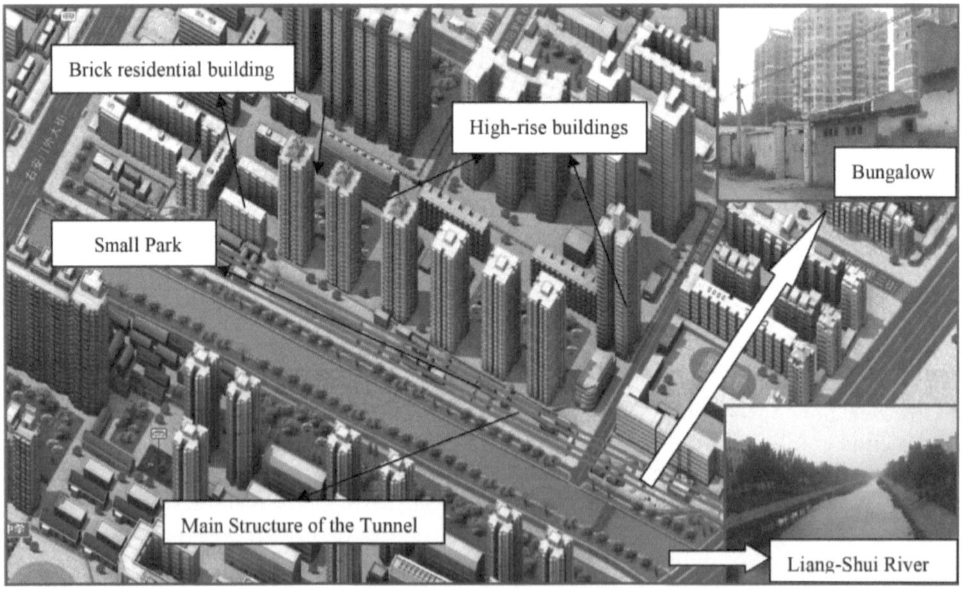

Figure 1. Sketch map of typical terrain features along the subway line.

Table 1. Physical and mechanical parameters of the soil strata.

Soil strata	Elastic modulus (MPa)	Poisson ratio	Severe ($kN \cdot m^{-3}$)	Cohesion (kPa)	Friction angle (°)
Filling	800	0.35	1700	5	20
Silty clay	1600	0.32	2000	20	23
Fine sand	1200	0.28	1800	10	21
Pebble	2000	0.30	2000	10	27
Conglomerate	2500	0.32	2200	40	35

The parts were mainly made up of thick sandy stratum and gravel stratum. The middle and middle-lower parts of the alluvial proluvial fan were mainly made up of interactive sediments consisting of clay, silt, sand, and gravel when crossing from west to east. The physical and mechanical parameters of soil are shown in Table 1.

The annual dynamic variation of the phreatic water level in this site was as follows: The groundwater level was generally high from November to March and modest in other months. The annual amplitude of variation was usually 1–2 m, and the maximum was 3 m. The hydrogeological investigation indicated that the surface water from the river in the site showed no obvious supplementation of seepage to the groundwater. This meant that in normal conditions with no exceptional cases such as the influence of construction, there was no distinct hydraulic connection between the river water and the phreatic water.

3 NUMERICAL CALCULATION

3.1 *Network model*

According to Saint-Venant's principle, the influence that the tunnel excavation had on the surrounding rock was inversely proportional to the distance between the surrounding rock and the center of the tunnel excavation (Sharma, 2001). In general, the variation in stresses was less than 5% when the distance from the edge of the model to the midline was three times larger than the diameter of the tunnel. When the distance was five times larger than the diameter, the variation was less than 1%. Taking into account all the factors including geological conditions, section size, and construction methods, the model dimension was designed to be five to seven times the diameter of the tunnel laterally, five to seven times the diameter of the tunnel under the surface, and 32 m longitudinally. The model was divided into 48960 elements and had 53907 nodes. It employed the eight-node hexahedral zone elements to simulate the surrounding rock, secondary lining, and temporary support, and the shell elements to simulate the primary support and the cable elements to simulate the feet-lock bolt. The model and mesh are shown in Figure 2.

3.2 *Model parameter*

Combined lining structure was applied in tunnel construction, and the supporting structure parameters are shown in Tables 2 and 3.

3.3 *Preliminary result*

The numerical calculation results showed that in the process of CRD excavation, after finishing the excavation of steps, the maximum principal stress of the surrounding rock around the free face was compressive, and the distribution of the minimum principal stress of the surrounding rock is shown in Figure 3. Stress concentration occurred in some small parts of the surrounding rock, which lay on the upper arch spring of the steps.

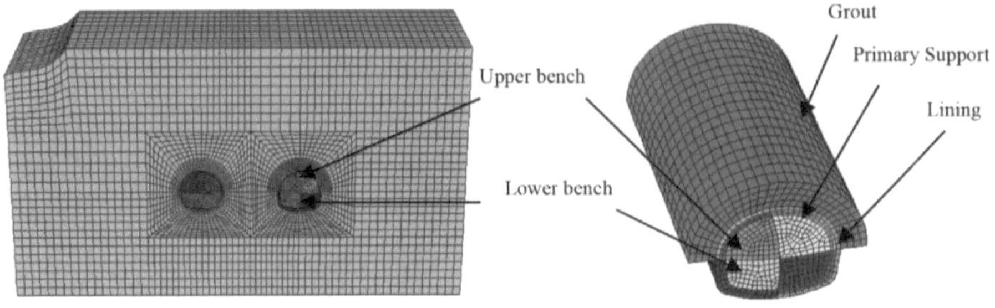

Figure 2. Numerical model in FLAC^{3D}.

Table 2. Supporting structure parameters.

Supporting structure	Elastic modulus (GPa)	Poisson ratio	Compressive strength (MPa)	Tensile strength (MPa)
Primary support	25.5	0.25	12.5	1.3
Lining	33.5	0.20	20	1.8

Table 3. Parameters of the feet-lock bolt.

Type	Diameter (mm)	Elastic modulus (GPa)	Compressive strength (MPa)	Tensile strength (MPa)
HRB335	42.3	210	410	540

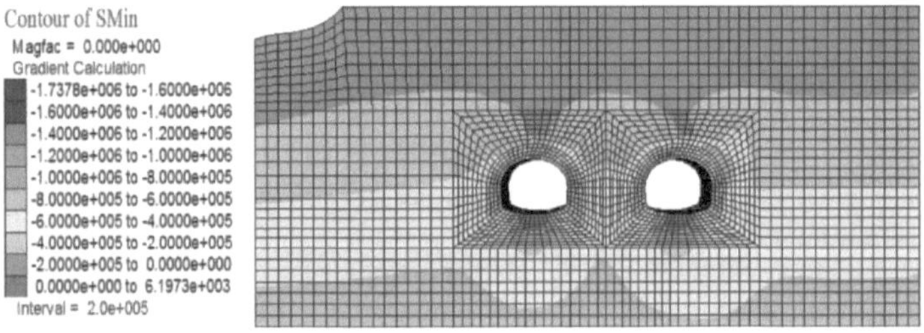

Figure 3. Contour of the minimum principal stress of the soil strata.

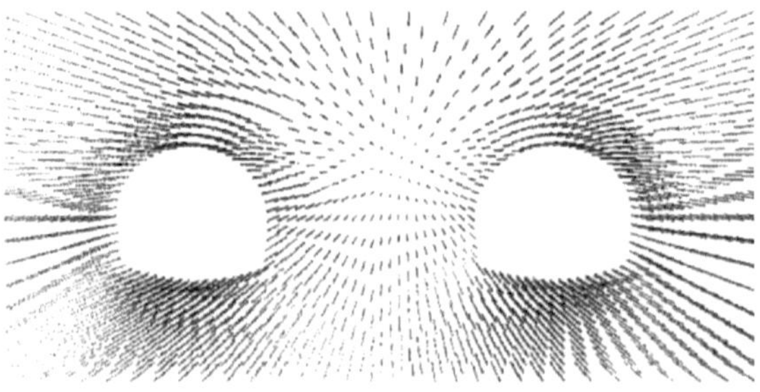

Figure 4. Vector diagram of the displacement of the surrounding rocks.

In excavation, the surrounding rock of the tunnel sprang back to the free face (Figure 4). In the process of construction, the surrounding rock close to the top part of arch emerged the trend of settlement, especially in the center of the vault. With the advance of the tunnel face, vault subsidence deformation increased and formed the cave arch gradually.

4 SUPPORT PARAMETER OPTIMIZATION ANALYSIS

4.1 *Design of the orthogonal testing scheme*

Orthogonal test was used for the optimization of support parameters of the tunnel structure and the secondary lining as safety stock in the process of analysis. The proposed optimal support parameters included anchor length, anchor spacing, row spacing, and the thickness of spray layer. First, a number of numerical calculation were done based on the orthogonal testing scheme, by employing range analysis and variance analysis, the influence and significance of the results could be tested, and then by comparing with the original design, the support parameters were modified. The flow diagram of optimization analysis is shown in Figure 5.

4.2 *Parameter optimization analysis*

On the basis of the statistic testing scheme and numerical simulation results, the surface settlement of the tunnel was counted as an evaluating index, and the results of orthogonal tests are shown in Table 4.

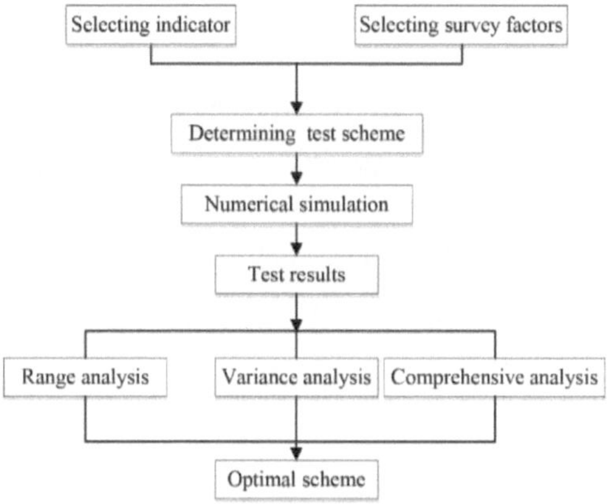

Figure 5. Flow diagram of optimization analysis.

Table 4. Results of orthogonal tests.

Number	Anchor length (m)	Anchor spacing (m)	Row spacing (m)	Thickness of spray layer (cm)	Displacement (mm)
–	1	2	3	4	–
1	3	0.8	0.8	20	11.323
2	3	1	1	25	12.113
3	3	1.2	1.2	30	13.421
4	3	1.5	1.5	35	13.936
5	4	0.8	1	35	8.783
6	4	1	0.8	30	9.245
7	4	1.2	0.5	25	12.720
8	4	1.5	1.2	20	14.252
9	4.5	0.5	1.2	25	11.210
10	4.5	1	1.5	20	11.165
11	4.5	1.2	0.8	35	10.863
12	4.5	1.5	1	30	10.838
13	5	0.8	1.5	30	10.809
14	5	1	1.2	35	10.472
15	5	1.2	1	20	11.161
16	5	1.5	0.8	25	10.951

Table 5. Results of range analysis.

Reasons	Kl	K2	K3	K4	k1	k3	k3	k4	Range
A	50.792	45.001	44.077	43.393	12.698	11.250	11.019	10.848	1.850
B	42.125	42.995	48.167	49.976	10.531	10.749	12.042	12.494	1963
C	42.384	42.894	49.354	48.630	10.596	10.724	12.339	12.158	1.743
D	47.901	46.993	44.313	44.056	11.975	11.748	11.078	11.014	0.961

Choosing the surface settlement as the evaluating index and observing four variables, including anchor length (F_A), anchor spacing (F_B), row spacing (F_C), and the thickness of spray layer (F_D), for range analysis and sensibility analysis, the results of range analysis are presented in Table 5.

The steps of variance analysis are as follows:

1. Calculate sum of squares of deviations:

$$T=\sum_{i=1}^{16} S_i = 11.232+12.113+\ldots+10.951=183.262 \tag{1}$$

$$P=T^2/n=2099.069 \tag{2}$$

$$Q=\sum_{i=1}^{16} S_i^2 = 11.232^2+12.113^2+\ldots+10.951^2=2133.58 \tag{3}$$

$$SS_T=Q-P=34.511 \tag{4}$$

$$\left(SS_A\ SS_B\ SS_C\ SS_D\ SS_e\right)=\begin{pmatrix} K_{A1}^2 & K_{A2}^2 & K_{A3}^2 & K_{A4}^2 \\ K_{B1}^2 & K_{B2}^2 & K_{B3}^2 & K_{B4}^2 \\ K_{C1}^2 & K_{C2}^2 & K_{C3}^2 & K_{C4}^2 \\ K_{D1}^2 & K_{D2}^2 & K_{D3}^2 & K_{D4}^2 \\ K_{e1}^2 & K_{e2}^2 & K_{e3}^2 & K_{e4}^2 \end{pmatrix}\begin{pmatrix} 0.25 \\ 0.25 \\ 0.25 \\ 0.25 \\ 0.25 \end{pmatrix}=(8.580\ 11.105\ 10.187\ 2.774\ 1.864) \tag{5}$$

2. Calculate mean square:

$$\left(MS_A\ MS_B\ MS_C\ MS_D\ MS_e\right)=\left(SS_A\ SS_B\ SS_C\ SS_D\ SS_e\right)\begin{pmatrix} df_A \\ df_B \\ df_C \\ df_D \\ df_e \end{pmatrix}=(2.867\ 3.702\ 3.396\ 0.925\ 0.621) \tag{6}$$

3. Calculate the value of F:

$$\left(F_A\ F_B\ F_C\ F_D\right)=\left(MS_A\ MS_B\ MS_C\ MS_D\right)\begin{pmatrix} \frac{1}{MS_e^{\Delta}} \\ \frac{1}{MS_e^{\Delta}} \\ \frac{1}{MS_e^{\Delta}} \\ \frac{1}{MS_e^{\Delta}} \end{pmatrix}=(4.605\ 5.961\ 5.469\ 1.490) \tag{7}$$

Table 6. Analysis of variance for orthogonal tests.

Reasons	SS	df	MS	F	Significance
A	8.580	3	2.860	4.605	–
B	11.105	3	3.702	5.961	**
C	10.181	3	3.396	5.469	*
D	2.774	3	0.925	1.490	–
Error	1.864	3	0.621	–	–

4. Test the value of F:
   The results of variance analysis are shown in Table 6, in which the critical values of $F_{0.05}$ (3, 6) and $F_{0.01}$ (3, 6) were 4.76 and 9.78, respectively. Because $F_{0.05} < F_C < F_B$, when $\alpha = 0.05$, both anchor spacing and row spacing had significant effects on the surface settlement.

It can be observed from the results of analysis above that:

1. The order of influence of four factors on the surface settlement is B > A > C > D.
2. On the basis of the results of sensibility analysis, it can be concluded that the influence of F_B on the surface settlement was significant. The influence of Fc was more significant than F_A. However, this result was inconsistent with the results of range analysis. By adjusting the optimal scheme, the results of sensibility analysis were more important. When F_A ranged from A_1 to A_2, the surface settlement was significantly changed. While F_A ranged from A_2 to A_3 and A_4, the surface settlement changed a little. Therefore, from an economic viewpoint, A_2 was the priority. The effect of F_B on the surface settlement was significant. Thus, reducing the anchor spacing contributed to forming bearing arches on excavation surface and limiting the surface settlement. By contrast, the effect of F_D on the surface settlement was not so significant.

In conclusion, when applying the surface settlement as the evaluating index, the optimal scheme was $A_2B_2C_2D_3$, which indicated that the anchor length was 4 m, the anchor spacing and the row spacing were both 1 m, and the thickness of spray layer was 30 cm.

5 CONCLUSION

Taking the tunnel of Beijing subway line 14 as an example, in this paper, we analyzed the optimization of support parameters of the tunnel by numerical simulation and orthogonal test. The optimal scheme shown by the results was as follows: the anchor length was 4 m, the anchor spacing and the row spacing were both 1 m, and the thickness of the spray layer was 30 cm.

ACKNOWLEDGMENT

This work was supported by the China Railway First Survey & Design Institute project (2014-93-08-01) and the National Key Research and Development Project (2016YFC0401801, 2016YFC0401804).

REFERENCES

Alessio C., Annamaria C., Glancarlo G. 2007. Numerical Evaluation of the Surface Displacements due to Soil Grouting and to Tunnel Excavation. *International Journal of Geomechanics* 7(3): 217–226.

Attewell P.B., Yeates J., Selby A.R. 1986. Soil movements induced by tunneling and their effects on pipelines and structures. *New York Chapman and Hall* 1986: 151–179.
Richard J., Finno, Frank T., et al. 2006. Evaluating damage potential in building affected by excavations. *Journal of Geotechnical and Geoenvironmental Engineering* 131(10): 1191–1210.
Sharma J.S., Hefny A.M., Zhao J., et al. 2001. Effect of large excavation on deformation of adjacent MRT tunnels. *Tunnelling and Underground Space Technology* 16(2): 93–98.
Shin H.S., Kim C.Y., Kimk Y. 2006. A new strategy for monitoring of adjacent structures to tunnel construction. *Tunnelling and Underground Space Technology* 5(21): 461–462.

Applied Mechanics and Civil Engineering VI – Xie (Ed.)
 ISBN 978-1-138-62631-7

Processes of storm-induced fluid mud formation in open muddy channels

Q.X. Pang, R.B. Zhang & C.P. Wen
Tianjin Research Institute of Water Transport Engineering, National Engineering Laboratory for Port Hydraulic Construction Technology, Key Laboratory of Engineering Sediment of Ministry of Transport, Tianjin, P.R. China

ABSTRACT: Fluid mud frequently forms in muddy channels after typhoon, but the processes of formation still need to be studied further, especially based on the field investigation. Hence, the processes of storm-induced fluid mud formation in the channel of Lianyungang Harbor, China were analyzed, mainly based on the experiment of fluid mud motion in flume and field surveyed data during typhoon Wipha. Based on the analysis, the results were obtained as follows. Strong waves last sufficient time, more than 18 hours during typhoon Wipha, for example, to lift enough sediments from shoals to form a high silt-laden layer with a height of 0.5 m near the bottom. Then, the sediment in the high silt-laden layer moves as density flows and pours directly into the channel to be fluid mud. In fact, parts of the suspended load settled in the channel also help in the formation of fluid mud.

1 INTRODUCTION

Fluid mud appears to be a common, perhaps even ubiquitous, feature of water bodies laden with fine-grained sediments (McAnally et al. 2007). In China, there are also many muddy harbors where fluid mud exists, such as Tianjin Harbor, Lianyungang Harbor, Guangzhou Harbor, etc. Fluid mud is a special existing status of mud, typically with a density ranging between 1030 and 1250 kg/m^3. Storm is often a main hydrodynamic factor that induces fluid mud in open muddy coasts, such as the Eel River continental shelf (Traykovski et al. 2002), the head of the Eel submarine canyon (Puig et al. 2004) and the Tai Shan port (Pang & Yang, 2012). However, the processes of fluid mud formation still need to be studied further, especially based on the field investigation during a storm. The field data surveyed during a storm are very rare. For example, the phenomenon of fluid mud had appeared many times in the channel of Lian Yungang harbor, China, but there was no corresponding hydrological observation. On the contrary, the hydrographic surveys in this area were also conducted several times, even lasting three months, but the channel did not form fluid mud during the entire survey.

This paper focuses on the processes of storm-induced fluid mud formation in the channel, mainly based on the experiment of fluid mud motion in laboratory and field surveyed data during typhoon Wipha, including the wave, current flow and Suspended Sediment Concentration (SSC) surveyed synchronously during the storm, and the thickness and density of fluid mud investigated after the storm.

2 METHODS

2.1 *Experiments of fluid mud motion*

The experiments were conducted in a long flume, focusing on the effect of fluid mud density, bed slope, and supply amount on the fluid mud motion. Three types of slopes of a muddy bed were built, namely the horizontal, 1:100 and 1:10, respectively, and several types of

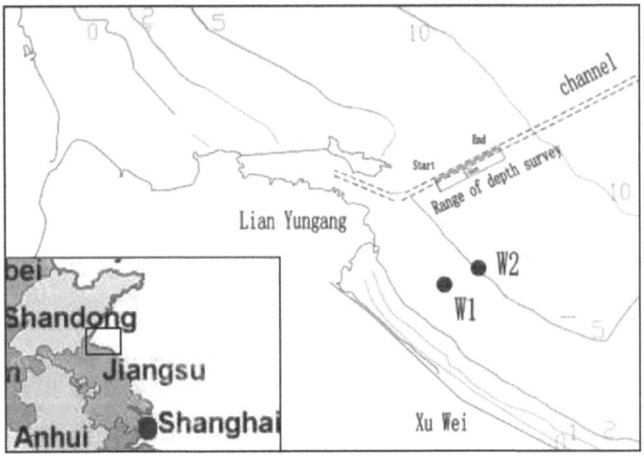

Figure 1. The bathymetric map of the Lian Yungang sea area showing the location of the channel and the range of depth survey before and after typhoon Wipha, and the stations W1 and W2 for the hydrological survey.

densities of fluid mud from 1050 kg/m³ to 1250 kg/m³ were prepared. Each time, fluid mud with a certain density was poured into the upstream container which could keep the same liquid level with a flow-off, and then it entered the flume through several pipes with valves imbedded in the fillister of bed and flowed down along the muddy bed under the static water.

2.2 *Field measurement during and after the storm*

Lian Yungang harbor is located at latitude 34°44' and longitude 119°27' in the Northeast coast of Jiangsu Province, China (see Figure 1). It belongs to typical muddy coast, the average sediment diameter (D_{50}) of which is approximately 0.006 mm.

No. 13 tropical storm WIPHA generated above the Northwest Pacific on September 16, 2007, and it reinforced to be super typhoon on September 18. The center of typhoon arrived at west Yellow Sea about 60 km from Lian Yungang, Jiangsu province, China on September 20 and the largest wind speed was 28.8 m/s, and the wind direction was NE or NW.

Before the typhoon arrived, the Tianjin Surveying & Designing Institute for Water Transport Engineering put two stations to conduct the field hydrologic survey on the isobaths of −3 m (W1) and −5 m (W2) of Lian Yungang shoals, as illustrated in Figure 1. Each station had a floater on the water surface and a steel frame mounted on the bed, connecting each other by steel wire. The self-contained instruments were used including optical turbidity meters such as RBR, COMPACT-CTD or OBS for suspended sediment concentration survey. Two turbidity meters were installed on the frame at a height of 0.5 m and 1.5 m above the bed, and one was placed at 0.5 m (and 1.5 m for W2 station) under the water surface. At the W1 (−3 m) station, a wave and tidal recorder was set on the bed to measure the wave height, period, direction and tidal level. At the W2 (−5 m) station, a Nortek AWAC sensor was anchored on the bed to survey the wave height, period, direction and current velocity and direction.

The water depth of the channel was surveyed by SWEDCC, before and after the typhoon, respectively, on September 13 and 21, 2007. The survey scope in the Lian Yungang channel was about 5.6 km, as illustrated in Figure 1.

3 RESULTS

3.1 *Time variation of wave heights and SSC*

The time variations of efficient wave heights and SSC at the height of 0.5 m above the bed surveyed at the W1 (−3 m) and W2 (−5 m) stations are plotted in Figure 2. In the first 18 hours, when the wind waves were still not strong enough, the SSC kept a low value of about 0.2 kg/m^3, meaning that the sediment on the bed surface had not been suspended. Afterward, during 12:00 o'clock on September 19 to 14:00 o'clock on September 20, the wind wave got stronger and the efficient wave heights were larger than 1.5 meters, and the SSC especially near the bottom increased significantly. At 23:00 o'clock on September 19, the wave got strongest, and the largest wave height *Hs* was 3.07 m, but the SSC at 0.5 m height above the bed was about 2.0 kg/m^3, which was smaller than the largest SSC of approximately 6.0 kg/m^3, registered at the moment when the second strongest wave appeared at 8:00 o'clock on September 20.

3.2 *Distributions of fluid mud in the channel*

The depths of the interested channel section are shown in Figure 3, surveyed respectively on September 13 and September 21. As shown in Figure 3, the channel bed before the storm was bumpy, and the depth along the channel varied significantly. The depth of the front 2.0 kilometer channel section was about 15.0 m under the local theoretical level, and the depth of the next 2.0 kilometer was about 0.5 m deeper. However, after the typhoon, the bed of this 4.0 kilometer channel was almost on the same level. So the thickness of siltation in the rear segment was larger than the one in the front part, which was approximately 1.0 m and

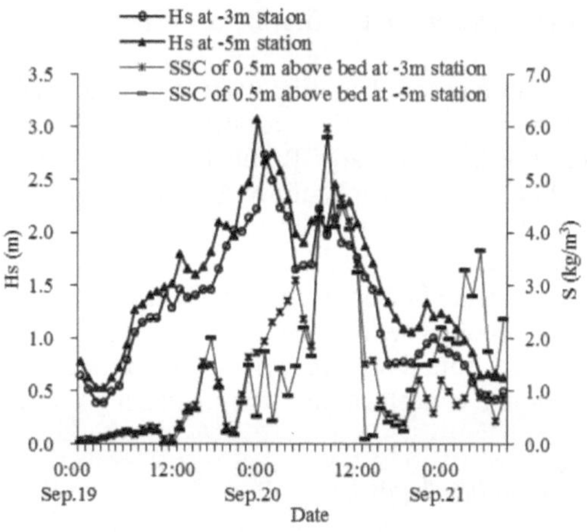

Figure 2. Time variation of efficient wave heights and SSC at the height of 0.5 m above the bed surveyed at –5 m station (W2) and –3 m station (W1) during typhoon Wipha.

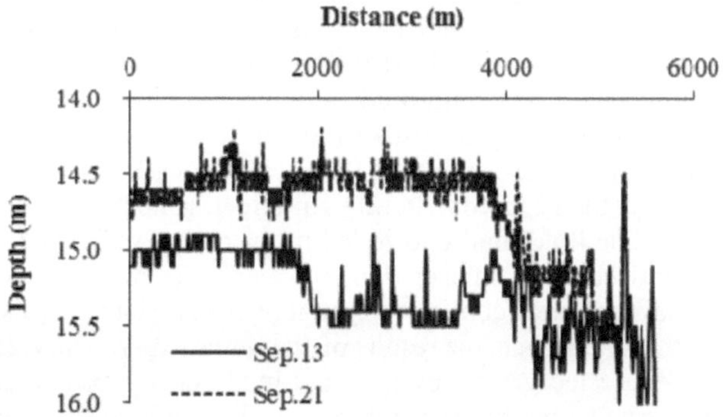

Figure 3. Water depths of the channel section of interest surveyed on September 13 and 21, respectively, represent before and after the typhoon. The depth variation is the thickness of the storm-induced fluid mud.

0.4~0.5 m, respectively. The rapid siltation in the channel should be induced by the storm, because the monthly variation of the water depth had never been so large like this under the normal sea conditions.

4 DISCUSSIONS

4.1 *Formation of a high silt-laden layer near the bottom*

Before the storm, the sediment on the muddy shoal is compacted firmly together. Waves acting on the over-consolidated sediment are likely to produce erosion of surface particles [Whitehouse et al. 2000]. With continuous strengthening of waves, under cyclic loading by the storm-induced waves and the shear forces by the current, the sediment structure has been destroyed gradually by excess pore pressure buildup. The upward moving pore fluid exerts a drag force on the sediment aggregates that exceeds the effective weight of the particles and/or the strength of higher order inter-aggregate bonds. When the upward pore fluid velocity exceeds a minimum value, aggregates are separated and the sediment becomes fluid supported, instead of grain supported, and the sediment is fluidized (McAnally et al. 2007, Whitehouse et al. 2000). During the process, the porosity or water content of the sediment changes continuously, which results in the sediment becoming more active to be eroded or entrained much more easily. This might be the reason that the largest SSC only appeared at the second strongest wave instead of the first largest wave height in typhoon Wipha, as shown in Figure 2.

Afterwards, the wave was strengthened continuously. When the wave height H_S was greater than 1.5 m, the SSC of 0.5 meters above the bed at the two stations rose up to 1.0 kg/m^3, meaning that the sediment was lifted to water. Particularly, after suffering the strongest wave turbulence, the maximum SSC at 0.5 m height above the bed was approximately 6.0 kg/m^3, but the maximum SSCs at 1.5 m height above the bed and 1.5 m under the water surface were only 1.3 kg/m^3, as illustrated in Figure 4. So, it can be concluded that a high silt-laden layer with a thickness of about 0.5 m near the bottom is formed during the stronger wave action. Even afterwards, when the storm becomes weaker, the SSC decreases quickly, and the suspended sediment settles into the bottom boundary layer, where horizontal and vertical trapping mechanisms can result in high enough concentrations to initiate a high silt-laden layer again.

4.2 *Fluid mud convergences in the channel*

Once the bed has formed fluid mud and/or a silt-laden layer, they will move more easily than a structured bed. Ogston et al. (2000) observed that suspended sediment concentrations at the two northern tripod sites following the large flood reached magnitudes typical of fluid mud (> 10 g/l) in a thin near-bed layer, and the fluid mud subsequently is transported seaward due to gravitational force. Fluid mud can flow down into the bottom slopes as a density current or horizontally as streaming under current or wave forcing (Traykovski et al. 2002, Puig et al. 2004, Hsu et al. 2007, Sequeiros et al. 2009, Cantelli, 2009).

Figure 5 is the picture of moving fluid mud in flume, particularly focusing on the head of density flow. Under the unchanged upstream conditions of fluid mud supply, namely unchanged upstream pressure, the fluid mud flowed down with small amplitude of mud wave, and the thickness of the head consistently got larger, which indicated that the turbulence induced by the mobile fluid mud had rolled up plenty of sediments from the muddy bed. Of course, the head becomes thicker due to water entrainment from above as well. Meanwhile, the head may accelerate continuously as it thickens, which had also been observed by Sequeiros et al. (2009). Meantime, the results of the flume experiments also show that the moving speed has been affected greatly by the fluid mud density, bed slope and upstream supply. Fluid mud with smaller density moves more quickly, and moves faster on the steeper slope. Hence, the sediment in the high silt-laden layers near the bottom can flow as a density current, which may be the main way of sediment transporting from shoals to the channel.

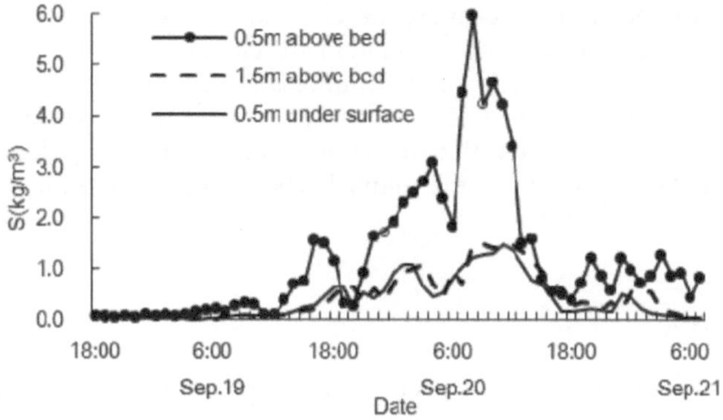

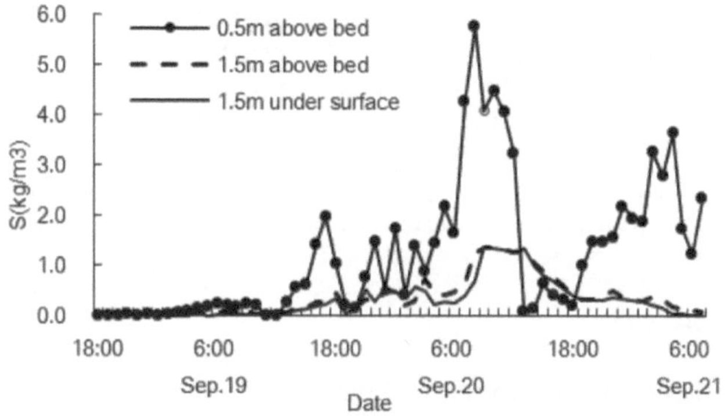

Figure 4. Time variation of SSC at different heights from the bed surveyed at W1 and W2 stations during typhoon Wipha.

Figure 5. Fluid mud moves on a muddy slope, particularly focusing on the head of the density flow.

But the fluid mud moving speed and the distance will decrease when the original density pressure decreases gradually caused by the upstream insufficient supply of fluid mud. Hence, in field, the wave should be strong and last a long time to supply sufficient sediments. In fact, the strong wave action lasted more than 18 hours and the high SSC maintained for about 10 hours during the storm. So the lasting time and supply amount of fluid mud are sufficient for the sediment moving to the channel from shoals.

In addition, parts of fluid mud pouring into the channel from the shoals will continue to flow into the deeper bottomlands, since it behaves on a slope as a liquid, resulting in the flat bottom of the channel after the typhoon. For example, as illustrated in Figure 5, the channel bed was bumpy before the storm, and the depth of the rear 2.0 kilometer segments was deeper of about 0.5 m than the depth of the front 2.0 kilometer channel section. But the bed surface of this 4.0 kilometer of the channel was almost on the same level after the typhoon. So, parts of fluid mud flow into the bottomland of the channel firstly.

Of course, suspended sediments is another important way for sediment transport to the channel. The sediment on the shoals is lifted to water under the action of wave and flow, which can transport with the current flow. When passing over the channel, a part of the suspended load will settle and deposit in the channel. The sediment settled in a short time also helps in the formation of fluid mud.

5 CONCLUSIONS

1. The vertical distribution of SSC indicates that a high silt-laden layer with a height of 0.5 m is formed near the bed during the storm.
2. Based on the analysis of the processes of storm-induced fluid mud formation in open channels, conclusions can be summarized as follows. Strong waves last sufficient time, more than 18 hours during typhoon Wipha, for example, to lift enough sediments from shoals to form a high silt-laden layer near the bottom. Then, the sediments move to the channel to form fluid mud by two ways. First, the sediment in the high silt-laden water layers near the bottom moves as density flows and pours directly into the channel; second, the suspended loads transport with the current flow and then settle into the channel.

ACKNOWLEDGMENTS

This research was funded by the Projects of Ministry of Transport (2012-329-A06-040) and the Central-level Nonprofit Research Institutes Fundamental Research Funds (TKS160201).

REFERENCES

Cantelli A. 2009. Uniform Flow of Modified Bingham Fluids in Narrow Cross Sections. Journal of Hydraulic Engineering 135(8): 640–650.

Hsu T.J., Traykovski P.A. & Kineke G.C. 2007. On modeling boundary layer and gravity-driven fluid mud transport. Journal of Geophysical Research Atmospheres 112(C4): 99–123.

McAnally W.H., Teeter A. & Schoellhamer D. 2007. Management of fluid mud in Estuaries, bays, and lakes (I). Journal of Hydraulic Engineering 133(1): 1–9.

Ogston A.S., Sternberg R.W., Kineke G.C. & Cacchione D.A. 2000. Observations of storm and river flood-driven sediment transport on the northern California continental shelf. Continental Shelf Research 20(16): 2141–2162.

Pang Q.X. & Yang H. 2012. Study of characteristics and forming conditions of sudden silting in Taidian Port. Journal of Sediment Research (2): 96–103.

Puig P., Ogston A.S., Mullenbach B.L., Nittrouer C.A., Parsons J.D. & Sternberg R.W. 2004. Storm-induced sediment gravity flows at the head of the Eel submarine canyon, northern California margin, Journal of Geophysical Research 109(C3): 1–10.

Sequeiros, O.E., Naruse H., Endo, N., Garcia & Parker G. 2009. Experimental study on self-accelerating turbidity currents, Journal of Geophysical Research, Oceans 114(C5): 1–26.

Traykovski P., Geyer W.R., Irish J.D. & Lynch J.F. 2002. The role of wave-induced density-driven fluid mud flows for cross-shelf transport on the Eel River continental shelf. Continental shelf research 20: 2113–2140.

Whitehouse R., Soulsby R., Roberts W. & Mitchener H. 2000. Dynamics of estuarine muds a manual for practical applications. London, Thomas Telford Publishing: 63–74.

Applied Mechanics and Civil Engineering VI – Xie (Ed.)

Analysis of runoff variation trend of the Dongting Lake in China

D.H. Mao & C. Feng
College of Resources and Environmental Science, Hunan Normal University, Changsha, China

H. Zhou & Z.Z. Li
Hydrology and Water Resources Survey Bureau of Hunan Province, Changsha, China

G.W. Hu
College of Architecture and Urban and Rural Plan, Hunan Industry University, Zhuzhou, China

R.Z. Guo
College of Mathematics and Computer Science, Hunan Normal University, Changsha, China

ABSTRACT: This paper analyzes the runoff variation trend of the Dongting Lake by using the Mann–Kendall trend test. The analysis showed that the runoff of four rivers in China had been changing during the period from 1951 to 2011, and their annual runoff trends fluctuated but did not exceed the confidence intervals. Thus, the reported change was not significant. The flow variation of Songzikou occurred in 1985; since then, the runoff trend had been showing significant downward changes. The flow variation of Taipingkou also took place in 1985, and exceeded the confidence interval threshold in 1990, showing a significant downward trend. The runoff trend of Ouchikou exceeded the confidence interval threshold in 1989, exhibiting a significant downward trend.

1 INTRODUCTION

Located in the northeast of Hunan Province and off the south bank of the Jingjiang River, a stem stream of the Yangtze River, Dongting Lake flows across the Hunan and Hubei Province and is closely connected to the middle reaches of the Yangtze River. Dongting Lake, the second largest freshwater lake in China, has a surface area of 2625 km^2 and plays an important role in the storage of flood water. The inflow to the Dongting Lake is contributed by four big rivers (Xiang River, Zi River, Yuan River, Li River), three bayous from the Yangtze River to the Dongting Lake (Songzikou, Ouchikou, Taipingkou), and several small rivers (Miluo River, Xingqiang River, etc.). These rivers flow into the Dongting Lake, are stored, and then discharge into the Yangtze River at Chenglingji (Fig. 1).

The temporal distribution of precipitation is uneven in the Dongting Lake region. The wet season (from April to September) accounts for about 65% of annual precipitation, increasing the risk of regional flood disaster. Because of the variation in inter-annual rainfall, the likelihood of drought and flood is very high. What's worse, even during flood years, the extreme weather like drought can occur in the Dongting Lake region. Flood and drought are the most severe disasters that occur in the Dongting Lake region (Li, et al, 2000; Mao, 1998,

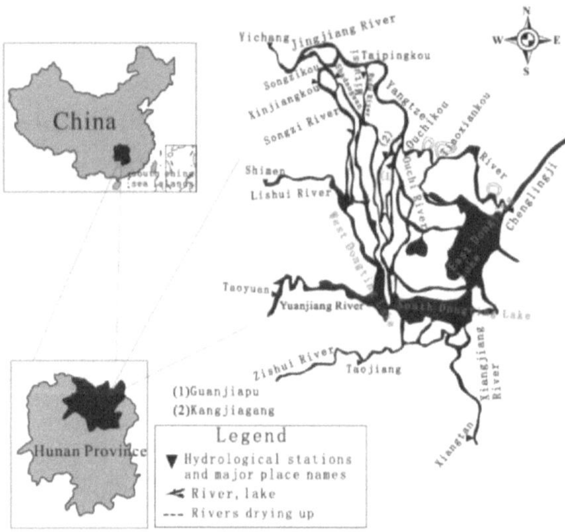

Figure 1. Schematic diagram of the study area.

2000). The main aim of the paper was to analyze the runoff variation trend of three bayous and four rivers of the Dongting Lake based on hydrological data.

2 DATA AND METHODOLOGY

This study used the hydrological data from 1951 to 2011 that were collected from hydrological stations located on the following rivers: Songzikou (Xinjiangkou and Shadaoguan), Taipingkou (Mituosi), Ouchikou (Kangjiagang and Guanjiangpu), Xiang River (Xiangtan), Zi River (Taojiang), Yuan River (Taoyuan), Li River (Shimen), and Chenglingji (Qilishan). Monthly data series were used for the analysis of annual runoff characteristics. All these data were obtained from Hunan Hydrology Bureau and *Bulletin of the Yangtze River sediment* published during 2000 and 2011.

The Mann–Kendall (M-K) trend test is usually used for the identification of the mutation characteristics of long time-series data (Wang, et al, 2011; Zhang, et al, 2008). The M-K trend test is a widely recognized test that does not require the original data to obey a certain probability distribution but requires only random and independent data. Let the time sequence $x_1, x_2, x_3, \ldots, x_n$, m_i represent the cumulative amount where the sample x_i is greater than x_j $(1 \leq j \leq i)$. The statistical magnitude is given by:

$$dk = \sum_{i}^{k} m_i \, (2 \leq k \leq n) \tag{1}$$

If the original sequence is random and independent, the mean and variance of dk are, respectively, $E[dk] = k(k-1)/4$, $var[dk] = k(k-1)(2k+5)/72$ $(2 \leq k \leq n)$. After further standardization of dk, the statistic can be defined as $UF_k = (dk - E[dk])/\sqrt{var[dk]}$ $(k = 1, 2, 3, \ldots, n)$. If $k > 10$, UF_k converges to the standard normal distribution. A bilateral trend test is adopted at a certain significant level a $(a = 0.05)$. The critical value of $U_{a/2}$ is determined in the normal distribution table. If $|UF_k| < U_{a/2}$, the trend is not obvious. If $|UF_k| > U_{a/2}$, the trend shows a significant change. If the UF_k value is positive, the long time series shows an increasing trend and vice versa. The inverse time series of x is calculated as above, while defining $UB_k = -UF_k$ $(k = n, n-1, \ldots, 1)$. The two curves of UB_k and UF_k cross at one point, and the crossover point is called the mutation point. If the mutation point is located between a given confidence level of 95%, it means that the mutation time is statistically significant.

3 RUNOFF TREND ANALYSIS OF THE DONGTING LAKE

There are many methods to analyze the continuous hydrological sequence. The most widely used and simple intuitive ways are the moving average method and the Mann–Kendall (M-K) trend test. This paper analyzes the inflow and outflow of the Dongting Lake by using the M-K trend test, which is widely accepted by hydrologists. This method does not assume the distribution form of the data under investigation, but requires that the data be random and independent with time. The statistical significance of the trend was assessed at the 5% level in this paper.

Fig. 2(a) shows four crosses of UF and UB within the confidence interval. Based on the statistical significance of the M-K mutation test, it can be seen that 1995, 1980, 1983, and 1987 are the possible runoff mutation time periods for Xiangtan. From 1957 to 1994, the UF curve is below the zero line within the confidence interval threshold, which explains a downward but not an obvious trend in its runoff. In Fig. 2(b), the UF and UB statistical curves within the confidence intervals show 1955, 1962, 1967, 1972, and 1978 as the potential mutation time periods for the Zi River. From 1956 to 1993, the runoff of Taojiang show a downward but not a significant trend while in the rest of the time period, it shows an upward trend. In Fig. 2(c), the UF and UB curves of Taoyuan have seven crossover points within the confidence interval. The abscissa value of these points is most likely to appear during mutation periods. An unexpected decreasing trend can be observed from 1956 to 1967, 1985 to 1995, and 2007 to 2011. In Fig. 2(d), runoff trend of Shimen is shown to be more complex than others. The UF and UB curves intersect significantly within the confidence interval and show fluctuation. Its runoff shows an upward trend from 1951 to 1975 and declines from 1976 to 2011. However, the trend is not significant as its crossover points are within the confidence interval.

In Fig. 3(a), UF and UB curves of Songzikou is shown to intersect in 1985 within the confidence interval. After the mutation, UF exceeds the confidence threshold and shows a

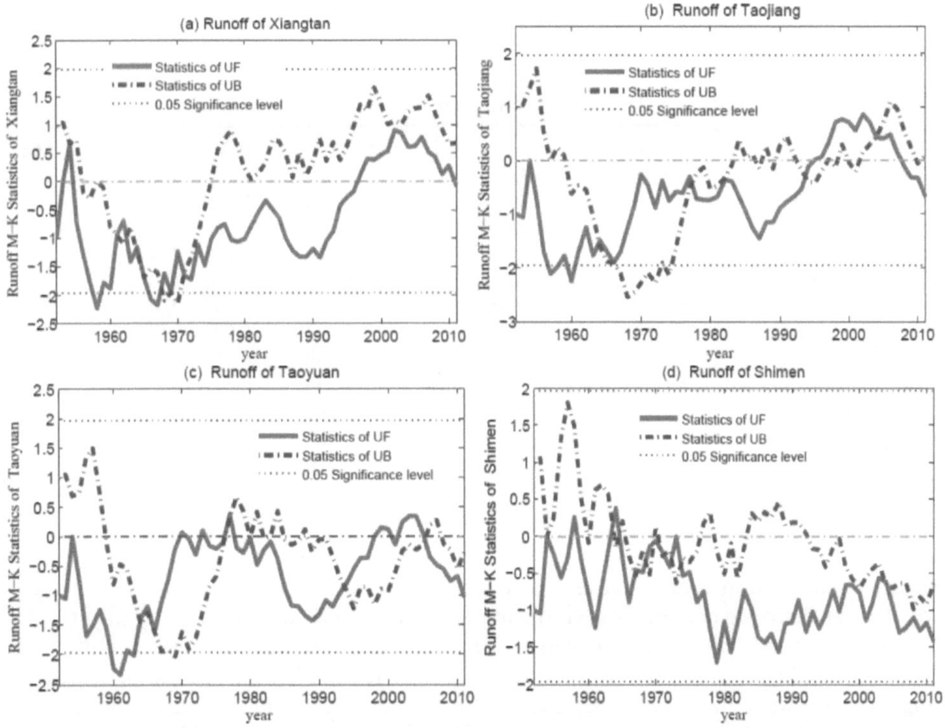

Figure 2. The M-K trend analysis of runoff of the four rivers.

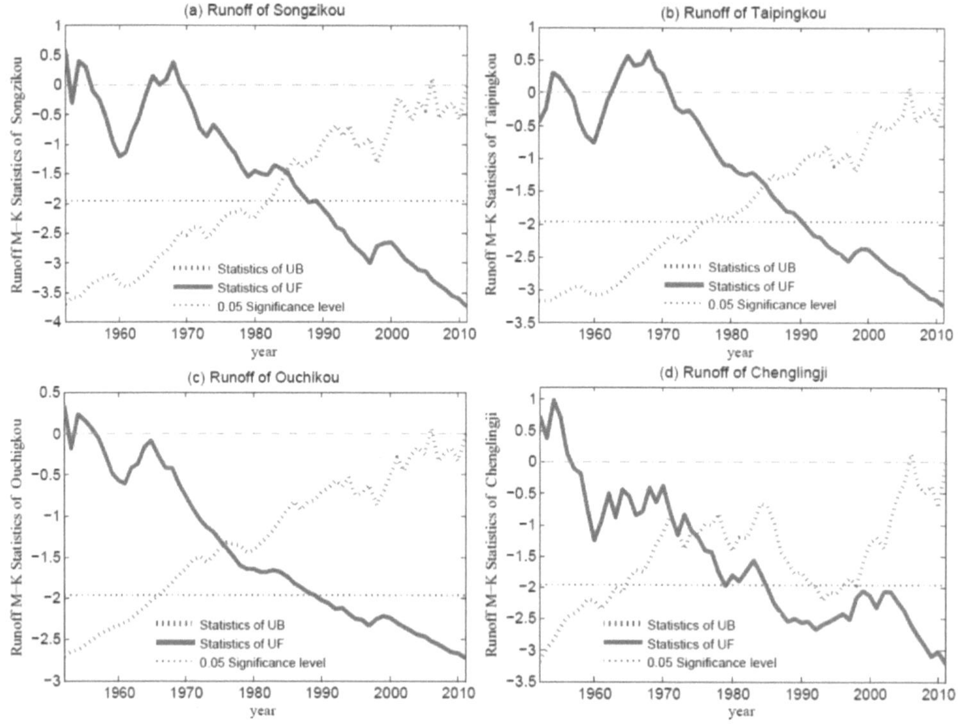

Figure 3. The M-K trend analysis of runoff of the three bayous and outflow at Chenglingji.

Table 1. Mean difference *t* test of the annual runoff of the three bayous, four rivers, and outflow at Chenglingji.

Station	M_j	Mutation index (A_{lj})	Statistic (t)	Significance level (a)	Thres-hold (ta)	Station	M_j	Mutation index (A_{lj})	Statistic (t)	Significance level (a)	Thres-hold (ta)
Songzikou	35	0.847	6.533	0.01	2.704	Tao-	14	0.02	0.143	0.01	2.704
Taipingkou	35	1.015	7.660	0.01	2.704	yuan	16	0.02	−0.148	0.01	2.704
Ouchikou	25	1.372	10.156	0.01	2.704		26	0.115	0.893	0.01	2.704
Chenglingji	25	0.559	4.366	0.01	2.704		27	0.175	1.366	0.01	2.704
Xiangtan	5	0.242	1.032	0.01	2.704		30	0.15	1.165	0.01	2.704
	30	0.085	−0.664	0.01	2.704		44	0.039	0.272	0.01	2.704
	33	0.012	−0.096	0.01	2.704		55	0.498	2.180	0.01	2.704
	38	0.127	−0.958	0.01	2.704		27	0.104	0.799	0.01	2.704
Taojiang	5	0.525	2.571	0.01	2.704	Shi-	37	0.099	0.750	0.01	2.704
	12	0.052	0.366	0.01	2.704	men	42	0.078	0.559	0.01	2.704
	17	0.050	−0.376	0.01	2.704		47	0.103	0.674	0.01	2.704
	22	0.006	−0.047	0.01	2.704		49	0.201	1.211	0.01	2.704
	28	0.001	0.012	0.01	2.704		53	0.346	1.576	0.01	2.704
							57	0.195	0.649	0.01	2.704

significant change. It indicates that after a continuous decreasing trend, its runoff undergoes an obvious and sudden change close to the mutation time 1985, which is exactly the 4th year after the completion of the Gezhou Water Control Project. The runoff shows a continuous downward trend after the mutation point. Similarly, the runoff of Taipingkou, illustrated in Fig. 3(b), shows mutation in 1985 and exceeds the confidence interval threshold in 1990, showing a significant downward trend. In Fig. 3(c), the mutation time of Ouchikou's runoff is shown to be earlier and its UF and UB curves intersect during the mutation times from

1975 to 1976 within the confidence interval threshold. Its UF exceed the confidence threshold and shows a significant downward trend in the mutation time 1989. The runoff of Ouchikou is affected from 1967 to 1972 due to the channel straightening project conducted in the lower Jingjiang River. Fig. 3(d) shows the mutation of the Dongting Lake's outflow between 1974 and 1975, after which the outflow shows a continuous downward trend and intersects with the confidence interval threshold in 1985, showing a significant downward trend.

To determine the accuracy of the mutation year, this paper used the mean difference *t* test. The results are summarized in Table 1.

As revealed by the mean difference *t* test, the statistical *t* values of the three bayous and Chenglingji were 6.533, 7.660, 10.156, and 4.366, respectively. All the values were found to be statistically significant (significance level $\alpha = 0.01$, critical value 2.704). In contrast, the runoff of the four rivers did not change, meaning that they were statistically non-significant.

4 CONCLUSIONS

River runoff can reflect the amount of water generated from precipitation, as well as hydrological phenomena such as watershed closure, soil infiltration, and evaporation. The change in watershed storage results from the comprehensive effects of climate change, hydraulic engineering construction, human activity, and underlying surface soil properties.

According to the results of the M-K test, the runoff of four rivers within the statistical time was changed and the annual runoff trends fluctuated, but did not exceed the confidence interval. Thus, this change was not significant. As observed, hydraulic engineering constructions, water and soil conservation, and other measures in the four watersheds showed a significant ecological effect. Runoff variation of Songzikou occurred in 1985, since then, it had been showing significant downward changes, which may be due to the effects of the completion of the Gezhou dam in 1981. Taipingkou's mutation time occurred in 1985, which is consistent with that of Songzikou, and exceeded the confidence interval threshold in 1990, showing a significant downward trend. The runoff of Ouchikou exceeded the confidence interval threshold in 1989, exhibiting a significant downward trend. From 1967 to 1972, after Jingjiang's channel straightening project, the runoff of Ouchikou was greatly affected, showing a sharp decreasing trend.

ACKNOWLEDGMENTS

This study was supported by the Key Research Project of Hunan Provincial Education Department (grant no. 15A113), the Key Research Project of Hunan Provincial Water Resources Department (grant no. [2015]13-22), and Construct Program of the Key Discipline in Hunan Province of China (2011001).

REFERENCES

Li J.B., Zheng Y.Y. & Gao C.H. 2000. A discussion on geographical regularity of flood and drought in Hunan Province. *Journal of natural disasters* 9(4): 115–120.

Mao D. 1998. Analysis of flood characteristics in Dongting Lake Region from 1471 to 1996. *Lake Science* 10(2): 85–91.

Mao D. 2000. Assessment and analysis of flood-waterlogging disaster condition in Dongting Lake Region. *Journal of natural disasters* 9(3): 46–53.

Wang C.H., Zhou S.W. & Tang X.P. 2011. Spatial and temporal distribution of precipitation in the past 48 years of Tibetan Plateau. *Geographical Science* 31(4): 471–477.

Zhang Q. Chen G.Y. & Jiang T. 2008. Water and sediment trends of Yangtze River and possible influence factors. *Resources and Environment in the Yangtze Basin* 17(2): 257–263.

Applied Mechanics and Civil Engineering VI – Xie (Ed.)

Numerical simulation of the lock release gravity current in a linearly stratified environment

Qian Yu & Bing Shi
College of Engineering, Ocean University of China, Qingdao, China

Zhiyong Zhang
Institute of Hydraulics and Estuary, Zhejiang, China

J.E.F. Baruch
School of Engineering and Informatics, University of Bradford, UK

ABSTRACT: A numerical simulation was used to investigate the movement of a lock released gravity current into stratified environments. A multiphase model was applied to simulate this phenomenon by solving the RANS equations and the RNG $\kappa-\varepsilon$ turbulence equations. The computational cases together with initial and boundary conditions and the numerical scheme are described. The time-dependent gravity current evolution and the head velocity were analyzed. The good agreement between the simulated and measured results indicates that the numerical model can be successfully applied to investigate the complex gravity current process moving into stratified environments.

1 INTRODUCTION

The density current, also called the gravity current, is a common effect in nature caused by fluid density differences, which can be caused by temperature, salinity, or other components. Experiments were carried out on lock release gravity currents in different width-depth ratio situations with modern Digital Image Processing (DIP) techniques (Hacker et al. 1996). They discovered that substance mixing mainly occurred in the early stages when the gravity current was growing. DIP techniques was also used to study the turbulent motion characteristic of gravity currents and investigated gravity currents with the RNG $\kappa-\varepsilon$ model numerically (Chen et al. 2002). Chen rebuilt two stages of the generation process of gravity currents: the initial collapse stage in which the velocity of current head was a constant and the later progressive self-similar stage in which the velocity of the current head was exponentially decreasing. It was found the loss of energy when the diffusion of the gravity current happened with a high Reynolds number was almost negligible (Shin et al. 2004), and they built a theoretical model used to predict the density of the flow velocity and thickness, based on this hypothesis.

Based on the Dynamic subgrid scale model in the two-dimensional Large Eddy Simulation (LES), A numerical study on the lock release gravity current was carried out to verify the accuracy of the forecasting results of the two-dimensional LES model's growing process, and the formation of the wave structure at the top of the gravity current head was precisely simulated (Ooi et al. 2007). With a multiphase flow model, Zhang (2008) carried out a numerical simulation of the diffusion of saltwater at the bottom of a sloping coast, and their results matched well with experiments. A numerical simulation was carried out for the diffusion of the circular symmetric gravity current with the RNG $\kappa-\varepsilon$ model (Zhang et al. 2010).

The lock release gravity current was firstly simulated in curved channels with a three-dimensional LES model, and compared with the results in rectangular channels (Mahdinia et al. 2012). Xu (2012) carried out a numerical simulation of the lock release gravity current with a multiphase flow model, and then demonstrated the feasibility of the model.

All researches mentioned-above are about lock release gravity currents in unstratified environments. The lock release gravity current in linearly stratified environments was observed by Maxworthy (Maxworthy et al. 2002), then he found a visible difference of the gravity currents in subcritical and supercritical conditions. Based on the energy parameter, Bolster (2011) built a model which could forecast the velocity of the gravity current in linearly stratified environments, and verified the accuracy of this model. But the lock release gravity current in linearly stratified environments has not been studied yet.

In this paper, a numerical simulation is used to investigate the movement process of the lock release gravity current into stratified environments. A multiphase model is applied to simulate this phenomenon by solving the RANS equations and the RNG $\kappa-\varepsilon$ turbulence equations. The time-dependent gravity current evolution and the head velocity is compared and analyzed using experimental data and numerically simulated data.

2 MULTIPHASE MODEL

2.1 *Governing equations*

The governing equations of every phase in multiphase flow are the momentum and continuity conservation equations.

$$\frac{\partial}{\partial x}(\alpha_q \rho_q) + \nabla \cdot (\alpha_q \rho_q \vec{v}_q) = 0 \tag{1}$$

$$\frac{\partial}{\partial t}(\alpha_q \rho_q \vec{v}_q) + \nabla \cdot (\alpha_q \rho_q \vec{v}_q \vec{v}_q) = -\alpha_q \nabla P + \nabla \cdot \overline{\overline{\tau}}_q + \alpha_q \rho_q \vec{g} + \sum_{p=1}^{4} K_{pq}(\vec{v}_p - \vec{v}_q) \tag{2}$$

where p and q are phase constituents; α_q is volume fraction of phase q; ρ_q is density of phase q; $\vec{v}$ is velocity vector; $\overline{\overline{\tau}}_q$ is stress-strain tensor of phase *q*; *P* is pressure shared by phases; $\vec{g}$ is gravity acceleration, and K_{qp} is momentum exchange coefficient between phases.

2.2 *Turbulence model*

In order to get an accurate results, the Reynolds number $\kappa-\varepsilon$ model is used to simulate the curved fluxion, because it relatively fits the case of meandering streamlines when the lock is opened and releases the gravity current into the stratified environment.

k equation:

$$\frac{\partial(\rho k)}{\partial t} + \frac{\partial(\rho k u_i)}{\partial x_i} = \frac{\partial}{\partial x_j}\left(\alpha_k \mu_{eff} \frac{\partial k}{\partial x_j}\right) + G_k + G_b - \rho\varepsilon \tag{3}$$

ε equation:

$$\frac{\partial(\rho\varepsilon)}{\partial t} + \frac{\partial(\rho\varepsilon u_i)}{\partial x_i} = \frac{\partial}{\partial x_j}\left(\alpha_\varepsilon \mu_{eff} \frac{\partial \varepsilon}{\partial x_j}\right) + \frac{C_{1\varepsilon}\varepsilon}{k}(G_k + C_{3\varepsilon} G_b) - C_{2\varepsilon}\rho\frac{\varepsilon^2}{k} \tag{4}$$

where u_i is the component of the velocity in the x_i-direction; ρ is the mixture density of all phases; G_k is the generation of *k* induced by the mean velocity gradients; α_k, α_ε are turbulence Prandtl number for *k* and *w*; μ_{eff} is effective dynamic viscosity of water; $C_{1\varepsilon}$, $C_{2\varepsilon}$, $C_{3\varepsilon}$ are empirical constants of the turbulence model.

3 MODEL VALIDATION

The model has been validated by the physical experimental results in reference 10. Fig. 1(a) shows the two-dimensional profile of the experiment. Water depth $H = 0.2\ m$, whole length

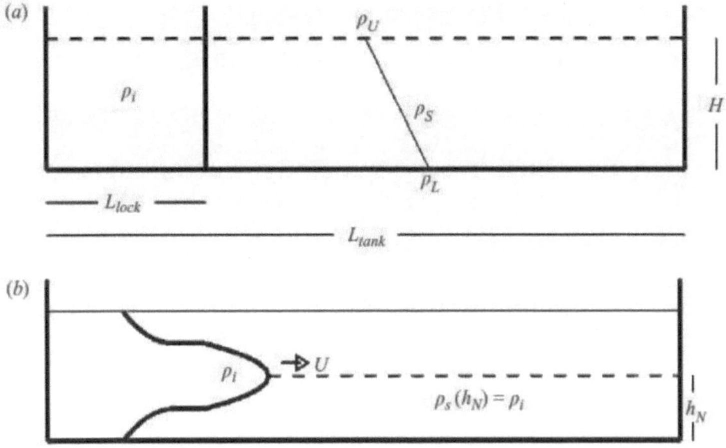

Figure 1. Sketch of experimental layout (a) and the diffusion of saltwater at $h_N = H/2$.

Table 1. Cases for calculating.

Cases	$\rho_U (Kg/m^3)$	$\rho_L (Kg/m^3)$	$\rho_i (Kg/m^3)$	N	h_N/H
1	1000.00	1020.50	1016.60	1	0.2
2	1000.00	1020.50	1012.30	1	0.4
3	1000.00	1020.50	1010.25	1	0.5
4	1000.00	1020.50	1008.20	1	0.6
5	1000.00	1020.50	1004.10	1	0.8

$L = 1.8$ *m*. The reach located the left side of the lock is 0.3 *m* long, which was filled with unstratified saltwater, and its density is ρ_i. The reach located the right side of the lock was filled with linearly stratified fluid, from the surface to the bottom, the density increased from ρ_U to ρ_L, and at the height h_N from the bottom the density is equal to ρ_i.

The buoyancy frequency of the stratified fluid can be expressed as:

$$N = (g/\rho_0)(\rho_L - \rho_U)/H \tag{5}$$

where $g = 9.81\ m/s^2$, represents the acceleration due to gravity, and $\rho_0 = 1000\ Kg/m^3$ is the reference density.

Fig. 1(b) shows the diffusion of saltwater when $h_N = H/2$, the velocity in the current head is U.

We started timing when the lock was abruptly removed, and observed the mixing of the saltwater on both sides. Table 1 shows the parameters in each case.

4 INITIAL, BOUNDARY CONDITIONS AND SOLUTION PROCEDURE

There are three different saltwater phases, of density:- phase 1 ρ_i, phase 2 ρ_U and phase 3 ρ_L respectively. The left side of the lock was filled with saltwater of phase 1, while the other side was filled with the mixture saltwater of phase 2 and phase 3 which were linearly stratified according to these rules. The distribution rules of the saltwater of phase 2 are as follows:

$$\alpha_2 = z/H \tag{6}$$

And the distribution rules of the saltwater of phase 3 are as follows:

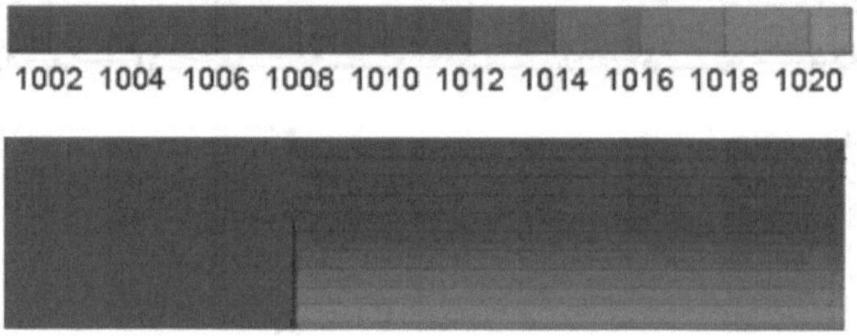

Figure 2. The initial density distribution in case 5 ($N = 1$, $h_N = 0.8H$).

$$\alpha_3 = 1 - z / H \tag{7}$$

Fig. 2 shows the initial density distribution in case 5, and demonstrates that the fluid density increases linearly with increasing water depth. The fluid density on the left side is equal to the fluid density at the $0.8H$ position on the right side, and a foundation for subsequent research is set up since the distribution of the density field matches the theoretical situation well.

The entire calculating region was discretized with structured meshes, and 180000 elements were used in simulation. The mesh size was 0.002 m in the flow direction and 0.001 m in the vectical direction.

All boundaries in the calculating domain are generated by the no slip condition. The governing equations were solved by the Finite-Volume Method (FVM). The pressure-velocity coupling is achieved by using the i algorithm.

The third-order QUICK scheme is applied for solving the momentum equation and the composition equation, the maximum residual for convergence is 10^{-5}, a constant time step is used 10^{-4} s.

5 RESULTS AND DISCUSSION

5.1 *Evolution of the gravity current*

Fig. 3 shows the diffusion of saltwater when $N = 1$, $h_N = 0.8H$.

It is found that the density of saltwater in the right side below $h_N = 0.8H$ position is greater than the density of the saltwater in the left side, the saltwater moves towards the left side under the influence of the density difference.

When t = 6 s, the stratified saltwater had moved to the edge of the left side, and then reflected back because of the obstruction of the end wall.

In this evolution, the stratified saltwater occupied the space of the left side, and then forced the initial saltwater on the left side to move towards the right.

Because the density of the stratified saltwater above $h_N = 0.8$ position is lower than the density of the saltwater in the left side, the saltwater on the left upper surface began to move and leave the upper surface by the action of gravity, and generated a current head which moves towards the right steadily.

The comparison shows good agreement between the simulated and measured results, this validated the proposal that the multiphase model can be used successfully to investigate the numerical simulation of lock released gravity currents in linearly stratified environments.

Fig. 4 and Fig. 5 show the evolution of the gravity current and the density diffusion when $h_N = 0.2H$ and $h_N = 0.5H$ respectively, and the evolution is opposite. When $h_N = 0.2H$, the saltwater in the left side diffused to the right side by the action of gravity. The density of the saltwater in the right side below the $0.2H$ position is greater than that in the left, and the left current head separated from the bottom and moved to the right at $0.2H$ when t = 5 s, 10 s and 15 s.

When $h_N = 0.5H$, the evolution of the left side saltwater shows symmetrical distribution along $0.5H$ level plane, the stratified saltwater at the top and the bottom moved to the left, the middle saltwater in the left side moved to the right, and the gravity current head was on the $0.5H$ level plane.

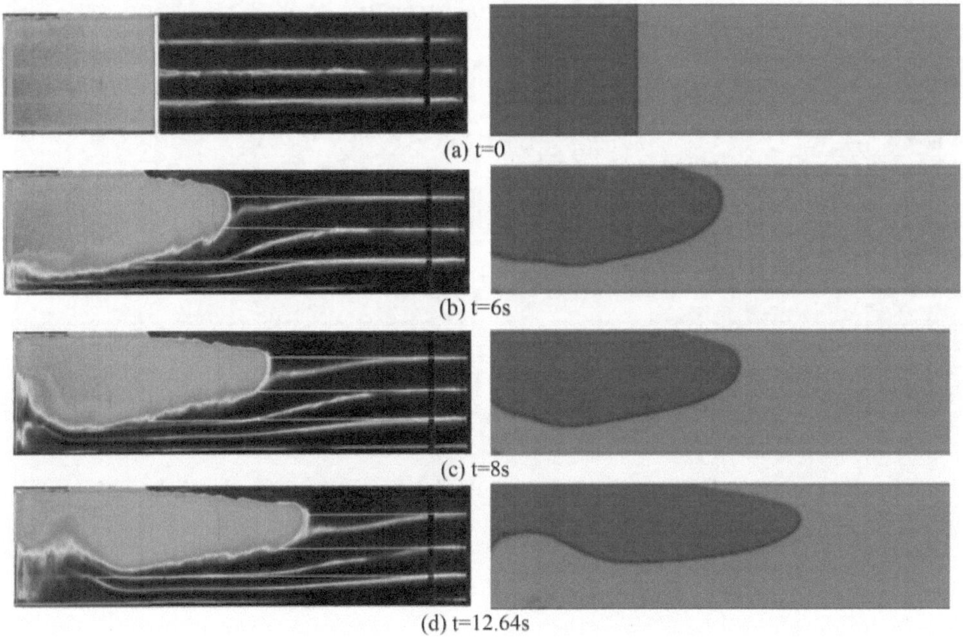

Figure 3. The evolution of saltwater when N = 1, $h_N = 0.8H$ (left: experimental result; right: simulated result).

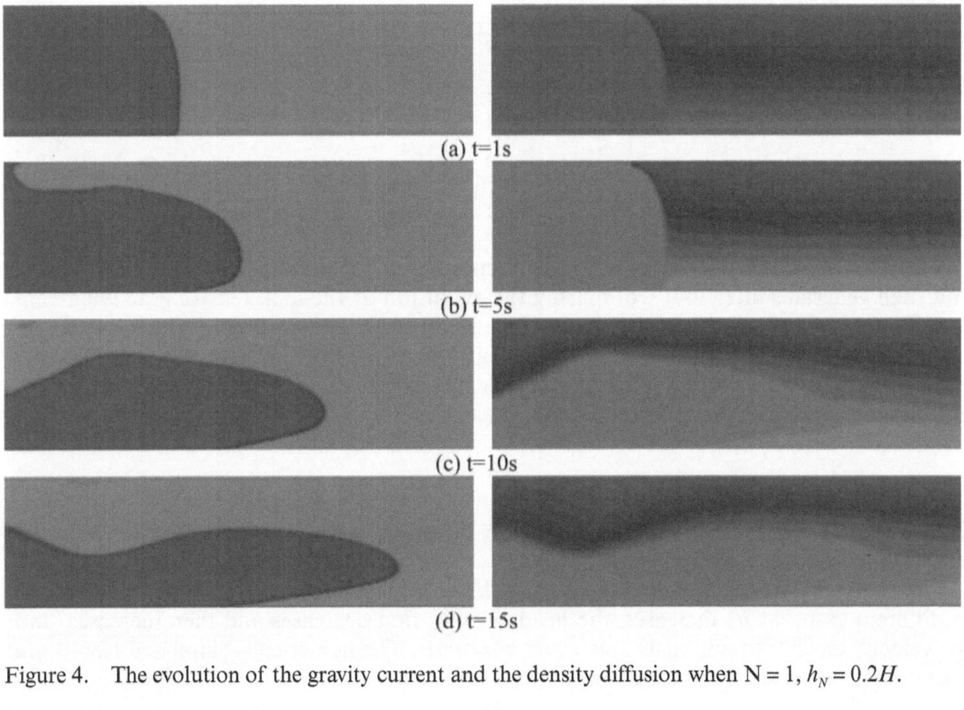

Figure 4. The evolution of the gravity current and the density diffusion when N = 1, $h_N = 0.2H$.

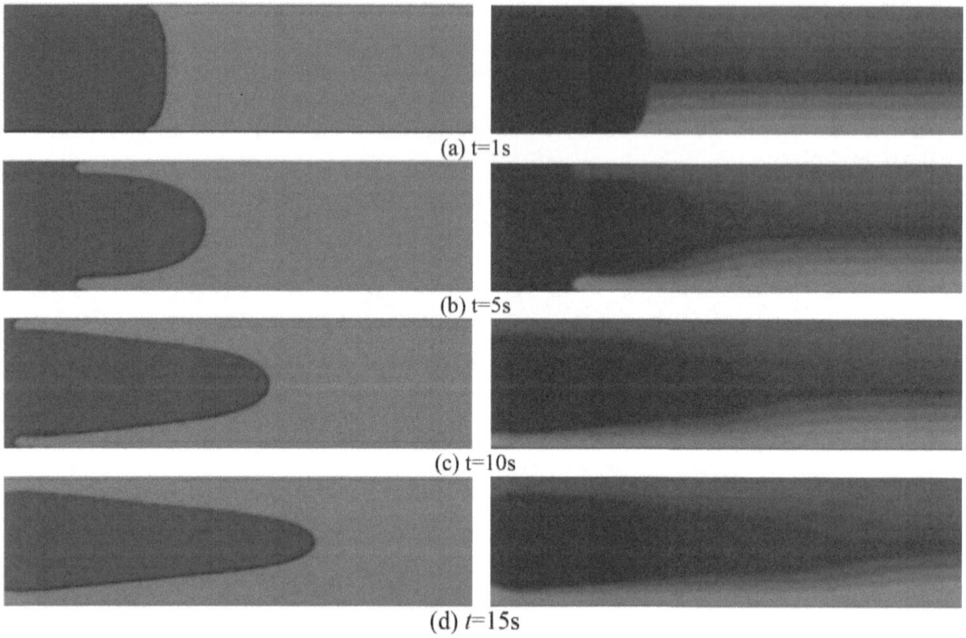

Figure 5. The evolution of the gravity current and the density diffusion when N = 1, $h_N = 0.5H$.

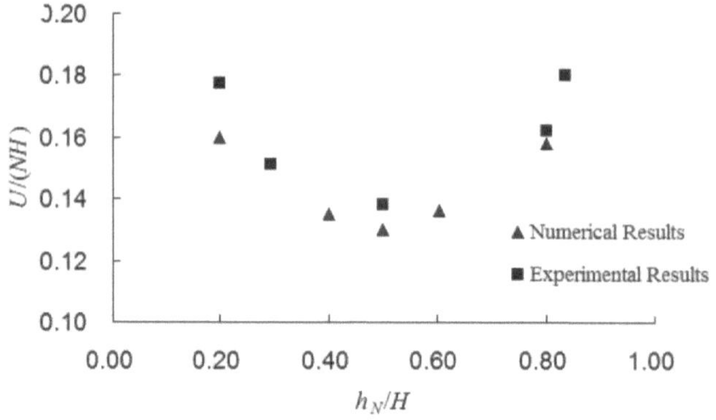

Figure 6. Comparing the dimensionless velocity of the current head with experimental data.

When t = 15 s, the saltwater on the left separates from both the top and the bottom, and then generates interflow. Comparing the evolution of the lock release gravity current in different hN conditions, when flowing into the stratified environment, the saltwater on the left moves to right neither close to the top nor to the bottom, but inserted the same density layer of the stratified environment, generated interflow, and diffused to the right gradually.

5.2 *Velocity of the current head*

The velocity of the current head keeps a nearly constant value moving to the right after the acceleration process.

Fig. 6 shows that changes between h_N and the dimensionless velocity of the current head in different cases: as h_N increases, the head velocity first decreases and then increases, and the velocity reaches to minimal value when $h_N = 0.5H$. The numerically simulated results are slightly less than the experimental ones.

6 CONCLUSION

A multiphase model was applied to simulate the evolution of the lock release gravity current in a stratified environment by solving the RNG $\kappa - \varepsilon$ turbulence equations. The evolution of the gravity current was simulated in different density cases, and the evolution of the current front was analyzed. Good agreement between the numerical simulations and the available laboratory measurements indicates that the model can accurately simulate the spreading and the motion of the gravity current in stratified environments.

REFERENCES

Bolster, Diogo & Hang, Alice & Linden, P.F. The front speed of intrusions into a continuously stratified medium. *Journal of Fluid Mechanics*, 2008, 594: 369–377.

Chen, G.Q., Li., X.W. & Li, Z. 2002. The turbulent gravity current of lock release type. *Science in China*, 2002, 32(6): 754–764.

Hacker, J., Linden, P.F. & Dalziel, S.B. 1996. Mixing in lock-release gravity currents. *Dynamics of Atmospheres and Oceans,* 1996, 24(1–4): 183–195.

Mahdinia, M., Firoozabadi, B. & Farshchi, M. 2012. Large Eddy Simulation of Lock-Exchange Flow in a Curved Channel. *Journal of Hydraulic Engineering*, 2012, 138(1): 57–70.

Maxworthy, T., Leilich, J. & Simpson, J.E. 2002. The propagation of a gravity current into a linearly stratified fluid. *Journal of Fluid Mechanics*, 2002, 453: 371–394.

Ooi, Seng Keat, Constantinescu, George & Weber, d Larry J. 2007. 2D Large-Eddy Simulation of Lock-Exchange Gravity Current Flows at High Grashof Numbers [J]. *Journal of Hydraulic Engineering*, 2007, 133(9): 1037–1047.

Shin, J.O., Dalziel S.B. & P.F. Linden 2004. Gravity currents produced by lock exchange. *Journal of Fluid Mechanics*, 2004, 521: 1–34.

Xu, X.z. 2012, Internal wave formation mechanism and 2-D numerical model simulation. *Harbin Institute of Technology*.

Zhang, L.Z., Zeng, L. & Zhang, B. 2010. Numerical study on head advancement of axisymmetric gravity currents. *Communications in Nonlinear Science and Numerical Simulation*, 2010, 15(7): 1893–1898.

Zhang, J., Guo, Y. & Shen, Y. 2008. Numerical Simulation of Flushing of Trapped Salt Water from a Bar-Blocked Estuary. *Journal of Hydraulic Engineering*, 2008, 134(11): 1671–1676.

Zhang, X.F., Yao, Z.J. & Lu, J.Q. 2011, Investigation on density currents in stratified reservoirs. *Engineering Journal of Wuhan University*, 2011, 44(4): 409–413.

Applied Mechanics and Civil Engineering VI – Xie (Ed.)
© 2017 Taylor & Francis Group, London, ISBN 978-1-138-62631-7

Simplified method to calculate buoyancy in multi-layered strata

M. Tang & C. Zhang
Guangzhou Institute of Building Science, Guangzhou, Guangdong Province, China

H. Chen
Guangzhou Institute of Building Science, Guangzhou, Guangdong Province, China
School of Civil Engineering, Guangzhou University, Guangzhou, Guangdong Province, China

ABSTRACT: It's complicated to calculate the buoyancy and anti-floating water level in multilayered strata. Based on the comparison and analysis of current progresses, both the generally accepted principles and inconsistent dispute are summarized, where the buoyancy reduction of underground structures in aquitards is highlighted. A simplified method based on local empirical reduction factors is proposed in this paper, where ten different models and corresponding formulae are given for various distributions of aquifers and aquitards. The applicability and safety of our proposed approach are verified by comparing the calculated results with reported experiments.

1 INTRODUCTION

The buoyancy for underground structures is generated from the water pressure difference between their bottom and top surfaces, where both hydrostatic and pore water pressure should be included. The underground water is rarely static but generally accompanied with seepage so that the porosities of the strata play the role of its passage as well as the storage space. The complexity of buoyancy calculation has attracted many representative discussions until now, especially for the determination of anti-floating water level. The recommended anti-floating water level with adequate safety should be given by survey engineers (claimed by GB50007-2011, Code for design of building foundation, national standard) after the comprehensive analysis of hydrology and geological conditions, the relationship among the recharge, run-off and discharge of underground water and the local dynamical maximum and minimum water levels. According to the rationality analysis, designers should decide whether to modify it before substituting it into the buoyancy related calculations, i.e. uplift bearing capacity and stability.

Neither the historical highest nor lowest water level can be directly employed as the anti-floating water level. Instead, long-term dynamical measurement and observation should be carried out. Zhang (2001), Zhang et al. (2001) and Men & Yan (2002) have proposed the classification of existence status of the underground water by the measured pore water pressure. They analyzed the main factors to influence the underground water level from the 50-year dynamical history, and then gave reliable predictions.

On the other hand, an aquitard (such as clay), where the transfer of hydrostatic pressure in clay mainly depends on the porosity (Huang, 2002), prevent the downward transfer of the hydrostatic pressure of interlayer phreatic (Huang et al., 2005). Therefore, the porosity is widely employed as the reduction factor during the calculation of water pressure, which, however, is inconsistent with the principle of effective stress as demonstrated by Li & Wu (2003).

In this paper, both the generally accepted principles and inconsistent dispute of buoyancy calculation are summarized, where our suggestions for modification of corresponding standards are highlighted. A simplified method based on local empirical reduction factors is also proposed in this paper, the applicability and safety of which are verified by the comparison with reported experiments.

2 THE BUOYANCY OF BASEMENT

2.1 *Hydrostatic pressure*

According to GB50007-2011, the hydrostatic resultant acting on the backplane of basement P is calculated as the product of the hydrostatic pressure p and the active area A, i.e. $P = pA$, where $p = \gamma_w H$, $H = h_w - h_0$ is the designed hydraulic head, and h_w and h_0 are the elevations of the anti-floating water level and the backplane of basement respectively. Indicated by above equation, it is of great importance to establish reasonable anti-floating water level, which in turn makes significant effects on the design of uplift piles or anchors, including their quantity and reinforcement ratio. Stochastic analytical and numerical mathematical models, based on regression, spectrum or time series analysis, have been established to predict the dynamic underground hydrostatic pressure over time (Zhang 2001, Men & Yan 2002). However, the results may deviate from the actual situation due to the randomness during the calculations.

2.2 *The clauses for buoyancy calculation*

The trade-off relation between the enhanced anti-floating stability to prevent fracture or exceeded floating displacement and the increasing costs for materials is the key factor to anti-floating design. The clauses for buoyancy related calculation in Chinese corresponding standards or technical codes are recapitulative, i.e. 1) in GB50021-2001, the anti-floating checking of basement or underground structures should be carried out in shallow underground water, the buoyancy calculation in the site of rocks with undeveloped joints and clay should be based on local experience or testing results and the seepage analysis should be performed if encountered; 2) in the design code for building foundation (DBJ15-31-2003, Guangdong Provincial standard), the fictional resistance acted on the side walls of the basement and the cohesion between the basement backplane and the strata are inadvisable to be involved into the buoyancy calculation, and the hydraulic head cannot be reduced except having long-term reliable controlling measures, which are obviously conservative; 3) although the reduction of buoyancy in clay foundation is allowed both in the code for geotechnical engineering investigation of high rise building (JGJ72-2004, industrial standard) and the technical code for construction of geotechnical engineering (DB22/46-2004, Guizhou provincial standard), the reduction local experience or testing results are additionally needed in the latter code; 4) a relative detail provision is found in the local standard of Shenzhen (SJG01-2010, code for design of survey and design of foundation), where, for coastal reclamation, deep basement embedded in sand, basement socketed into sandy/gravelly destructured layer, and shallow basement in back-filing silty clay, earthy destructured layer, gravelly destructured, distinctly weathered rock, hills, tableland or partially weathered rocks with undeveloped fissure, buoyancy reduction should be considered. However, the descriptions in these clauses are general terms, and thereby should be modified in the three aspects below:

1. It is difficult to define the underground water as shallow or not. Here, we suggest a clear classification of five kinds, i.e. 1) the water 0~2 m under the ground with annual fluctuation <0.5 m, 2) the water 2~5 m under the ground with annual fluctuation 0.5~2.0 m, 3) the water 5~10 m under the ground with annual fluctuation 2.0~5.0 m; 4) the elevation is larger than 10 m with annual fluctuation <2.0 m due to the negligible effects of surface evaporation; and 5) the elevation is larger than 10 m with annual fluctuation >2.0 m. On the other hand, if the recharge of underground water is larger than its exploitation

or drainage, a higher value of the anti-floating water level should be assigned. On the contrary, if the exploitation is larger than the surface penetration form surface or direct recharge in long-term, it is safe enough to check the anti-floating performance according to the dynamic difference during last 50 or 100 years.

2. The local anti-floating stability is not required in existing clauses. For single building without podium, the anti-floating stability is satisfied if the load from the superstructure is greater than the buoyancy of underground water. However, if the basement is extended outward or the integral basement connects several building towers, local fracture or large floating displacement maybe encountered.
3. The depth, distribution and flowing of underground water are complex so that, to ensure enough safety of the anti-floating performance, the regional geological structure, hydrogeological condition, and the underground water/space exploitation scale and intensity should be also considered when providing geotechnical investigation report, where the buoyancy reduction in each stratum should be clearly given.

2.3 *Buoyancy reduction factor*

In the provision explanation of the code for geotechnical engineering investigation of high rise building (JGJ72-2004, industrial standard), for the temporal high water level induced by rainstorm, the buoyancy can be reduced by a factor from 0.6 to 0.8 m if the basement is socketed in clay with well drainage conditions. The rainstorm is frequent in northern China so that the waterlogging phenomenon is commonly happened in the cities in the Pearl River Delta and along Yangtze River. The low frequency of rain in the northern and northwestern China results in the long-term decline of underground water level. However, the anti-floating calculation in waterlogging is independent on its frequency in our opinion. For safety, the buoyancy cannot be reduced, even though the worst flooding in 50 years, because the immediately eliminating the effects of rainstorm cannot be guaranteed even with well drainage conditions and it is difficult to define the duration of the high water level.

In Handbook of Engineering Geology (2001), the calculated buoyancy should be based on 100% of the designed water level for the buildings in foundation of sand, gravelly soil or rocks with developed joint, while 50% for the rocks with undeveloped joint foundation. Only the empirical method could be employed for the cases with clay foundation. In Shenzhen, an approach based on the reduction of active area is proposed by Yang & Zhu (2001), which is suitable for phreatic and perched water, i.e.

$$(W/F)/(\gamma_w \psi h) \geq 1.2 \tag{1}$$

where W is the load from superstructure, F is the nominal contact area between foundation and strata, h is the anti-floating water level related to bottle of the bearing stratum, and $\psi(=0.5\sim1.0)$ is the empirical reduction factor for buoyancy. The effective safety factor in Eq. (1) is 1.2, which is large than that in other technical codes. Zhang (2004) discussed the buoyancy reduction in general strata structure, where the buoyancy p depends on both the upper and lower aquifers. The hydraulic gradient consumption by an aquitard leads to the decrease of hydrostatic pressure. The water level at the the basement backplane can be approximated according to the initial hydraulic gradient, which gives the pressure acting on the basement backplane as

$$p = \gamma_w\left[h_1 - d(h_1 - h)/m\right] \tag{2}$$

where h_1 and m are the thickness of the upper aquifer and aquitard respectively, d is the socketed depth of the basement into the aquitard and h is water level in the aquitard. However, Eq. (2) is independent on porosity and permeability of the clay (Zhang 2004, Li 2014). It is also indicated by Eq. (2) that the decrease of hydraulic head is more severe in the thinner aquitard and the hydraulic pressure attenuates to unphysical zero at the lower boundary of the aquitard.

3 THE SIMPLIFIED METHOD OF BUOYANCY CALCULATION

In fact, the hydraulic head can be measured in aquifer, while it is difficult to do such measurement in aquitards. It is also demonstrated by Ma (2007) that the actual hydraulic pressure is far less than the theoretical hydrostatic pressure. On the other hand, the buoyancy calculation becomes more complex due to the fluctuation of geotechnical conditions, especially the thickness of strata. Therefore, a simplified method based on reduction factor is reasonable.

Flowing through aquitards, the seepage of underground water leads to the loss of hydraulic head Δh, then the decrease of pressure Δp presents as

$$\Delta p = \gamma_w \Delta h = \xi \gamma_w d \tag{3}$$

where ξ is the empirical reduction factor related to permeability of the strata given by local experiments or experiences, the typical values of which for different strata are listed in Table 1.

Various combination of the foundation backplane elevation and the aquifers/aquitards distributions are schematically shown in subfigures 1(a)~(j), the corresponding formulae to get the buoyancy and their applicable conditions are listed below:

$$p = \gamma_w h \tag{4a}$$

$$p = \gamma_w \left[h - (1+\xi) d \right], h \geq (1+\xi) d \tag{4b}$$

$$p = \gamma_w \left(h + h_z - \xi d \right), h_z \geq \xi d \tag{4c}$$

$$p = \gamma_w \left(h_z - \xi d \right), h_z \geq \xi d \tag{4d}$$

$$p = \gamma_w \left(h_z + d \right) \tag{4e}$$

$$p = \gamma_w \left[h - (1+\xi) d \right], h \geq (1+\xi) d \tag{4f}$$

$$p = \gamma_w \left[h - (1+\xi) d \right], h \geq (1+\xi) d, h_z < m - d \tag{4g}$$

$$p = \gamma_w \left(h - d + h_z - \xi m \right), h \geq (1+\xi) d, h_z \geq \xi (m - d) \tag{4h}$$

$$p = \gamma_w d, h \geq (1+\xi) m \tag{4i}$$

$$p = \gamma_w \left[h - (1+\xi) m + h_z \right], h \geq (1+\xi) m \tag{4j}$$

4 COMPARISON WITH EXPERIMENTS

Our simplified method to calculate the buoyancy is compared with the example introduced by Zhang et al. (2001), where the three-floor basement with the depth of 13.4 m is buried into the strata with three aquifers up to 30 m below the ground surface. The shallowest aquifer

Table 1. Reduction factor of water pressure.

District	Stratum	Reduction factor
Beijing	Fine silty sand	0.65
	Tableland silt	0.25
	Clay	0.19
Shen-zhen	Sand, coastal reclamation Sandy/gravelly destructured layer	1.0
	Earthy destructured weathered layer	0.8
	Thick back-filing and silty clay, gravelly destructured/distinctly weathered rocks	0.7~0.8
	Hills, tableland, partially weathered rocks with undeveloped fissure	0.5~0.6
Guang-zhou	Sand	1.0
	Gravelly distinctly/partially weathered rocks	0.8~1.0
	Sand clay, destructured and reworked rocks	0.7~0.8
	Silty clay and silt	0.5~0.6
	Mucky soil	0.4~0.5

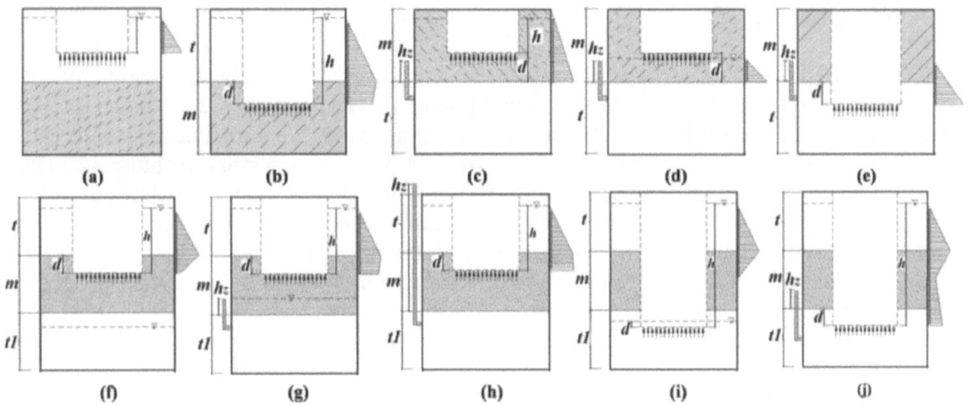

Figure 1. Models of buoyancy calculation under different design and environmental conditions.

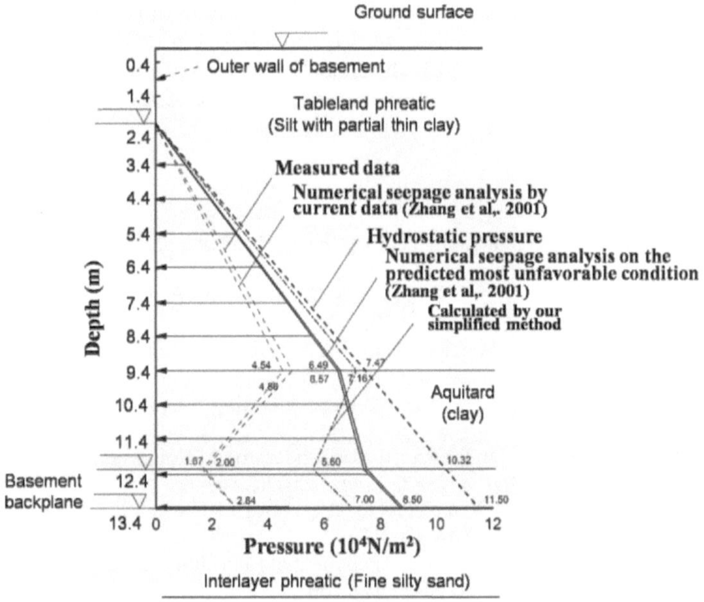

Figure 2. Water pressure distribution in underground structure.

is tableland phreatic with the water level −2.24~−4.65 m. Then, the interlayer phreatic is followed with the water level −10.94~−12.02 m and the predicted long-term maximum value −4.99 m. The water level of the last confined aquifer is about −18.07~−19.37 m below the ground, but the predicted average annual value reaches −10.00 m due to the south-to-north water diversion program.

Substituting the measured maximum water value into Eq. (4j), the calculated pressure at the bottle of the tableland phreatic aquifer, clay layer and basement backplane is 71.6 kN/m^2, 56.0 kN/m^2 and 70.0 kN/m^2, where the water pressure reduction factor of the clay aquitard is assigned as 0.6. The seepage results based on both the measured actual and long-term predicted most unfavorable water levels are conducted by Zhang et al. (2001), where our result for the pressure on casement backplane is between them and our pressure distribution over depth is consistent with the measured results, as shown in Fig. 2. Therefore, it is indicated that our theoretical method not only presents simpler process other than finite element seepage analysis, but also ganrantees enough safety.

5 CONCLUSION

In this paper, the determination of anti-floating water level and the calculation of the buoyancy are discussed. None of the perennial maximum, *in-situ* measured value, historical maximum value of the water level can be directly employed as the anti-floating water level. Instead, the comprehensive analysis of the natural conditions (i.e. history, recharge, seepage and drainage of the underground water) and the underground water/space exploiting should be carried out to satisfy both the safety and costs requirements. In aquitards, the transmission of the hydrostatic pressure mainly relies on the pores of the clay, which leads to the loss of hydraulic head. A simplified approach based on local empirical reduction factors is proposed in this paper. By comparasion with reported experiments, our theoretical method not only presents the simpler process other than finite element seepage analysis, but also keeps enough safety affluence.

ACKNOWLEDGEMENT

Author to whom correspondence should be addressed: chenhang.thu@gmail.com (Chen, H.). The authors acknowledge the financial support from national NSFC (Grants No. 51608139 and 51678171), Guangdong Science and Technology Department (Grant No. 2004B36001028), and China Postdoctoral Science Foundation (Grant No. 2016M592471).

REFERENCES

Editorial board of handbook of engineering geology. 2001. *Handbook of engineering geology*. Beijing: China Architecture & Building Press.

Huang, Z. 2002. Analysis of Ground Water's Up lift Pressure on Underground Construction. *Geotechnical Engineering Technique*. 2002(5): 273–275.

Huang, Z., Ma, J. & Li, C. 2005. The Water Level for Prevention of Up-floating under the Condition of Multi-level Ground water. *Geotechnical Engineering Technique*. 2005, 19(4): 182–217.

Li, C. 2014. Reasonable value of water level for building anti-floating design. *Geotechnical investigation & surveying*. 2014(4): 49–54.

Li, G. & Wu, J. 2003. Calculation of Up lift Pressure on Underground Construction and Effective Stress Principle in Clay. *Geotechnical Engineering Technique*. 2003(2): 63–66.

Ma, B. 2007. Ground Subsidence and Water Level for Prevention of Up-floating of Beijing Shahe Area. *Geotechnical Engineering Technique*. 21(4): 202–203.

Men, Y. & Yan, C. 2002. Application of seasonal exponential smoothing method to ground water level prediction. *Geotechnical Investigation & Surveying*. 2002(2): 13–18.

Shen, X., Zhou, X., Wang, J. & Han, X. 2013. *Anti-floating of underground water and structure*. Beijing: China Architecture & Building Press.

Yang, R. & Zhu, L. 2001. Discussion on design and construction of underground structure design and construction design of water level of selection and anti-floating [J]. *Geotechnical Investigation & Surveying*. 2001(1): 43–46.

Zhang, X. 2004. Analysis of Defence Water Level and Reduction Factor of Buoyancy for Underground Buildings in Shenzhen Area. *Site Investigation Science and Technology*, 2004(2): 12–20.

Zhang, Z. 2001. *Groundwater and building foundation engineering*. Beijing: China Architecture & Building Press.

Zhang, Z., Sun, B. & Xu, H. 2001. Effect of characteristics of ground water distribution and seepage on anti-uplift analysis of building foundations. *China civil engineering journal*, 34(1): 73–78.

Applied Mechanics and Civil Engineering VI – Xie (Ed.)
 ISBN 978-1-138-62631-7

Effect of dissolved oxygen on nitrate removal from drinking groundwater using rice wine

Q.L. Tong, Y.D. Chen, L.Y. Li & Y.P. Jiang
Guilin University of Technology, Guilin, Guangxi, China

ABSTRACT: To better understand the effect of the influencing factors of ethanol on nitrate removal from drinking groundwater, this study conducted batch experiments using rice wine as a carbon source aimed at the problem of nitrate removal from drinking groundwater in rural settings. The study focused on the effect of Dissolved Oxygen (DO), and used four types of microcosms: aseptic control group (M0), anaerobic group (M1), micro-aerobic group (M2), and aerobic group (M3). The results indicated that the presence of DO slightly delayed the initiation of denitrification. Moreover, the denitrification process was not significantly inhibited by DO, but the denitrification rate was the most ideal one under strictly anaerobic conditions. The formation of NO_2^- was obviously inhibited and the accumulation of NO_2^- was not observed at high oxygen content. This study can provide a reference for the practical application of nitrate removal from drinking groundwater in rural areas.

1 INTRODUCTION

Due to the application of nitrogen fertilizers in agriculture, the discharge of domestic sewage and industrial wastewater that contain nitrogen, as well as the leaching of solid wastes (Jiang et al. 2009, Zhang et al. 2014, Pastén-Zapata et al. 2014), nitrate has become one of the most common pollutants of groundwater (especially shallow groundwater). Nitrogen pollution will not only cause the decline in groundwater quality and applicability, but also harm human health directly or indirectly (Li 2011, Qu 2007). Therefore, the study of groundwater nitrate pollution and its control has become a hot research topic at home and abroad. Biological denitrification is a common method of nitrate removal from groundwater. The key influencing factors of this process are Dissolved Oxygen (DO), Temperature (T), mass ratio of Carbon to Nitrogen (C/N), etc.

In our previous study, a simple sand bucket device was used to remove nitrate from drinking groundwater, in which rice wine brewed in a rural area was proved to be a powerful carbon source due to the presence of ethanol (Liu et al. 2015). To better understand the effects of DO on nitrate removal using rice wine, a batch experiment was conducted under aerobic, micro-aerobic, and anaerobic conditions. Rice wine was used as a sole carbon source in the experiment.

2 MATERIALS AND METHODS

2.1 *Material handling*

For the experiment, groundwater was collected from the Yanshan campus at the Guilin University of Technology, the chemical indicators of which are listed in Table 1. Sand material with a particle size of 2~10 mm was collected from a construction site. Because denitrifying bacteria were present in the environment, groundwater and sand material samples were used as the source of microbial strains without special cultivation. The nitrate solution used in the experiment consisted of $NaNO_3$ (analytical reagent) and groundwater. According to the

Table 1. Background values of groundwater chemical indicators.

Items	NO_3^- (mg/L)	NO_2^- (mg/L)	CH_3COO^- (mg/L)	Cl^- (mg/L)	DO (mg/L)	Ec (μs/cm)	pH	T (°C)
Concentration	6.6	–	–	6.3	7.8	255	7.9	25.8

* "–" means not detected.

Table 2. Types of microcosms and concentrations of additives.

Experimental group	Type	Solid (g)	GW (ml)	NO_3^- (mg/L)	EtOH (ml)	$HgCl_2$ (ml)
M0	Aseptic + sealed	400	800	100	0.355	5
M1	Strictly anaerobic	400	800	100	0.355	0
M2	Micro-aerobic + sealed	400	800	100	0.355	0
M3	Aerobic	400	800	100	0.355	0

groundwater quality standard of China, the nitrate concentration in drinking groundwater (calculated by N) should not exceed the limit of 20 mg/L, about 88.55 mg/L as NO_3^-. The actual pollution level was greater than the threshold, so the target concentration was set to 100 mg/L. Rice wine (alcohol contents about 20%) was selected as an external carbon source.

There were four experimental groups: M0, M1, M2, and M3. For each group, experiments were conducted in triplicate, and the average value for each group was taken for analysis. A total of 12 transparent glass bottles (capacity 1000 ml) were used for making the microcosms. M0 represents an aseptic control group, which was used to confirm the effects of biological action on denitrification. The materials used in this group were treated at high temperatures and high pressures (sterilizing pot, daily sterilization 60 min, for 3 days). Then, 5 ml $HgCl_2$ (concentration 500 mg/L) was added for enhanced sterilization. M1 represents the anaerobic group: nitrogen was filled into the bottle containing the solution until the DO concentration was less than 1 mg/L, and then the microcosms were cultivated in the anaerobic glove box. M2 represents the micro-aerobic group: nitrogen was filled into the bottle containing the solution until the DO concentration was less than 4 mg/L, and the bottle was sealed and placed in a laboratory. M3 represents the aerobic group: not treatment was applied.

2.2 *Microcosm production*

Four types of microcosms and concentrations of the additives are described in Table 2. The batch experiments had a total of 12 microcosms: M0 aseptic control group (M0-1#, M0-2#, M0-3#), M1 anaerobic group (M1-1#, M1-2#, M1-3#), M2 micro-aerobic group (M2-1#, M2-2#, M2-3#), and M3 aerobic group (M3-1#, M3-2#, M3-3#).

2.3 *Sampling and analysis methods*

Samples were taken every 12 hours (8:00 am or pm) in the initial stage of the experiment. The frequency of sampling was properly released when the start time of denitrification was stable. The sample size was 10 ml, and each sample was transferred into a borosilicate glass bottle (capacity 30 ml). DO, pH, T, and Ec were determined by using a German WTW handheld multiparameter tester (multi 3420). The concentrations of NO_3^-, NO_2^-, and CH_3COO^- were analyzed by using an ion chromatograph (DionexICS 2100) under the following chromatographic conditions: column temperature 30°C; pool temperature 35°C; suppressor (ASRS) 4 mm; suppressor current 33 mA; eluent (3.5 mmoL Na_2CO_3, 1 mmoL $NaHCO_3$) flow rate 1 ml/min; oxygen intake 25 μL; minimum detection 0.1 mg/L.

Ethanol was not the main measured parameter in this study; therefore, few samplings and measurements were done to determine its degradation trend. A 5 ml solution was taken from each microcosm and transferred into a headspace bottle (capacity 10 ml) lined with teflon gasket and containing 1 g NaCl. Ethanol concentration in each bottle was analyzed by gas chromatography (Agilent 6890 N) under the following chromatographic conditions: column temperature 50°C (10 min), 20°C/min, 150°C/min; sample inlet temperature 200°C; detector temperature 250°C; nitrogen carrier gas flow rate 1.0 ml/min. It was also analyzed under the following headspace sampling conditions: bottle area temperature 60°C; bottle balancing time 10 min; shaking time 5 min; bottle pressurization time 0.13 min; sampling time 1 min; minimum detection 0.1 mg/L.

3 RESULTS AND DISCUSSION

After the initiation of cultivation, the variation in the chemical composition for each microcosm was determined by the analysis of the water samples. Except in the aseptic control group, the nitrate concentration in the other microcosms had obvious changes, followed by the production of nitrite. A very similar concentration was found between the microcosm duplicates, and their average values obtained are graphically represented in the below figures. Through the comparison of the microcosm experiments under different DO conditions, important findings were obtained and are discussed below.

3.1 *Change in dissolved oxygen concentration*

DO has a strong oxidation ability and is the main electron acceptor in the oxidation of groundwater (Cortez et al. 2009). The results showed that the DO concentration in the aseptic control group (M0) was relatively higher and stable (5.5~6.7 mg/L), indicating that the microcosms under the aseptic conditions without oxidation did not consume DO as an electron acceptor. In the anaerobic group (M1), the initial DO concentration was 0.5 mg/L, and then after 24 h, its concentration was below 0.1 mg/L, i.e. there was almost no DO. In the micro-aerobic group (M2) and the aerobic group (M3), the DO concentration was significantly decreased from the start of the cultivation (2 mg/L), and then there was an upward trend (Fig. 1). Under the micro-aerobic and aerobic conditions, the concentration of DO decreased significantly, which indicates that DO as the most important electron acceptor is first used by microorganisms for the oxidation of groundwater. The increase in the DO concentration in both M2 and M3 groups at the late stage may be due to the presence of oxygen in the air dissolved in water and the absence of electron donors at the late stage of the reaction.

3.2 *Nitrate removal effect*

By injecting ethanol as the external carbon source can promote denitrification in groundwater (Magram et al. 2010). The effect of nitrate removal in four types of microcosms is shown in Figure 2. There was no significant change in the nitrate concentration in the M0 group,

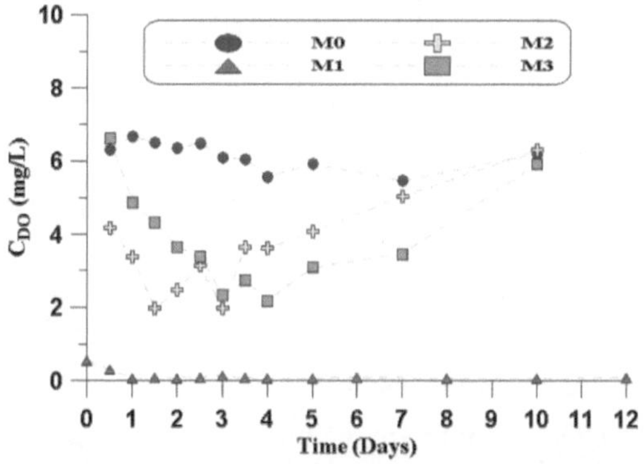

Figure 1. Curves showing the change in the dissolved oxygen concentration in each microcosm.

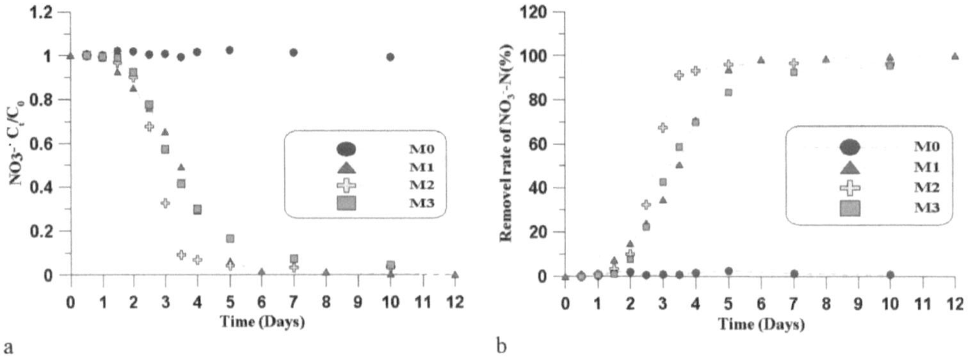

Figure 2. Curves showing the change in the nitrate concentration and nitrate removal rate in each microcosm.

i.e. the nitrate removal rate was almost 0, which confirms the biological effect on denitrification. Nitrate concentration in the M1~M3 groups showed a significant decreasing trend, which indicates that the use of rice wine as a sole carbon source can accelerate the initiation of denitrification, and has strong adaptability to microorganisms. This result is consistent with that of Jiang Yaping et al. (2011) showing that the nitrate in groundwater can be effectively removed by using indigenous denitrification bacteria in aqueous media with ethanol as an electron donor. From Figure 2b, the final removal rate of nitrate in the M1~M3 groups was found to be more than 95%, which shows that the degradation of nitrate by microorganisms can be carried out, and DO was found to have no significant inhibitory effect on the denitrification process under different DO conditions. The removal rate of nitrate in the M1 group was 100%, which was slightly higher than that found in the micro-aerobic and aerobic groups. Although the difference was not significant, it is consistent with the study of Desimone & Howes (1998) who monitored and analyzed NO_3^-–N pollution in groundwater and found that at a DO concentration <0.2 mg/L, the removal rate of nitrate was the best.

3.3 *Accumulation of nitrite*

Denitrification is the process of reducing NO_3^-–N or NO_2^-–N to NO_2 or N_2 by denitrifying microorganism under the catalysis of a series of enzymes, where nitrite produced as an intermediate product can cause secondary pollution, and even cancer when its concentration is high. The denitrification process consists of four steps (Philippot 2002): $NO_3^- \rightarrow NO_2^- \rightarrow N_2O \rightarrow N_2$.

The presence of intermediate products will affect the water treatment to a certain extent. For example, the limiting value of nitrite concentration is 1 mg/L according to China's current drinking water sanitary standard (GB5749-2006). In this experiment, except the M0 group in which NO_2^- was absent, the other groups (M1~M3) had NO_2^- in different degrees. As can be seen from Figure 3, the proportion of NO_2^- in the M3 group was less than that in the M1 and M2 groups. The maximum value in the former group was 1.8 mg/L, while that of the latter groups was 11.5 mg/L and 14.2 mg/L, respectively, and the retention time of NO_2^- in the M3 group was significantly shorter than that in the M1 and M2 groups. It can be observed that at high oxygen content, the formation of nitrite was inhibited and the accumulation of nitrite was not observed. It was possible to have a strong effect on denitrification and weak nitrification under the low DO conditions.

3.4 *Effect of dissolved oxygen on denitrification*

Nitrate is only second to DO, and both are the most powerful electron acceptors in natural metabolism. In the denitrification process, the presence of DO will compete with the nitrate for an electron donor (or a carbon source), which will affect the denitrification process and

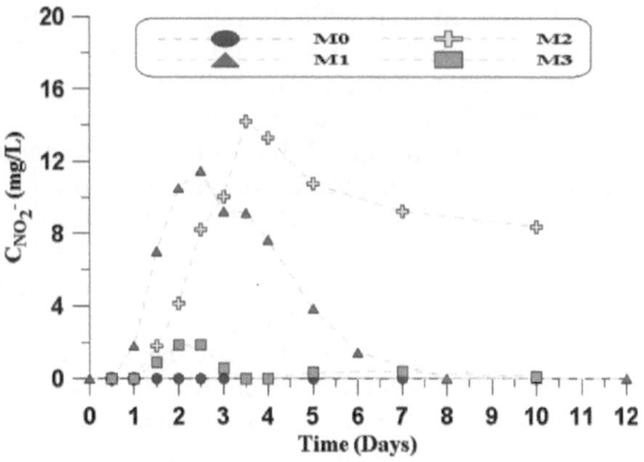

Figure 3. Curves showing the change in the nitrite concentration in each microcosm.

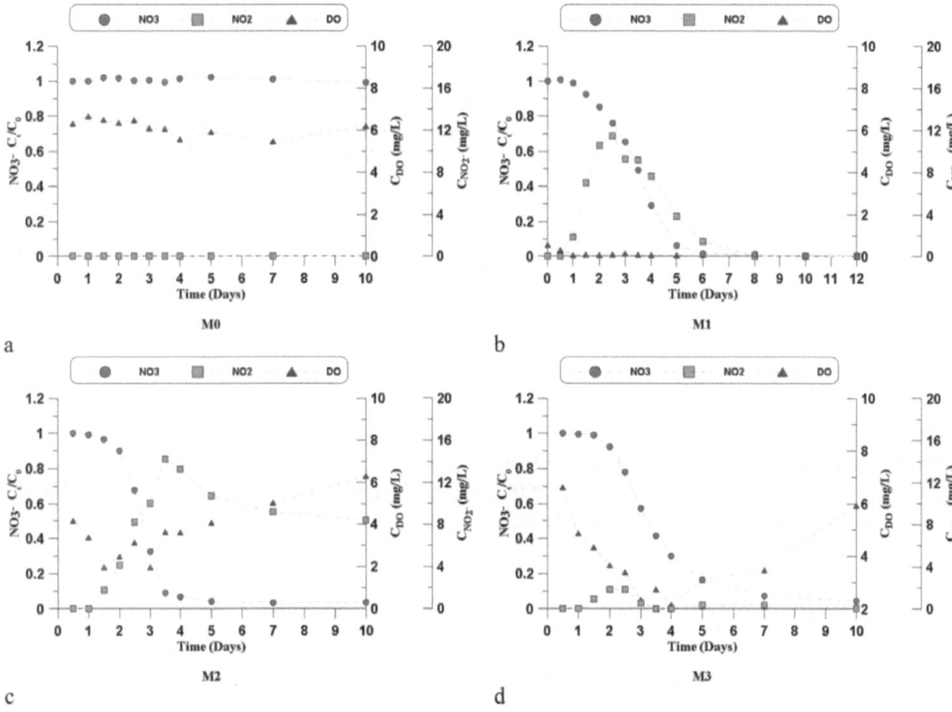

Figure 4. Effect of dissolved oxygen on denitrification in each microcosm.

nitrate removal. However, some researchers also speculated that since denitrifying bacteria are facultative anaerobes, denitrification can be carried out under aerobic conditions. A study by Gillham & Cherry (1978) performed a statistical analysis and found that the upper limit of DO for denitrification in groundwater was 2 mg/L. Another study by Deng Kang et al. (2010) found that when the concentration of DO was higher than 3.5 mg/L, the denitrification rate decreased while the concentration of DO increased. However, in another study (Jin Zanfang et al. 2004), it was found that the denitrification process was less affected by DO. When the value of the influencing factor DO was varied from 2 to 6 mg/L, the removal efficiency of the reactor did not change. In the case of higher DO concentration, it was supposed that denitrification was performed by aerobic denitrifying microorganisms.

In this experiment, there were no significant changes found in the water chemical indicators of the M0 group (Fig. 4a), which shows that the sterilization effect was relatively complete, with no biological effect. A significant decrease in the nitrate concentration in the M1 group requires 1.5 days (36 h), which corresponds to 0.1 mg/L of DO, as well as the generation and disappearance of NO_2^- essentially correspond to the initiation and termination time of the reduction process of the NO_3^- concentration (Fig. 4b). A significant decrease in the nitrate concentration in the M2 group requires 2 days (48 h), corresponding to 2.5 mg/L of DO (Fig. 4c). A significant decrease in the nitrate concentration in the M3 group requires 2 days (48 h), which corresponds to 3.7 mg/L of DO (Fig. 4d). Based on these findings, the starting time of denitrification in the M1 group was found to be 12 hours ahead of that in the M2 and M3 groups, and the presence of DO slightly delayed the initiation of denitrification. Gómez et al. (2002) found that the effects of DO concentration on nitrate removal in groundwater depended on the selected carbon source, and the effects were the least affected when ethanol was used as the carbon source, but there was an overall negative correlation between the concentration of DO and the denitrification rate. Therefore, it was speculated that the slight inhibitory effect of DO on denitrification in this experiment may be related to the carbon source.

4 CONCLUSIONS

This paper focused on the problem of nitrate removal from drinking groundwater in rural settings. It not only considered the feasibility of individual applications, but also tried to understand the influencing factors in the process of implementation. The microcosm experiments were conducted to study the effects of DO on nitrate removal from drinking groundwater using rice wine. Based on the results, the following conclusions were drawn:

1. Rice wine as a sole carbon source accelerated the denitrification process rapidly, and had strong adaptability to microorganisms. The degradation of nitrate by microorganisms could be carried out, and DO had no significant inhibitory effect on the denitrification process under different DO conditions. However, the denitrification rate was the most ideal under strictly anaerobic conditions.
2. At high oxygen content, the formation of NO_2^- was inhibited and the accumulation of NO_2^- was not observed. This may be attributed to a strong effect on denitrification and weak nitrification under low DO conditions.
3. The presence of DO slightly delayed the initiation of denitrification, which may be related to the presence of different carbon sources.

REFERENCES

Cortez, S., Teixeira, P. & Oliveira, R. et al. 2009. Effect of operating parameters on denitrification in an anoxic rotating biological contactor. *Environmental Technology*, 30(13): 1381–1389.

Deng, K., Huang, S.B. & Hu, T. 2010. Study on Aerobic Denitrification in BAF. *Environmental Science*, 12: 2945–2949.

Desimone, L.A. & Howes, B.L. 1998. Nitrogen transport and transformation in a shallow aquifer receiving wastewater discharge: a mass balance approach. *Water Resour Res*, 34(2): 271–285.

Gillham, R.W. & Cherry, J.A. 1978. Field evidence of denitrification in shallow groundwater flow systems. *Water Pollut. Res. Can*, 13(1): 53–71.

Gómez, M.A., Hontoria, E. & González-López, J. 2002. Effect of dissolved oxygen concentration on nitrate removal from groundwater using a denitrifying submerged filter. *Journal of Hazardous Materials*, 90, 267–278.

Jiang, Y., Wu, Y. & Groves, C. et al. 2009. Natural and anthropogenic factors affecting the groundwater quality in the Nandong karst underground river system in Yunan, China. *Journal of Contaminant Hydrology*, 109(1): 49–61.

Jiang, Y.P., Chen, Y.D., Liu, H.L., Huang, Y.Q. & Sui, H.S. 2011. Availability of Ethanol for In-situ Nitrate Bioremediation in Polluted Groundwater. *Journal of Natural Resources*, 08: 1446–1452.

Jin, Z.F. & Chen, Y.X. 2004. Denitrification of Groundwater Using Cotton as Energy Source. *Journal of Agro-Environment Science*, 23(3): 512–515.
Li, D. 2011. The study on the immobilized-cell membrane bioreactor for removing nitrate from groundwater. *East China University of Science and Technology*.
Liu, Z.T., Chen Y.D. & Jiang Y.P. 2015. A preliminary study on removing nitrate in groundwater by using rice wine for the rural drinking water. *Ground Water*, 04: 44–47.
Magram, S.F. 2010. Drinking water denitrification in a packed bed anoxic reactor: effect of carbon source and reactor depth. *J Appl Sci*, 10(7): 558–563.
Pastén-Zapata, E., Ledesma-Ruiz, R. & Harter, T. et al. 2014. Assessment of sources and fate of nitrate in shallow groundwater of an agricultural area by using a multi-tracer approach. *Science of the Total Environment*, 470: 855–864.
Philippot, L. 2002. Denitrifying genes in bacterial and Archaeal genomes. *Biochimica et BiophysicaActa*, 1577(3): 355–376.
Qu, J.H. 2007. Pollution and health effects of drinking water [J]. *Scientific Chinese*, 06: 69–72.
Zhang, Y., Li, F. & Zhang, Q. et al. 2014. Tracing nitrate pollution sources and transformation in surface- and ground-waters using environmental isotopes. *Science of the Total Environment*, 490: 213–222.

Applied Mechanics and Civil Engineering VI – Xie (Ed.)

Low-frequency variability of sea surface height in the tropical Pacific Ocean

Y.R. Jia, M.X. Chen & J.C. Zuo
College of Oceanography, Hohai University, Nanjing, China

ABSTRACT: In this paper, low-frequency variability of Sea Surface Height (SSH) in the tropical Pacific Ocean during 1950–2014 is analyzed based on two combined SSH anomaly data. Then, the influence of decadal variations on the sea level trend since 1993 is mainly discussed. The results indicate that the interannual and decadal variations of SSH in the tropical Pacific Ocean are related to the phenomenon of ENSO (El Niño-Southern Oscillation) and PDO (Pacific Decadal Oscillation), respectively. It also indicates the influence of decadal variations on the sea level trend from 1993 to 2014, with a rising trend of SSH in the western tropical Pacific and a declining trend in the eastern tropical Pacific over the last 22 years of decadal oscillations.

1 INTRODUCTION

Sea level change has become one of the hot spots of recent climate change research. Since the tropical Pacific Ocean is the most important source of global atmospheric movement and the sensitive area of global climate anomaly, changes in sea level have attracted much attention. The altimeter data show that the rate of sea level rise in the western tropical Pacific over the last 20 years is more than 10 mm/a, which is about 3–4 times of the global average rise rate (~3 mm/a) (Merrifield 2011, Qiu & Chen 2012). From the view of climate change, the long-term trend of sea level change is the most important threat to human survival. Previous studies have focused on the influence of interannual and decadal variations on long-term trend (Becker et al. 2012, Si et al. 2014, Chen et al. 2014). For example, Nidheesh et al. (2013) found the area with the fastest rate of sea level rise in the western tropical Pacific, as well as with the largest amplitude of interannual and decadal sea level variations. Therefore, the estimation of the long-term trend of the regional sea level change in the tropical Pacific Ocean is influenced by the interannual and decadal signals. In order to accurately estimate the long-term trend of sea level change, the interannual and decadal signals must be estimated accurately.

Satellite altimetry has been providing broad and accurate measurements of near-global Sea Surface Height (SSH) since 1993, which has improved our understanding of sea level change regionally and globally. However, the data are too short to accurately study the influence of SSH variability on the decadal timescales, so the reconstructed data with longer time series based on tide gauge and altimeter data are used to study the low-frequency variability of SSH (Church et al. 2004, Hamlington et al. 2011, Cheng et al. 2015). Taking into account the time length and the application, in this paper, two reconstructed SSH data, one based on EOF from Church et al. (2004) and another based on CSEOF from Hamlington et al. (2011), are combined with satellite altimetry to construct two SSH Anomaly (SSHA) fields from 1950 to 2014, respectively. The two new sets of SSHA data are used to verify each other and analyze the low-frequency variability of SSH in the tropical Pacific Ocean. Based on this analysis, the influence of decadal variations on sea level trend since 1993 is discussed.

2 DATA

Monthly SSH data in Ssalto/Duacs multimission altimetry products were obtained from Archiving, Validation, and Interpretation of Satellite Oceanographic (AVISO) for the period from 1993 to 2014, with the resolution of 1/4° × 1/4°.

Two sets of reconstructed SSH data were used in this paper. One was the monthly reconstructed SSH data developed by Church (hereafter, CH01) with the resolution of 1°× 1° spanning from January 1950 to December 2001 without the seasonal signals. The other was the weekly reconstructed SSH data developed by Hamlington (hereafter, HL09) with the resolution of 0.5° × 0.5°, covering the period of January 1950 to June 2009. The data were interpolated at monthly intervals and 1°× 1° grids before analysis. For more details, see Church et al. (2004) and Hamlington et al. (2011).

The PDO index and the SOI index are available online at http://jisao.washington.edu/pdo/PDO.latest and http://www.cpc.noaa.gov/data/indices/soi, respectively.

3 INTERANNUAL AND DECADAL VARIABILITY DURING 1950–2014

The domain of the tropical Pacific Ocean is defined as 20°S–20°N, 120°E–80°W. Two new sets of SSHA data (CWTP and HLTP) covering the period from 1950 to 2014 were reconstructed by combining altimeter data with CW01 and HL09, respectively. Prior to this, the annual and semiannual signals were removed from the altimeter data (Church et al. 2004). Furthermore, a 14-month low-pass filter was applied to all the data as our current research interest was in the low-frequency variability of SSH. Figure 1 shows the regional mean SSHA in the tropical Pacific Ocean, in which both the reconstructed data and altimeter data show a good agreement over the common time period in 1993–2001 and in 1993–2009 with the correlation coefficient being larger than 0.8 or 0.9, respectively. Thus, CWTP based on CW01 and HLTP based on HL09 were combined with the altimeter data for analysis, with the average value being used in the common period.

The interannual and decadal SSHA variations in the tropical Pacific Ocean were inferred using an Empirical Orthogonal Function (EOF) analysis. Previously, the long-term trend was removed from CWTP and HLTP, and the data were filtered with a 2- to 8-year band-pass filter and an 8-year low-pass filter, respectively, to study the interannual and decadal variations of SSH. The contributions of the first five EOFs computed by CWTP and HLTP are given in Table 1. The first EOF mode (EOF1) explain more than 50% of the total variance, which is also far more than other modes on the interannual and decadal timescales. Therefore, we will only analyze the EOF1 of interannual or decadal SSH variations.

Since the EOF1 as well as the corresponding Principal Component (PC1) computed from CWTP and HLTP for both interannual and decadal timescales are very similar, for the sake of illustration, only the results computed from CWTP are reported (Fig. 2).

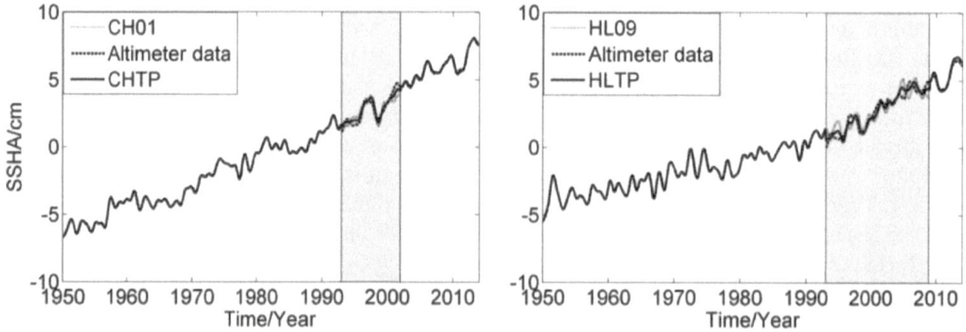

Figure 1. Regional mean SSH anomaly in the tropical Pacific Ocean derived from the altimeter data and CW01 (left) and HL09 (right). CWTP and HLTP are also displayed in the map. The shaded part is the common period of the altimeter data and the two reconstructed data, respectively.

Table 1. Contribution of the first five EOFs calculated from CWTP and HLTP (%).

EOF mode	CWTP		HLTP	
	Interannual	Decadal	Interannual	Decadal
1	72.2	63.9	67.2	54.0
2	13.8	13.3	15.4	21.1
3	3.6	9.5	5.0	8.4
4	2.4	5.1	2.7	6.7
5	1.5	3.3	2.2	3.4

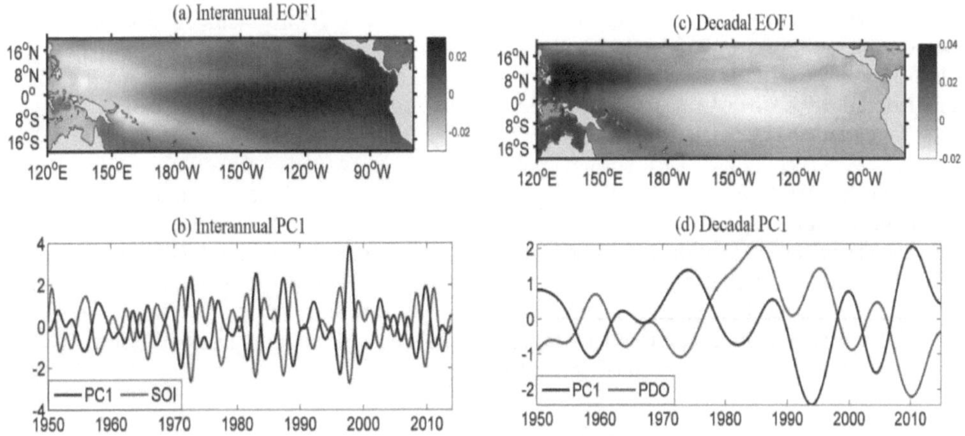

Figure 2. EOF1 patterns of SSH anomalies in the tropical Pacific Ocean (top line) and their corresponding normalized Principal Component (PC) (bottom line) on interannual timescales (a, b) and decadal timescales (c, d).

On interannual timescales, the spatial distribution of the EOF1 is dominated by opposite signs, with negative anomalies in the western tropical Pacific and positive anomalies in the central/eastern tropical Pacific (Fig. 2a). Several previous studies have described that this mode is associated with ENSO events in the tropical Pacific Ocean (e.g. Nidheesh et al. 2013). Furthermore, the maximum correlation coefficient between the SOI and PC1 is −0.89 (Fig. 2b). A negative SOI corresponds to El Niño events, while a positive SOI corresponds to La Niña events. These correlations confirm a lower/higher-than-average sea level during El Niño/La Niña events in the western tropical Pacific, and the opposite in the central/eastern tropical Pacific. This phenomenon is mainly linked to ENSO-related trade wind anomalies in the tropical Pacific Ocean, which is not discussed in this paper.

On decadal timescales, the spatial distribution of the EOF1 dominated by positive anomalies in the western tropical Pacific and negative anomalies in the central/eastern tropical Pacific (Fig. 2c). The most significant interannual and decadal variations of SSHA in the western tropical Pacific are both located on the north and south sides of the equator (5°N–15°N, 130°E–150°E, 6°S–11°S, 140°E–160°E) (Nidheesh et al. 2013). The maximum correlation coefficient between the PC1 and the PDO index is −0.67 when SSH leads PDO up to 4 months (Fig. 2d). These results agree well with those of Chang et al. (2015) indicating that the decadal variability of SSH in the tropical Pacific Ocean is significantly modulated by PDO. In addition, since the spatial distribution of this mode is similar to the distribution of sea level rise after 1993 (Figs. 2c, 3b, d), there must be some relationships between this mode and the distribution of sea level trend (Chang et al. 2015); in other words, the decadal variations may have an influence on the sea level trend. This will be further discussed in the following section.

4 INFLUENCE OF DECADAL VARIATIONS ON THE SEA LEVEL TREND AFTER 1993

Sea level trend computed from CWTP and HLTP since 1993 (Figs. 3b, d) reveals the rapid rise in sea level (maximum 10 mm/a) in the Western Tropical Pacific (WTP) while negligible rise (or even decreasing) in the Eastern Tropical Pacific (ETP), which is consistent with previous studies (e.g. Zhang & Chruch 2012). This makes researchers think that the sea level rise is accelerating; however, this inference is drawn by considering only the long-term trend of SSH changes but ignoring the influence of low-frequency variations on sea level trend. In the long term, the spatial distribution of sea level trend will be different. As shown in Figures 3a, c, sea level trends computed from CWTP and HLTP for the period between 1950 to 2014 differ significantly, which could be due to some differences between CW01 and HL09 (Hamlington et al. 2011), but both the values indicate that the rate of sea level rise since 1950 is not as fast as that observed since 1993. To explore the reason, we study the influence of decadal variations on the sea level trend since 1993.

Before analysis, the long-term trend was removed from CWTP and HLTP, and the data were filtered with an 8-year low-pass filter so that they contained only decadal fluctuations. Decadal SSHA variations in the WTP (15°S–15°N, 140°E–180°E) and in the ETP (15°S–15°N, 80°W–120°W) are shown in Figure 4. The two sets of data show the rising trend over the last 22 years in the WTP with the average rate of 3.4 mm/a and 3.5 mm/a from CHTP and HLTP, respectively. Compared with the WTP, the overall declining trend over the last 22 years in the ETP with the rate of –1.0 mm/a and –0.7 mm/a from CHTP and HLTP, especially decadal fluctuations, is also more obvious in the ETP. As these decadal variations influence the sea level trend from 1993, they lead to a rapid rate of sea level rise as seen in the WTP. We must realize that the decadal variations have a direct and great influence on the prediction of the long-term trend of sea level.

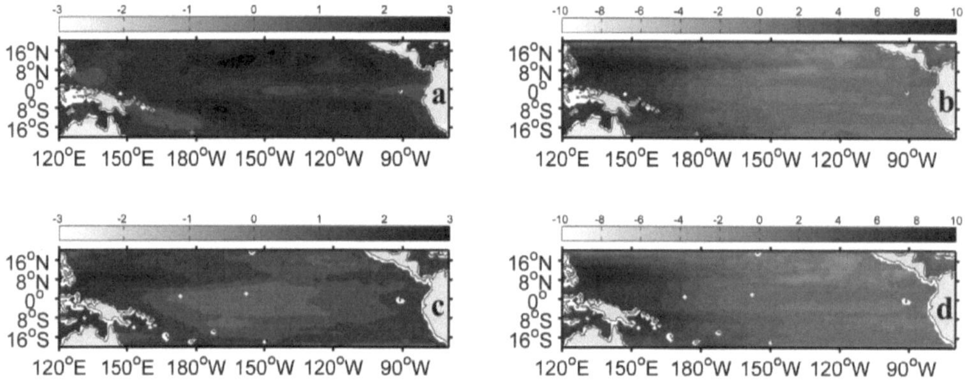

Figure 3. Long-term trend of SSHA derived from CWTP and HLTP for the period between 1950 and 2014 (a, c) and between 1993 and 2014 (b, d) (unit:mm/a).

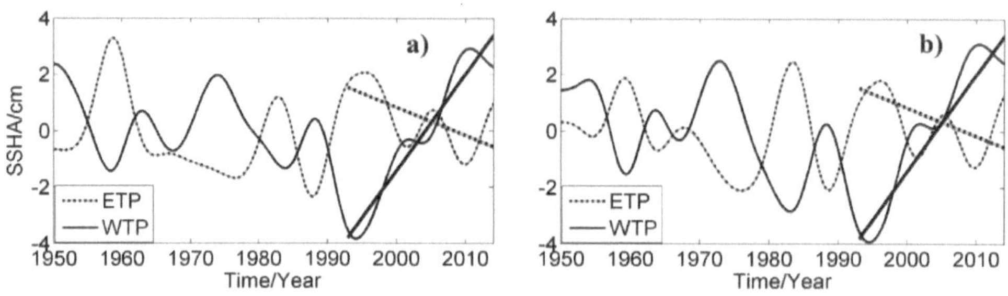

Figure 4. Decadal SSHA variations in the WTP and ETP for the period from 1950 to 2014 with the linear trend after 1993: (a) CWTP and (b) HLTP.

5 CONCLUSIONS

Two sets of SSH anomaly data based on reconstructed SSH and satellite altimeter data were combined to study the low-frequency variability of SSH in the tropical Pacific Ocean over the period of 1950–2014. The results indicate that the EOF1 of SSHA computed from CWTP and HLTP are very similar on both interannual and decadal timescales, and that the interannual and decadal variations of SSH are related to the ENSO and PDO events, respectively.

The sea level trends since 1993 are consistent with each other and with previous studies. However, in the long term, the spatial distribution of the sea level trend derived from CWTP and HLTP for the period from 1950 to 2014 will have a significant difference and the rate of the sea level trend is not as large as that observed since 1993. Therefore, we conclude that SSHA trends have been rising in the WTP but declining in the ETP over the last 22 years of the decadal oscillations. The rapid rise in the sea level observed since 1993 must be influenced by decadal variations. So we must realize that the decadal variations have a direct and great influence on the prediction of the long-term trend of sea level change.

ACKNOWLEDGMENTS

This study was supported by the NSFC project (41506006, 41376028, 41576020, and 41276018), the Ocean Public Welfare Scientific Research Project (201005019 and 20115009), the NSSFC project of No. 15ZDB170, the Jiangsu Province Nature Science Foundation of China (BK20140846), the Foundation of the Key Laboratory of Data Analysis and Application of the State Oceanic Administration (LDAA-2013-01), and the National Program on Global Change and Air-Sea Interaction (GASI-IPOVAI-04).

REFERENCES

Becker, M., Meyssignac, B., Letetrel, C., Llovel, W., Cazenave, A. & Delcroix, T. 2012. Sea level variations at tropical pacific islands since 1950. *Global & Planetary Change,* 80–81, 85–98.

Chang, J., Zhu, W., Ren, Z., Mei, H. & Kuang, X. 2015. Low-frequency variation of sea surface height of north pacific and its factors. *Jiefangjun Ligong Daxue Xuebao/Journal of Pla University of Science & Technology,* 16(4), 374–380.

Chen, X., Feng, Y. & Huang, N.E. 2014. Global sea level trend during 1993–2012. *Global & Planetary Change,* 112(1), 26–32.

Cheng, X., Xie, S.P., Du, Y., Wang, J., Chen, X. & Wang, J. 2015. Interannual-to-decadal variability and trends of sea level in the south china sea. *Climate Dynamics,* 46(9–10), 1–14.

Church, J.A., White, N.J., Coleman, R., Lambeck, K. & Mitrovica, J.X. 2004. Estimates of the regional distribution of sea level rise over the 1950–2000 period. *Journal of Climate,* 17(13), 2609–2625.

Hamlington, B.D., Leben, R.R., Nerem, R.S., Han, W. & Kim, K.Y. 2011. Reconstructing sea level using cyclostationary empirical orthogonal functions. *Journal of Geophysical Research Oceans,* 116(C12), 90–99.

Merrifield, M.A. 2011. A shift in western tropical pacific sea level trends during the 1990s. *Journal of Climate,* 24(15), 4126–4138.

Nidheesh, A.G., Lengaigne, M., Vialard, J., Unnikrishnan, A.S. & Dayan, H. 2013. Decadal and long-term sea level variability in the tropical indo-pacific ocean. *Climate Dynamics,* 41(2), 381–402.

Qiu, B. & Chen, S. 2012. Multidecadal sea level and gyre circulation variability in the northwestern tropical pacific ocean. *Journal of Physical Oceanography,* 42(1), 193–206.

Si, Z.S. & Xu, Y.S. 2014. Influence of the pacific decadal oscillation on regional sea level rise in the pacific ocean from 1993 to 2012. *Chinese Journal of Oceanology & Limnology,* 32(6), 1414–1420.

Zhang, X., & Church, J.A. 2012. Sea level trends, interannual and decadal variability in the pacific ocean. *Geophysical Research Letters,* 39(21), 653–660.

Applied Mechanics and Civil Engineering VI – Xie (Ed.)
 ISBN 978-1-138-62631-7

Study on the effect of water content on the stability of gypsum pillars

Ai-bao Chen
School of Resources and Safety Engineering, Central South University, Changsha, Hunan, China
Administration of Hunan Work Safety, Changsha, Hunan, China

Chun-de Ma
School of Resources and Safety Engineering, Central South University, Changsha, Hunan, China
Center for Advanced Study, Central South University, Changsha, Hunan, China

Xiao-ming Liu & Ya-nan Zhou
School of Resources and Safety Engineering, Central South University, Changsha, Hunan, China

ABSTRACT: Gypsum pillars left in the gob area of gypsum mines are subject to the physical and chemical actions of water, which degrades their mechanical properties and eventually leads to gob collapse. To investigate the effect of water content on the stability of gypsum pillars, laboratory tests were conducted on gypsum specimens collected from a gypsum mine in Changde. Meanwhile, the tributary area method and the equation of pillar strength proposed by Bieniawski were applied to analyze the stability of gypsum pillars with different water contents based on the test results, and the safe depth of mining under different moisture conditions was determined. The experimental results indicated that the uniaxial compressive strength and elastic modulus of gypsum specimen decreased with increasing water content. The calculated results indicated that, at the same depth, the stability of gypsum pillars decreased with the increase in water content.

1 INTRODUCTION

Gypsum is widely used in medicine, ceramic, decoration, and construction industries as a very important raw material worldwide. China has the largest reserve of gypsum resource with a total proven reserve of more than 60 billion tons (Wang, 2007), and is the fifth largest producer of gypsum in the world, with about 500 mines distributed in 24 provinces (Zhang, 2007). With the development of China, the gypsum industry has broad prospects as the demand for gypsum minerals increases. According to statistics in 2005, the current exploitation of gypsum is more than 32 million tons, and it has accelerated at a rate of 10% each year (Huang, 2008).

Gypsum mines not only bring huge economic benefits, but also would lead to massive potential security risks to facilities, houses, and people. On the one hand, room-and-pillar mining method is predominant in gypsum underground mines in China, and a large number of chambers are left without taking any precaution against their collapse. With the progressive exploitation of minerals, the area and volume of gob are larger, which contributes to the increasing pillar stress. The inappropriate mining method and insufficient administration and monitoring further lead to the deterioration and damage of the pillars. On the other hand, because of the hydrolysis of mineral contained in gypsum (Yilmaz, 2001), water or even air humidity could dissolve gypsum pillars gradually, thereby decreasing the strength of gypsum pillars under moisture condition. The two main reasons mentioned above lead to frequent gypsum mine accidents with fatalities and incalculable economic losses. From 2000 to 2015, more than 100 disaster events of gypsum mine had occurred in China, which indicates that, in order to ensure safe exploitation of mines underground and safeguard the life and property of people, it is necessary to analyze the stability of gob. In general, pillars play a crucial role in maintaining gob stability, thereby addressing the need to investigate the effect of water on the stability of gypsum pillars.

The Changde gypsum mine, located in Hunan Province of China, has been extensively influenced by mining activities in the gypsum deposits since the late 1990s. The annual production of gypsum in this mine is about 100,000 tons, and the total area of gob is more than 160000 m^2 after nearly 20 years of mining. It has a monoclinic structure, and the dip angle of ore body is about 15° in accordance with rock occurrence. The range of mining depth was 60 to 220 m from free surface. The excavation was carried out using the room-and-pillar method, which used long chambers and rib pillars with width ranging from 4 to 5 m. The chamber span ranged from 10 to 14 m, and the chamber height ranged from 6 to 8 m. Water could easily flow into the mine located close to Dongting Lake, where water table is high.

This paper aims to conduct a safety analysis of a gypsum mine in Changde. In this paper, uniaxial compressive tests were conducted by using the MTS-815 machine to determine the uniaxial compressive strength, elastic modulus, and Poisson ratio of gypsum specimen with different water contents. The stability of gypsum pillars with different water contents was also analyzed with the tributary area method and the equation of pillar strength proposed by Bieniawski.

2 COMPRESSIVE TESTS ON GYPSUM SPECIMENS

2.1 *Specimen preparation*

The test specimens were obtained from the same gypsum block retrieved from a site with high geometrical integrity and petro-graphic uniformity. They were manufactured in accordance with the standards in the ISRM specification. The water content of well-saturated gypsum specimens is approximately 8.2%; thus, five sets of water contents levels, 0%, 2%, 4%, 6%, and 8.1%, were designed for tests, and four specimens were prepared for each test set.

2.2 *Test results and analysis*

The uniaxial compressive tests were carried out on MTS815 in the Advanced Research Center at Central South University. All tests were conducted on the specimens by computerized control at a constant loading rate of 0.15 mm/min. Table 1 shows the test results of gypsum specimens with different water contents.

To demonstrate definite trends of mechanical properties with increasing water content, their relationship should be expressed by exact mathematical equations. Hawkins and McConnell (Hawkins, 1992) conducted several tests to determine the effect of water content on the strength of 15 different British sandstones, and found that the relationship between water content and uniaxial compressive strength could be described by an exponential function:

$$\sigma_c(\omega) = ae^{-b\omega} + c \tag{1}$$

where σ_c is the compressive strength; ω is the water content; and a, b, and c are constants.

With this analysis method, the best-fit lines shown in Fig. 1 were developed and the equations of relationship between water content and three mechanical properties with their coefficients of determination were listed.

Fig. 1(a)–(b) presents the variation of compressive strength and elastic modulus with different water contents, respectively. It can be seen that, with increase in water content, both the compressive strength and elastic modulus decrease gradually. When a specimen became saturated from dry, the compressive strength was reduced by around 38.5%, and the elastic modulus far more, by 48.4%. Fig. 1(c) shows that the Poisson ratio of gypsum increases with the increase in water content. The Poisson ratio of saturated specimen was 0.298, which is approximately 110.4% of that of the dry specimen.

Table 1. Test results of gypsum specimens with different water contents.

Specimen No.	Water content/%		Specific weigtht /kN•m^{-3}		Strength/MPa		Elastic modulus/GPa		Poisson ratio	
	Designed	Real	Test	Average	Test	Average	Test	Average	Test	Average
SC1-1	0.0	0.00	24.30	24.35	31.46	32.42	7.68	7.74	0.269	0.270
SC1-2		0.00	24.56		32.60		7.83		0.272	
SC1-3		0.00	24.37		34.40		7.74		0.270	
SC1-4		0.00	24.17		31.22		7.69		0.271	
SC2-1	2.0	1.97	24.65	24.66	26.02	25.86	6.05	6.08	0.281	0.283
SC2-2		2.05	24.67		24.80		5.89		0.283	
SC2-3		2.12	24.68		26.71		6.17		0.285	
SC2-4		1.94	24.64		25.91		6.19		0.283	
SC3-1	4.0	4.01	25.14	25.13	22.34	22.78	5.31	5.19	0.290	0.289
SC3-2		3.88	25.11		23.15		5.04		0.289	
SC3-3		3.94	25.12		22.65		5.27		0.287	
SC3-4		4.06	25.15		22.96		5.16		0.292	
SC4-1	6.0	6.03	25.63	25.62	20.77	20.27	4.29	4.35	0.294	0.295
SC4-2		5.94	25.61		21.03		4.44		0.294	
SC4-3		6.01	25.62		19.34		4.31		0.297	
SC4-4		5.97	25.61		19.96		4.35		0.296	
SC5-1	8.1	8.18	26.15	26.12	19.84	19.93	4.17	4.15	0.297	0.298
SC5-2		8.10	26.13		19.47		4.22		0.298	
SC5-3		8.05	26.12		19.93		4.09		0.299	
SC5-4		7.96	26.09		20,48		4.13		0.297	

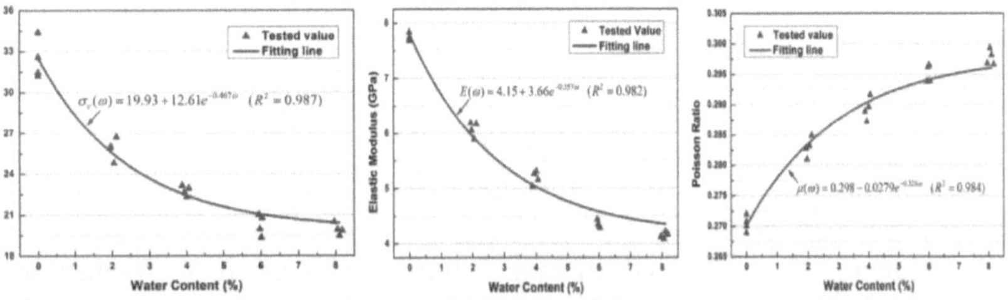

(a) Uni-axial compressive strength (b) Elastic modulus (c) Poisson ratio

Figure 1. Variation of uniaxial compressive properties with different water contents.

3 STABILITY ANALYSIS OF PILLARS WITH DIFFERENT WATER CONTENTS

In a mine excavated with room-and-pillar method, the pillars show that the global stability of gob supports the overlying strata up to the surface. Once a pillar is loaded beyond its resistance, failure of a single pillar would occur. Load redistribution caused by the single pillar failure could overload adjacent pillars, and would result in wide-area failure of pillars (Zipf, 1997). This wide-area failure could further lead to a catastrophic gob collapse. Thus, to study the stability of pillar in different moisture conditions is crucial. The key is the estimation of the pillar stress and strength.

3.1 *Determination of pillar stress*

The average pillar stress in regular layouts of pillars can be estimated by the tributary area method (Zipf, 1997), which considers that, if the directions of all stress components in the stress field of pillar are parallel to that of pillar axis, the load undertaken by the pillar is the gravity of overlying "rock pillar" up to the surface; the sum of pillar area and the mining area shared by pillar is often described as the area tributary to the respective pillar. The Changde gypsum mine was excavated with long rooms and rib pillars, and the average pillar stress can be expressed by:

$$\sigma_p = \gamma z\left(1+\frac{W_o}{W_p}\right) \tag{2}$$

where σ_p is the average pillar stress, γ is the specific weight of the overlying rocks, z is the depth of cover, and W_o and W_p are the chamber span and pillar width, respectively.

3.2 *Determination of pillar strength*

Pillar strength can be defined as the maximum resistance of a pillar to axial compression, which reflects the bearing capacity of the pillar to support the overlying rock mass directly; determination of pillar strength is the basis for analysis of pillar stability. Since the 19th century, several researchers have studied the problem of predicting the strength of coal mine pillars based on uniaxial compressive tests. From these studies, many different equations of pillar strength relating compressive strength and specimen size and shape factors have been proposed (Brady, 2013), and some representatives of pillar strength equations are listed in Table 2.

The equation of pillar strength proposed by Bieniawski has been widely used in many countries (Esterhuizen, 2011), and has proved its practical applicability throughout time. Thus, this equation was applied to estimate the pillar strength in Changde gypsum mine.

3.3 *Safety factor of pillars with different water contents*

The safety factor (F_s) is defined as:

$$F_s = \frac{S_p}{\sigma_p} \tag{3}$$

Substituting (2) and the pillar strength formula proposed into (3), the factor of safety can be expressed as:

$$F_s = \frac{S_l[0.64+0.36(W_p/h)]^{\alpha}}{\gamma z(1+W_o/W_p)} \tag{4}$$

where S_p is the pillar strength; S_l is the uniaxial compressive strength of standard specimen; W_p and h are the width and height of the pillar, respectively; and α is a constant.

When judging whether a system of pillars is in stable condition, the factor of safety must be considered, because it must compensate for the variability and uncertainty related to pillar strength and stress and mining inconsistencies (Hustrulid, 1976). With factor of safety decreasing, the probability of failure of the pillars can be expected to increase. According to Bieniawski's suggestion, the factor of safety of pillars in the Changde gypsum mine can select 1.5.

Table 2. Some equations of pillar strength in common use.

Proposer	Expression of equation	Appendix
Bunting	$S_p = S_l[0.7+0.3(W_p/h)]$	The earliest empirical equation of estimating pillar strength
Zern	$S_p = S_l(W_p/h)^{0.5}$	The standard specimen whose uniaxial compressive strength ranged from 4.8 to 7.0 MPa is suggested
Holland & Gaddy	$S_p = kW_p^{0.5}/h$	The width-to-height ratio ranged from 2 to 8 of coal pillar is suggested
Salamon & Munro	$S_p = 7.2W_p^{0.46}/h^{0.66}$	The recommended factor of safety ranges from 1.31 to 1.88
Bieniawski	$S_p = S_l[0.64+0.36(W_p/h)]\alpha$	If $W_p/h < 5$, then $\alpha = 1.0$; otherwise $\alpha = 1.4$

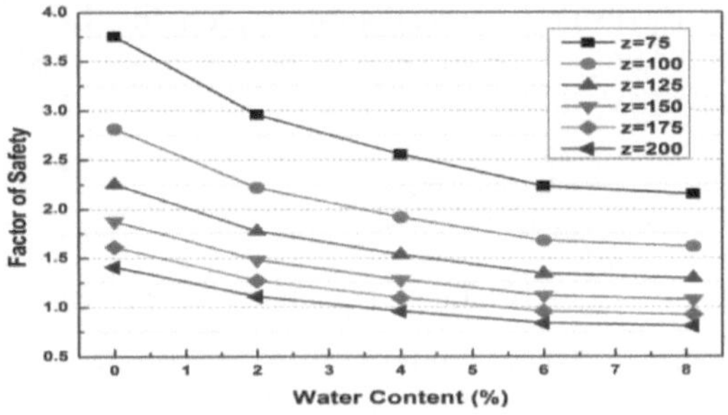

Figure 2. Safety factor of pillars with different water contents.

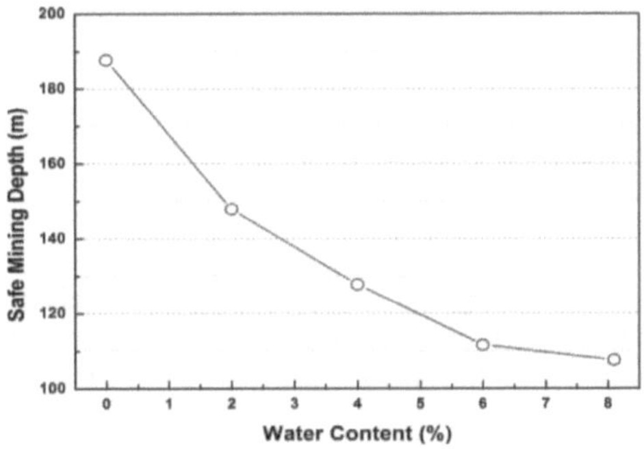

Figure 3. Safe mining depth under different moisture conditions.

According to the structural parameters and the current situation of the Changde gypsum mine, the calculation parameters are: chamber span is 12 m and the width and height of pillars are 4 and 7 m, respectively. The depth of cover ranges from 60 to 220 m. The safety factor of pillars with different water contents was calculated using Equation (4) and the schematic is shown in Fig. 2.

Fig. 2 further shows the safety factor of pillars with different water contents under different depths of cover. As shown in Fig. 2, on the same depth of cover, the pillar's factor of safety decreases gradually with the increase in water content. It is also observed that, when the cover depths of gypsum pillar are on 75 and 100 m, the factor of safety of all pillars is greater than 1.5, which indicates that the strength of gypsum pillar is enough to support the upper loads, and the phenomenon of instability will not happen, regardless of the moisture conditions of the pillars. When the cover depths of gypsum pillar are on the order of 125 m, losing stability will occur in pillar with water content of more than 6.0%, whose factor of safety is 1.34 less than 1.5. The failure would occur when water content of pillars reaches 2.0% and 0.75% on the cover depth of 150 and 175 m, respectively. When the cover depth reaches 200 m, failure occurs even in the dry gypsum pillars, let alone pillars in moisture conditions.

3.4 *Determination of safe mining depth*

Considering the effect of water content on gypsum pillars only, the safe mining depths with different water contents were calculated as:

$$Z_s = \frac{S_l[0.64 + 0.36(W_p / h)]^{\alpha}}{F_s\gamma(1 + W_o / W_p)} \tag{5}$$

Fig. 3 shows that according to the parameters of pillar, when water content is 0, the safe mining depth is around 188 m, that is, losing stability does not occur in the pillar that is above 188 m as long as the pillar remains dry. When the gypsum pillar is saturated, the safe mining depth is 108 m, almost a 42.6% reduction. It is difficult to keep the gob in dry state all the time, because the Changde gypsum mine is adjacent to Dongting Lake. Thus, determination of safe mining depth according to the stability of pillars in saturated state can ensure the safety of gob to the greatest extent.

4 CONCLUSION

In this paper, a case study of a gypsum mine in Changde was presented.

1. Water content has a pronounced effect on the mechanical properties of gypsum. The laboratory test results indicated that, with increasing water content, the uniaxial compressive strength and elastic modulus decreased, and the reductions were 38.5% and 46.4% when gypsum specimens reached the saturated state. On the contrary, the Poisson ratio increased gradually; the Poisson ratio of saturated specimen was 110.4% of that of the dry specimen.
2. The safety factor of gypsum pillars with different water contents was calculated. The results indicated that, on the same depth of cover, the pillar's factor of safety decreased gradually with the increase in water content, even pillars could not support the upper loads. Thus, water is the main factor that leads to gob collapse in gypsum mines.
3. According to the parameters of pillar design in the Changde gypsum mine, when the depth of cover was less than 108 m, even though a pillar became saturated, it would not fail. When the depth of cover was greater than 108 m, the instability would occur in damp pillars; it cannot ensure the safety of pillars designed with the original parameters at that time, and then increasing the pillar width and decreasing the chamber span or height appropriately should be recommended.
4. As for gypsum gobs, the water inside should be drained before leaving, and then appropriate measures of plugging and draining should be applied to prevent gypsum pillars from deterioration of water. For the gypsum pillars in moisture conditions, reinforcement must be adopted to make the pillar strength high enough to support upper loads and further ensure the stability of gobs.

ACKNOWLEDGMENTS

The authors would like to acknowledge the financial support from the National Natural Science Foundation of China (51322403, 51274254) and the Hunan province Natural Science Foundation of China (2015JJ2172). The corresponding author (Chun-de MA) has made an important contribution to this research.

REFERENCES

Brady, B.H., & Brown, E.T. *Rock mechanics: for underground mining* [M]. Springer Science & Business Media. 2013.

Esterhuizen, G.S., Dolinar, D.R., & Ellenberger, J.L. Pillar strength in underground stone mines in the United States [J]. *International Journal of Rock Mechanics and Mining Science*. 2011, 48(1): 42–50.

Hawkins, A. B., & McConnell, B. J. Sensitivity of sandstone strength and deformability to changes in moisture content [J]. *Quarterly Journal of Engineering Geology and Hydrogeology*. 1992, 25(2): 115–130.

Huang, Y. H. The mechanisms of gypsum gob instability mined with room-and-pillar method [D]. *Master's dissertation, Changsha Institute of Mining Research*, 2008.

Hustrulid, W.A. A review of coal pillar strength formulas [J]. *Rock Mechanics*. 1976, 8(2): 115–145.

Wang, G., You, G., & Xu, Y. Investigation on the Nanjing gypsum mine flooding [J]. *Geotechnical Engineering for Disaster Mitigation & Rehabilitation*, 2007: 920–930.

Yilmaz, I. Gypsum/anhydrite: some engineering problems [J]. *Bulletin of Engineering Geology & The Environment*. 2001, 60(3): 227–230.

Zhang, X.F. The method of dispose gob in gypsum mine [D]. *Doctoral dissertation Wuhan university of technology*, 2007.

Zipf, R.K., & Mark, C. Design methods to control violent pillar failures in room-and-pillar mines [J]. *Transactions of the Institution of Mining and Metallurgy*. 1997, 106(1): 124–131.

Applied Mechanics and Civil Engineering VI – Xie (Ed.)
© 2017 Taylor & Francis Group, London, ISBN 978-1-138-62631-7

Study of the material parameters of self-anchored suspension bridges

C.H. Kou, L. Chen & S.W. Ma
Department of Civil Engineering, Chung-Hua University, Taiwan

M.L. Yang
Department of Architecture and Urban Planning, Chung-Hua University, Taiwan

ABSTRACT: This paper will analyze and explore when a self-anchored suspension bridge's main cables, main girders, towers, hangers, and other structural elements' material elastic modulus is individually altered, what the effect on the bridge's vertical earthquake responses would be in order to clearly understand the self-anchored suspension bridge's dynamic characteristics. The results of this research indicate that the effect of a change in the main cable's elastic modulus on the maximum internal forces and displacements produced by an earthquake in the vertical direction is the largest, followed by the effect of a change in the elastic modulus of the main girder, towers, and hangers.

1 INTRODUCTION

Because the main cables of self-anchored suspension bridges are already directly anchored at the two ends of the main girder, this structure saves the cost of installing other anchoring devices; furthermore, the axial pressure is transferred from the main cable to the main girder and used as a source of pre-stressed force that indirectly enhances the bending resistance capability of the main girder, which then renders the self-anchored suspension bridge worthy of further research and widespread use as a water-spanning structure. Therefore, this paper, upon completing a stochastic static analysis of the self-anchored suspension bridge (Wollmann, G.P. 2001a, b, Kao, C.S. et al. 2006), further explores a self-anchored suspension bridge's earthquake responses in order to clearly understand the dynamic characteristics of the self-anchored suspension bridge in addition to the impact that different material characteristics may have on the earthquake responses of this very structure.

2 THEORETICAL ANALYSIS OF THE DYNAMIC CHARACTERISTICS OF SELF-ANCHORED SUSPENSION BRIDGES

In considering the self-anchored suspension bridge's large displacement generalized potential energy function, this paper will be based on the following assumptions (Zhang, H.J. 2004):

a. All the materials must comply with Hook's law.
b. At the complete stage, the statical load is evenly distributed along the span, and the cable forms a parabolic curve.
c. At its points of connection with the towers, the main girder receives vertical support.

d. Given that the hangers are distributed densely, they are treated as even membranes with only vertical resistance and malleability is not taken into account.
e. Without considering the bending resistance rigidity and axial compression deformation in the direction along the main girder of the tower, the paper will assume that the rigidity in the transverse direction is infinitely large.
f. The paper considers only the warping deformation of the twist in the main girder but does not consider the cross-section's distorted deformation and will assume the diaphragm is distributed densely and its shearing rigidity is infinitely large.

Assume the displacements along the main girder are u, vertical displacements are v, displacements in the transverse direction are w, twisted angles are θ, displacements along the direction of the main girder, in the vertical direction, and transverse direction of the left side cables of the main girder are u_l, v_l, w_l, displacements along the direction of the main girder, in the vertical direction, and transverse direction of the right side cables of the main girder are u_r, v_r, w_r, $H_l(t)$, $H_r(t)$ represent the left and right side increment horizontal forces instigated by the vibration inertial force, E_c, A_{cl} are the cross-sectional areas and the elastic modulus of the main cables, q represents total load density, $q = q_s + q_c$, q_s is the statical load density of the main girder, q_c is the statical load density of the main cables, H_q is the horizontal force of the two main cables under statical load conditions, L_c is the length of each span, λ_l, λ_r are the Lagrange Multipliers, f_l, f_r are the forces constraining the extension of the left and right hangers, g is gravity acceleration, h' is the cable length along the x-axis, A_s, I_y, I_z are the cross-sectional area, lateral, and vertical flexural inertial moments of the main girder, respectively, E is the elastic modulus of the main girder, μ is the shearing influence coefficient, $\beta E_l J_w$ is the main girder's warping rigidity, $\beta = J_t/(J_p - J_t)$, $E_l = E/(1 - \nu^2)$, J_p is the pole moment inertia of the cross section, J_t is St. Venant's torsional constant.

The large displacement generalized potential energy function (Zhang, H.J. 2004, Fan, J.J. 2007):

$$\begin{aligned}\Pi^* = \int_{t_1}^{t_2} \frac{1}{2}\sum_{i=1}^{3}\int_0^{L_i} \frac{q_s}{gA_s}\left[I_y\left(\frac{\partial^2 w}{\partial x \partial t}\right)^2 + I_z\left(\frac{\partial^2 v}{\partial x \partial t}\right)^2 + (I_y + I_z)\left(\frac{\partial \theta}{\partial t}\right)^2 + A_s\left(\frac{\partial u}{\partial t}\right)^2\right.\\
\left. + A_s\left(\frac{\partial v}{\partial t}\right)^2 + A_s\left(\frac{\partial w}{\partial t}\right)^2\right] dx_i + \frac{1}{2}\sum_{i=1}^{3}\int_0^{L_i} \frac{q_c}{2g}\sqrt{1+h'^2}\left[\left(\frac{\partial u_1}{\partial t}\right)^2 + \left(\frac{\partial v_1}{\partial t}\right)^2 + \left(\frac{\partial w_1}{\partial t}\right)^2\right.\\
\left. + \left(\frac{\partial u_r}{\partial t}\right)^2 + \left(\frac{\partial v_r}{\partial t}\right)^2 + \left(\frac{\partial w_r}{\partial t}\right)^2\right] dx_i - \sum_{i=1}^{3}\frac{L_c}{E_c A_{cl}}\left[\frac{1}{2}H_q(H_l + H_r) + \frac{1}{2}(H_1^2 + H_r^2)\right]\\
- \frac{1}{2}\sum_{i=1}^{3}\left[\int_0^{L_i} \beta E_1 J_w\left(\frac{\partial^2 \theta}{\partial x^2}\right)^2 dx_i + \int_0^{L_i} GJ_t\left(\frac{\partial \theta}{\partial x}\right)^2 dx_i + \int_0^{L_i} EI_y\left(\frac{\partial^2 w}{\partial x^2}\right)^2 dx_i\right.\\
+ \int_0^{L_i} EI_z\left(\frac{\partial^2 v}{\partial x^2}\right)^2 dx_i + \int_0^{L_i} EA_s\left(\frac{\partial^2 u}{\partial x^2}\right)^2 dx_i + \int_0^{L_i} N_x\left(\frac{\partial v}{\partial x}\right)^2 dx_i\\
\left. + \int_0^{L_i} \mu GA_s\left(\frac{\partial u}{\partial y} + \frac{\partial v}{\partial x}\right)^2 dx_i + \int_0^{L_i} \mu GA_s\left(\frac{\partial u}{\partial z} + \frac{\partial v}{\partial x}\right)^2 dx_i\right] - \sum_{i=1}^{3}\left[\int_0^{L_i} q_s v dx_i\right.\\
\left. + \int_0^{L_i} \frac{1}{2} q_c (v_l + v_r) dx_i\right] - \sum_{i=1}^{3}\int_0^{L_i} \lambda_l f_l dx_i - \sum_{i=1}^{3}\int_0^{L_i} \lambda_r f_r dx_i \Bigg\} dt\end{aligned} \qquad (1)$$

where H_q is the sum of horizontal forces triggered by the statical load of the two main cables.

The vibration equations of the self-anchored suspension bridges can be obtained from following operation

$$\delta\Pi^* = \frac{\partial \Pi^*}{\partial \phi_i}\delta\phi_i = 0 \quad \phi_i = u, v, w, \theta \ldots \qquad (2)$$

3 BASIC DATA ANALYSIS

In this paper's analysis, the choice of bridge corresponds to that in reference (Fan, J.J. 2007), namely a double tower, suspension bridge with a main span of 160 meters, two side spans of 76 meters, an entire length of 312 meters, tower height of 42.4 meters, rise-span ratio of 1/6, hanger to hanger distance of 5 meters, a reinforced concrete box girder as the main girder with a standard cross-section of 5 single box cells, girder width of 41 meters, height of girder's geometric center of 2.5 meters, and main girder height-span ratio of 1:64. Furthermore, the construction site that this paper considers is located in a strong earthquake zone where the site's underlying geological plates are of category 3. The earthquake response spectrum is displayed in Figure 1.

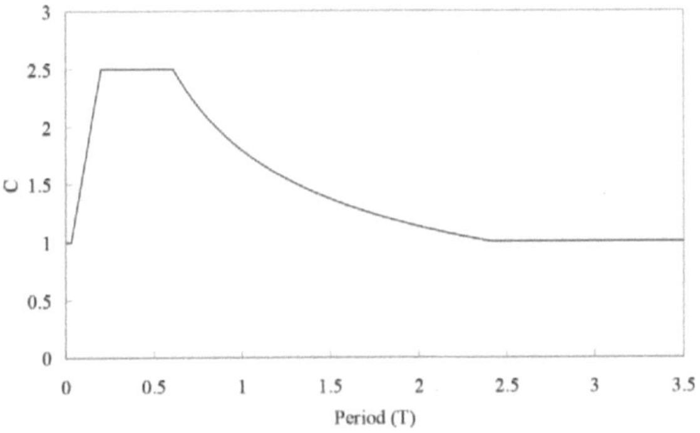

Figure 1. Category 3 plates earthquake response spectrum.

Table 1. The effects of a change in the material elastic modulus on the structure's maximum internal force during an earthquake in the vertical direction.

Item		Main cable's maximum hanger force (N)	Hanger's maximum hanger force (N)	Tower base's maximum shear force (N)	Tower base's maximum bending moment (N-m)	Main girder's maximum bending moment (N-m)
Main cable E value	Reduce to 0.7 times the original value	–11.92%	–11.27%	+12.34%	+16.77%	+3.87%
	Increase to 1.3 times the original value	+10.09%	+11.45%	–12.47%	–15.61%	–3.84%
Main girder E value	Reduce to 0.7 times the original value	+8.03%	+6.64%	+4.13%	+8.97%	–13.95%
	Increase to 1.3 times the original value	–7.63%	–5.77%	–3.39%	–6.95%	+12.85%
Tower E value	Reduce to 0.7 times the original value	–0.31%	–0.13%	–25.45%	–25.25%	+0.74%
	Increase to 1.3 times the original value	+0.30%	+1.16%	+27.83%	+27.41%	–0.69%
Hanger E value	Reduce to 0.7 times the original value	–0.98%	+3.46%	–3.29%	+1.06%	+0.23%
	Increase to 1.3 times the original value	+0.59%	–2.99%	+4.11%	–1.28%	–0.17%

Table 2. The effects of a change in the material elastic modulus on the structure's displacement during an earthquake in the vertical direction.

Item		Main girder vertical displacement			Tower roof horizontal displacement (m)
		Left side span mid-point (m)	Main span midpoint (m)	Right side span midpoint (m)	
Main cable E value	Reduce to 0.7 times the original value	+5.46%	+22.23%	+5.51%	+17.72%
	Increase to 1.3 times the original value	-4.82%	-21.59%	-4.89%	-15.43%
Main girder E value	Reduce to 0.7 times the original value	+19.32%	+8.74%	+19.26%	+10.00%
	Increase to 1.3 times the original value	-18.78%	-7.00%	-18.76%	-8.23%
Tower E value	Reduce to 0.7 times the original value	+0.95%	+1.81%	+0.95%	+3.42%
	Increase to 1.3 times the original value	-0.90%	-1.70%	-0.90%	-3.33%
Hanger E value	Reduce to 0.7 times the original value	+0.39%	+1.55%	+0.40%	+1.22%
	Increase to 1.3 times the original value	-0.24%	-0.91%	-0.24%	-0.79%

4 THE EFFECT OF A CHANGE IN THE MATERIAL ELASTIC MODULUS ON THE EARTHQUAKE RESPONSE IN THE VERTICAL DIRECTION

Tables 1 and 2 both indicate that a change in the main cable's elastic modulus has the greatest effect on the structure's internal forces and displacements generated by an earthquake in the vertical direction followed by a change the elastic modulus in the main girder, towers, and hangers.

5 CONCLUSION

A double tower self-anchored suspension bridge with bridge length 312 m, main span 160 m, two side spans of 76 m, tower height 42.4 m, and hanger to hanger distance of 5 m, this paper's analysis yields the following conclusions:

a. During an earthquake in the vertical direction, a reduction in the main cables' elastic modulus will cause the tower base's maximum internal force and the main girder's maximum bending moment to increase; a reduction in the main girder's elastic modulus will cause the maximum hanger force of the main cables and hangers and the tower base's maximum internal force to increase. Also, a reduction in the towers' elastic modulus will cause the main girder's maximum bending moment to increase; finally, reducing the hangers' elastic modulus will cause the hanger's maximum hanger force, tower base's maximum internal force and the main girder's maximum bending moment to increase.
b. During an earthquake in the vertical direction, a reduction in the main cables' elastic modulus will cause the main girder's vertical displacement and tower roof's horizontal displacements to increase; a reduction in the main girder's elastic modulus will cause the main girder's vertical displacement and tower roof's horizontal displacements to increase. Also, a reduction in the towers' elastic modulus will cause the main girder's vertical displacement and the tower roof's horizontal displacements to increase; finally, reducing the

hangers' elastic modulus will cause the main girder's vertical displacement and the tower roof's horizontal displacements to increase.

c. A change in the main cables' elastic modulus has the largest effect on the structure's internal forces and displacements generated during an earthquake in the vertical direction followed by a change the elastic modulus in the main girder, tower, and hanger.

REFERENCES

Fan, J.J. 2007. *Study of the Dynamical Response of Self-Anchored Suspension Bridge*. Master Thesis, Department of Civil Engineering, Chung-Hua University, Hsinchu.

Kao, C.S. et al. 2006. Nonlinear Stochastic Static Analysis of Self-Anchored Suspension Bridge. *Asia Pacific Review of Engineering Science and Technology* 4(1): 653–668.

Wollmann, G.P. 2001a. Self-Anchored Suspension Bridges. *Journal of Bridge Engineering, ASCE* 4: 156–158.

Wollmann, G.P. 2001b. Preliminary Analysis of Suspension Bridges. *Journal of Bridge Engineering, ASCE* 6: 227–233.

Zhang, H.J. 2004. *Mechanical Performance Analysis and Experimental Research on Concrete Self-Anchored Suspension Bridge*. Master Thesis, Department of Bridge Engineering, School of Civil & Hydraulic Engineering, Dalian University of Technology, Dalian.

Applied Mechanics and Civil Engineering VI – Xie (Ed.)
© 2017 Taylor & Francis Group, London, ISBN 978-1-138-62631-7

Author index